Advancing Sustainability through Land-Related Approaches: Theoretical Foundations and Empirical Investigations

Advancing Sustainability through Land-Related Approaches: Theoretical Foundations and Empirical Investigations

Guest Editors

Bing-Bing Zhou
Taiyang Zhong
Yujia Zhang
Ligang Lv
Yuzhu Zang

MDPI

Basel • Beijing • Wuhan • Barcelona • Belgrade • Novi Sad • Cluj • Manchester

Guest Editors

Bing-Bing Zhou
School of International Affairs and Public Administration
Ocean University of China
Qingdao
China

Taiyang Zhong
School of Geography and Ocean Science
Nanjing University
Nanjing
China

Yujia Zhang
Environmental Studies Program
Oberlin College
Oberlin
USA

Ligang Lv
School of Public Administration
Nanjing University of Finance and Economics
Nanjing
China

Yuzhu Zang
School of Public Administration
China University of Geosciences
Wuhan
China

Editorial Office
MDPI AG
Grosspeteranlage 5
4052 Basel, Switzerland

This is a reprint of the Special Issue, published open access by the journal *Land* (ISSN 2073-445X), freely accessible at: https://www.mdpi.com/journal/land/special_issues/79Q89Z37E4.

For citation purposes, cite each article independently as indicated on the article page online and as indicated below:

Lastname, A.A.; Lastname, B.B. Article Title. *Journal Name* **Year**, *Volume Number*, Page Range.

ISBN 978-3-7258-4557-6 (Hbk)
ISBN 978-3-7258-4558-3 (PDF)
https://doi.org/10.3390/books978-3-7258-4558-3

Cover image courtesy of Bing-Bing Zhou

Contents

Preface . . . . **vii**

Bing-Bing Zhou, Jingyuan Liu and Xiaoke Wang
Advancing Sustainability Through Land-Related Approaches: Insights from NRC (1999) and a Bold Call to Action
Reprinted from: *Land* **2025**, *14*, 756, https://doi.org/10.3390/land14040756 . . . . **1**

Qiqi Yang, Lijie Pu and Sihua Huang
Review of Urbanization-Associated Farmland Research in China: A Sustainability Perspective
Reprinted from: *Land* **2024**, *13*, 534, https://doi.org/10.3390/land13040534 . . . . **12**

Xinyu Hu, Chun Dong and Yu Zhang
Impacts of Cropland Utilization Patterns on the Sustainable Use Efficiency of Cropland Based on the Human–Land Perspective
Reprinted from: *Land* **2024**, *13*, 863, https://doi.org/10.3390/land13060863 . . . . **28**

Yawen Jin, Rongxiao He, Jingke Hong, Dan Luo and Guoling Xiong
Assessing the Accessibility and Equity of Urban Green Spaces from Supply and Demand Perspectives: A Case Study of a Mountainous City in China
Reprinted from: *Land* **2023**, *12*, 1793, https://doi.org/10.3390/land12091793 . . . . **54**

Lu Han, Yanbo Qu, Shufeng Liang, Luyan Shi, Min Zhang and Haiyan Jia
Spatiotemporal Differentiation of Land Ecological Security and Optimization Based on GeoSOS-FLUS Model: A Case Study of the Yellow River Delta in China Toward Sustainability
Reprinted from: *Land* **2024**, *13*, 1870, https://doi.org/10.3390/land13111870 . . . . **71**

Xinwei Pei, Xinger Zheng and Cong Wu
How Part-Time Farming Affects Cultivated Land Use Sustainability: Survey-Based Assessment in China
Reprinted from: *Land* **2024**, *13*, 1242, https://doi.org/10.3390/land13081242 . . . . **92**

Xue Wang, Mingyu Wang, Lianbing Gong and Chuangchuang Yu
Understanding the Rise and Fall of Rural Specialty Agriculture from Social–Ecological Land System Perspective: A Longitudinal Case Study in China
Reprinted from: *Land* **2025**, *14*, 254, https://doi.org/10.3390/land14020254 . . . . **111**

Yang Yang, Zhe Dong, Bing-Bing Zhou and Yang Liu
Smart Growth and Smart Shrinkage: A Comparative Review for Advancing Urban Sustainability
Reprinted from: *Land* **2024**, *13*, 660, https://doi.org/10.3390/land13050660 . . . . **137**

Zhi Li, Lixin Zhang, Tian Hu and Yifei Wu
History, Progress, and Prospects of Urban Fringe Research in China: An Idiosyncratic Synthesis from a Spatial Perspective
Reprinted from: *Land* **2025**, *14*, 248, https://doi.org/10.3390/land14020248 . . . . **158**

Lin Tan, Qinyu Cui, Lan Chen and Lan Wang
An Exploratory Study on Spatial Governance Toward Urban–Rural Integration: Theoretical Analysis with Case Demonstration
Reprinted from: *Land* **2024**, *13*, 2035, https://doi.org/10.3390/land13122035 . . . . **176**

Jinpeng Wang, Zhengkai Mao and Zhijun Zhang
A Leverage Points Perspective on China's Governance of Marine Protected Areas: Current State and Ways Forward
Reprinted from: *Land* **2025**, *14*, 425, https://doi.org/10.3390/land14020425 . . . . **199**

Preface

This is a reprint containing a particular selection of peer-reviewed articles that were originally published in the *Land* journal under the Special Issue "Advancing Sustainability through Land-Related Approaches: Theoretical Foundations and Empirical Investigations". This volume reflects the growing recognition that land is not merely a material resource but a rich governance toolbox for transformative sustainability endeavors at multiple domains and scales.

This reprint not only showcases the latest scholarly advancements from various land-related sciences such as land system science, landscape ecology, natural resource management, geography, and public administration in mobilizing land for advancing sustainability, but also illustrates how seemingly distinct land-related sustainability approaches should form a coherent set of governance levers at four complementary intervention domains. The covered topics range widely from farmland protection, farmland use efficiency, the spatial justice of urban green spaces, landscape ecological security, agro-farmland sustainability, sustainability dynamics of rural land systems, challenges amid coupled urban growth and shrinkage, changing paradigms of sustainable urban fringe systems, the spatial governance of rural–urban systems, and the under-researched governance of marine space. However, from a leverage point perspective on sustainability, they all fall into either of the four land-related sustainability governance levers, i.e., rebuilding capital stocks, redirecting interaction flows, reforming governance architectures, and re-/co-piloting sustainability transitions. With the bigger picture in mind, transdisciplinary land stakeholders including scholars, practitioners, and policymakers should work collectively to harness the full potential of land-related approaches to advancing sustainability in various contexts and at multiple scales. It is our sincere hope that this collection of studies can inspire new thinking, transdisciplinary collaboration, and more real-world sustainability transformations.

We are deeply grateful to all contributing authors for their intellectual engagement. Special thanks to the editorial team at *Land* and the MDPI production staff for their continued support and professionalism throughout the publication process.

Bing-Bing Zhou, Taiyang Zhong, Yujia Zhang, Ligang Lv, and Yuzhu Zang
Guest Editors

land

MDPI

Perspective

Advancing Sustainability Through Land-Related Approaches: Insights from NRC (1999) and a Bold Call to Action

Bing-Bing Zhou [1,2], Jingyuan Liu [1] and Xiaoke Wang [1,*]

1 School of International Affairs and Public Administration, Ocean University of China, Qingdao 266100, China; bbz@ouc.edu.cn (B.-B.Z.); liujingyuan6963@stu.ouc.edu.cn (J.L.)
2 Maxwell School of Citizenship and Public Affairs, Syracuse University, Syracuse, NY 13244, USA
* Correspondence: wangxiaoke@stu.ouc.edu.cn

Abstract: This paper investigates the critical role of land in advancing sustainability, drawing insights from the landmark report by the U.S. National Research Council, *Our Common Journey: A Transition Toward Sustainability* (hereafter referred to as NRC (1999)), and aligning them with the leverage points perspective on sustainability. Four key problem entries—land as a resource, land use and ecosystem services, land systems, and landscapes or regional scales—are identified as pivotal framings for addressing sustainability challenges, and are further elaborated with practical examples. Regretfully, despite decades of multidisciplinary research progress, land-related approaches remain fragmented. This paper contributes to the existing research by illustrating, for the first of time, how these multidisciplinary research traditions can be integrated cohesively using the four nested realms of sustainability leverage points—rebuilding capital stocks, redirecting interaction flows, reforming governance architectures, and re-/co-piloting sustainability transitions—to achieve nested transformations across varying timeframes. We hope that this hierarchical perspective fosters top-down and bottom-up collaborations among researchers, policymakers, and practitioners to take transformative actions. To leave a legacy of sustainability for future generations, we must act collectively, boldly, and without delay to harness the transformative potential of all land-related approaches.

Keywords: sustainability science; land system science; landscape sustainability; social-environmental systems; sustainability transition; leverage points; actionable knowledge

1. Introduction

Sustainability—meeting the fundamental needs of humanity more equitably without threatening Earth's life support systems [1]—has been increasingly recognized as the theme of our time [2]. Yet, with the severely off-track progress we have achieved [3], bolder efforts are urgently needed to navigate our "World-Earth Spaceship" [4] onto more sustainable tracks, more than ever. To a certain degree, finding the effective approaches to advancing sustainability has become the holy grail of the science community of the 21st century [5]. Therefore, arguably one of the most critical research missions is to advance sustainability through land-related approaches—given the pivotal role of land as the spatial interface of the World and Earth subsystems.

As a matter of fact, scholarly efforts to advance sustainability through land-related approaches date back to the late 1980s (e.g., [6–9]), about same time that the Brundtland Report popularized the idea of sustainable development in 1987 [1]. That said, global political recognition of the land's critical roles in addressing sustainability can trace far earlier, to the 1972 United Nations Conference on the Human Environment in Stockholm [10]. After several decades of multidisciplinary efforts involving agriculture, forestry, geography, landscape architecture, landscape ecology, and many other related fields, fruitful studies have been conducted (e.g., [11–13]), creating two main fields of research which seem to begin converging: land system science originated from geography [14,15] versus the more niche and contested landscape sustainability science emerging from landscape ecology [16,17].

Notwithstanding the encouraging research progresses, land-based practices to address sustainability are rather piecemeal and unsatisfying, likely due to the complex nature of land and the multidisciplinary tradition of research. Lacking the big picture of land-related approaches to advancing sustainability hinders policymakers and practitioners from taking systematic and integrated land-based actions. In this context, in the present paper (along with the other papers in this Special Issue), we attempt to map out the big picture by drawing on insights from the landmark publication that played a historic role in catalyzing a burgeoning science of sustainable development—*Our Common Journey: A Transition Toward Sustainability* (hereafter NRC (1999) for short) [18]—and further, try proposing a more holistic and action-oriented agenda by integrating the leverage points perspective, in a hope of stimulating the much needed real-world transformations.

2. Insights from NRC (1999): The Four Problem Entries of Land When Addressing Sustainability

To draw on the insights from NRC (1999) [18], we searched the main text of the report (i.e., from the Executive Summary to Chapter 6, excluding references and bibliography) for the word "land", and identified a total of 146 sentences containing "land" and/or "landscape" (excluding words like Brundtland and Thailand), with four pairs of duplicated sentences (in the Executive Summary). The first author then conducted a content analysis and a grounded coding of these sentences [19] (like [20]). A 100% consistency was reached while coding the four pairs of duplicated sentences, hinting at a reliable internal validity of coding (see Supplementary Table S1 for details). Afterwards, we elaborate on the four problem entries that emerged from the land-related text corpus in NRC (1999) [18] (Table 1), with examples from relevant publications or real-world practices. As the start point of our discussion, a problem entry here means how land is framed into a sustainability context (i.e., the way by which land relates to a sustainability problem).

Table 1. The four problem entries of land when addressing sustainability based on NRC (1999) [18].

Problem Entry	Code Example
Land resource per se or ecosystem	• *Example 1*: Meeting even the most basic needs of a stabilizing population at least half as large as today's implies greater production and consumption of goods and services, increased demand for land, energy, and materials, and intensified pressures on the environment and living resources (page 1). • *Example 2*: Productivity must be increased in robust areas and restored to degraded lands, while the damage to fragile land areas must be reduced (page 13).
Land use and ecosystem services	• *Example 1*: In fact, most new housing in the cities in developing countries has been built on illegally occupied or subdivided land (page 35). • *Example 2*: For example, the Forest Principles set forth at the 1992 World Conference on Environment and Development do not set quantitative targets for forest protection; but they establish in Article 2b that "forest resources and forest lands should be sustainably managed to meet the social, economic, ecological, cultural and spiritual needs of present and future generations" for forest products and services (page 45).

Table 1. *Cont.*

Problem Entry	Code Example
Land system or social-ecological system	• *Example 1*: Moreover, sustainable communities have not been realized across the U.S. landscape (page 14). • *Example 2*: Coastal ecosystems are often affected by activities that occur far inland, through changes in the delivery of water, nutrients, and chemical contaminants from rivers and atmospheric deposition (page 88).
Landscape or regional scale	• *Example 1*: Greater understanding is needed of how biological systems work, how to stem the continued loss of habitats, and how ecosystems can be restored and managed at the landscape or regional scale (page 13). • *Example 2*: It is therefore necessary to work at the scale of whole landscapes, with explicit attention both to the preservation of critical ecosystems and to the interactions between human activities and the managed and unmanaged ecosystems among them (page 99).

2.1. Problem Entry 1: As Land Resource per se or Ecosystem

The most frequent sustainability problem entry regarding land in NRC (1999) [18] is the narrow framing that conceptualizes land as land per se, as a development resource, or as an ecosystem type of development potential (e.g., wetland, forest land), with a frequency of 51.37% (i.e., a total of 75 counts). This "stock" framing of land is the original sustainability problem entry, with societal concerns either on the loss of productive ecosystems and land degradation following a development perspective, or regarding ecosystem conservation and restoration in case of future development impacts following an environmental perspective, or both, emphasizing balancing the two sides of environment–development from an integrated perspective on sustainability. NRC (1999) [18] devoted a standalone sub-subsection in Chapter 5 on "Conserving Productive Landscapes" (page 254 therein). Studies with such a problem entry were popular in the past (e.g., [7,21]), yet they would be hard to get published now (but see [22–24]). That said, land-related approaches following this problem entry are perhaps the most actionable in practice.

As corroboration, China has established vast areas of ecological conservation areas nationwide in various forms (e.g., national parks of particular importance or fragility, theme parks of specific ecosystems, ecological protection areas in biodiversity hotspots). Such efforts safeguard the bottom-line of China's ecological sustainability, which is perhaps why it is better known as China's "Ecological Redline Policy" [25,26]. Moreover, China routinely implements restoration/construction programs for productive landscapes like wetlands, forests, and high-quality farmlands, with dedicated government agencies and specialized government funding. Perhaps even more fundamental than ensuring no net loss of natural habitats, a recent policy known as biodiversity offsets has gained significant popularity worldwide over the past decade, as it aims to achieve no net loss of biodiversity from development activities through various measures [27,28]. In addition, to reduce farmland loss and contain the sprawl of construction land, China has implemented a comprehensive set of policies such as "Farmland Redline Policy," "Farmland Dynamic Balance Policy" (i.e., zero net loss of farmland), "Urban-Rural Construction Land Balance Policy" (i.e., zero net increase in urban–rural construction land) [29].

2.2. Problem Entry 2: As Land Use and Ecosystem Services

The second frequent problem entry of land in sustainability in NRC (1999) [18] is a somewhat broadened framing of land as land use, with the real interests lying in the land's functions or ecosystem services (41.40%, 60 counts). With this "flow" framing of land, the societal concerns would be directed onto land transformations and the associated impacts of specific land use changes to meet human needs. The NRC (1999) [18] highlights in its standalone subsection on "Land" in Chapter 2 (page 93 therein) three long-term global trends in land use change, i.e., the increasing intensification of agriculture, clearing of tropical forests, and regrowing of temperate zone forests—all with high agro- and ecological stakes. Studies with such a problem entry are still popular nowadays. For example, Hu et al. [30], in this Special Issue, attempt to optimize the cropland utilization patterns to increase cropland use efficiency and, thus, enhance sustainability in general sense. Similarly, the study by Jin et al. [31] in this Special Issue attempts to understand the determinants of the accessibility of urban green spaces, so as to improve their patterns and thus reduce the disparities in their multifaceted services.

This flow-thinking land-related approaches to addressing sustainability can be applied to two action domains in practice. The upstream domain is to control land conversions (e.g., urban encroachment of farmland), whereas the downstream domain is to come up with ways of regulating the land-related processes causing a sustainability concern (e.g., food security, carbon emission, economic output). Apparently, the former is way easier than the latter in real-world actions, as the latter gets decoupled from land per se to varying degrees. For example, China has developed a relatively mature land resource management system, featured by subsystems like spatial planning, top-down built-up land quota allocation, land conversion regulation, territorial spatial use inventory, as well as the supervision and enforcement of illegal land use [32,33]. However, the downstream governance is beyond the political scope of the land management department, causing the need for coordination and collaboration across multiple departments and even agents of the market and society. As is often the case, it is when emergencies of wide societal concern happen that such non-institutionalized, collaborative governance arrangements would occur, due to high transaction costs. In this sense, the downstream land-related sustainability governance would most likely be reactive, occasional, and focusing on either one or few ecosystem services. For instances, the "Economical and Intensive Construction Land Use Policy" has long been implemented in China to improve the economic efficiency of construction land [29], and a regime known as "Payment for Ecosystem Services" has been practiced in many jurisdictions to provide financial incentives to landowners, communities, or resource managers to sustainably manage ecosystems to provide targeted ecosystem services [34].

2.3. Problem Entry 3: As Land System or Social-Ecological System

A further broadened framing of land in the NRC (1999) [18], as relevant to sustainability, is the problem entry of land as land systems, either as places of cultural landscapes in the general sense, or as archetypes of social-ecological systems. This problem entry can be referred to as a social-ecological land system perspective, or a land system perspective for short. With an occurrence frequency of 2.74% (four counts) in the NRC (1999) [18], it appears to be the least applied sustainability entry regarding land, while, in fact, there is extensive content on degradation syndromes therein that can be interpreted as unsustainability archetypes of social-ecological systems, each with characteristic land uses (see the list of syndromes on page 287). Overall, most such studies adopt the general land system framing, treating a socio-ecological system by focusing on its dominant land use, e.g., urban land system [35] and agri-food system [36]. Contrastingly, archetypal land system studies have received little interest since the original work by the German Advisory Council on Global Change (WBGU) [37] in 1997, but seem to be gaining popularity in the recent ten years [38–40].

In practice, the general land system problem entry requires systematic thinking, and often also an integrated perspective on sustainability that explicitly accounts for the multidimensional tradeoffs and synergies of land-dependent ecosystem services. A case study of Australia shows that the land-sector alone is insufficient for achieving all SDGs, and must focus on priorities like food, energy, water, and biodiversity by actively collaborating with other sectors such as clean energy, food systems, and water resource management [41]. In this case, the operational challenges would be daunting, as the general land system approaches to sustainability necessitate institutionalizing a place-based, adaptive, collaborative land-related/spatial governance. Alternatively, archetypal land system approaches seem more promising in practice, as the epistemic complexity of land systems is reduced in one aspect or another with archetypes. In this regard, China has established several kinds of thematic pilot or demonstration zones (e.g., National Pilot Zone for Ecological Civilization [42] and Demonstration Zone for the Transformation and Upgrading of Resource-Based Cities [43]). These initiatives aim to provide the testing ground for new governance approaches and policy frameworks, with practical successes reported. Yet, regretfully, the efforts in researching generalizable insights from these zones that could inform broader land-related sustainability governance are lacking (c.f., Ostrom's design principles for governing the commons [44]). As the ultimate end of sustainability is sustaining and improving inclusive human wellbeing (with sustaining and improving the life-support system as the means) [45], the land system framing to address sustainability, be it general or archetypal, ultimately directs sustainability actions toward deep systemwide changes in the administrative, political, and legal structures.

2.4. Problem Entry 4: As Landscape Scale or Interacting Land Systems

The broadest framing of land as relevant to sustainability in NRC (1999) [18] is the problem entry of land as a landscape or a regional scale, which essentially means an interacting mosaic of heterogeneous social-ecological land systems. This problem entry, though not the least occurring (4.79%, seven counts), has not been systematically elaborated on in NRC (1999) [18] regarding how to operationalize it. It is likely due to the fact that with such a framing, land/landscape has only a nominal relevance to sustainability. Though this framing is mostly used by landscape ecologists [46–48], it actually involves many fields that may seem lacking substantial relevance to land, such as international relations, geopolitics, international trade, supply chain management, new economic geography, and political economy. As of today, the science and practice of upscaling from local (un)sustainability to regional (un)sustainability and to global sustainability remains arguably the most challenging. Studies with such a problem entry are either highly conceptual [49–51] or necessitate an epistemic simplicity by focusing on spatial tradeoffs and synergies of relatively limited social-ecological dimensions [52–54].

This upscaling framing of land into real-world sustainability contexts can spell out an increasing concern among the global community on the spatial displacement of land use and illegal land occupation in a telecoupled world, calling for better governance of coupled archetypal land systems (e.g., [55,56]). Alternatively, with implicit relation to land, such a framing can also translate into an issue of spatial collective action [57] or governing spatially coupled infrastructure [4], such as transboundary governance [58] and polycentric governance [59]. For example, China has achieved widespread success with its Accountability-to-Chief system that designates government officials as the person-in-charge for managing and protecting specific ecosystems like rivers, lakes, and forests within their jurisdictions (e.g., [60]). China has also shown success in using relational or polycentric governance approaches to coupling spatially unconnected social-ecological systems for paired assistance [61] or pan-regional cooperation [62]. Nonetheless, these practical approaches often fail to explicitly account for the social-cultural plurality critical when attempting to achieve the collective sustainability of inclusive human wellbeing.

After all, one man's food can be another man's poison; similarly, societies in different cultural landscapes can have strikingly divergent sustainability visions and preferences.

3. An Extended Leverage Points Perspective on Land-Related Approaches to Advancing Sustainability

Drawing upon the insights from the foundational report of NRC (1999) [18], this paper has uncovered the four major problem entries of land when addressing sustainability, each with its own historical, theoretical, and practical nuances. Although these problem entries have produced a wide range of land-related sustainability efforts as noted above, such efforts are notably fragmented due to disciplinary silos and conceptual plurality. Here, we argue that the leverage points perspective on sustainability [5,63] holds substantially underappreciated promise for integrating the conceptual plurality of various land-related sustainability framings—by which a more holistic and actionable agenda to harness the full potential of land-related approaches can be catalyzed (Figure 1).

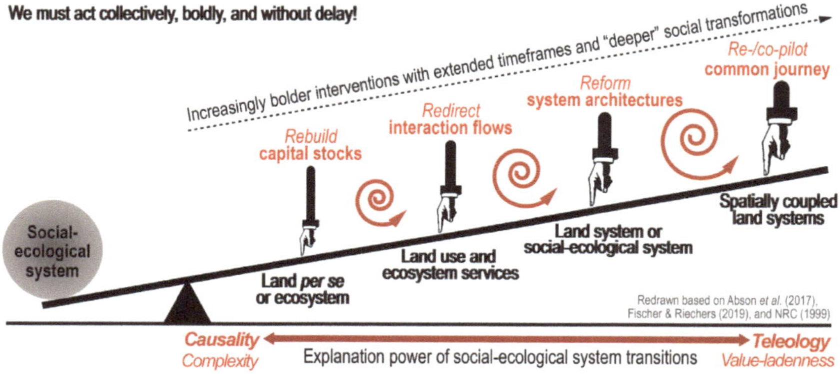

Figure 1. An extended leverage points perspective on sustainability, specifically applied to land-relevant contexts. The four problem entries of land, as identified from NRC (1999) [18], are interpreted as the four realms of leverage points following Abson et al. [63] and Fischer and Riechers [5]. The spirals represent interactions between the realms of leverage points. A key difference between the redrawn figure and the original lies in our extension of the deepest realm of leverage points. This extension aims to re-/co-pilot the common journey toward sustainability in coupled social-ecological systems. It encompasses not only the original focus on changing the mindsets and paradigms of the focal social-ecological systems but also extends to changing those of the intercoupled social-ecological systems when necessary. By fostering social consensus on common interests and coordinated interventions (i.e., shared future blueprints), local-scale (un)sustainability can potentially scale up to achieve regional-scale sustainability.

The original leverage points perspective on sustainability, drawing on an analogy to the principles of levers in physics, highlights that the points of sustainability interventions can be conceptualized as a continuum of four realms of levers along a leverage system, with the realm of levers positioned at the furthest end of the lever arm, exerting a stronger transformative influence when external pressure is applied, due to the extended leverage [5,63]. In increasing the order of transformative potential, the four realms of sustainability levers are system parameters, system feedbacks, system structures, and system intents (Box 2 in [63]), corresponding, respectively, to sustainability interventions involving altering materials, changing processes, designing institutions, and shaping values, goals, and worldviews (Figure 1 in [5]). In this regard, the four increasingly broadened problem entries of land in addressing sustainability align well with the four realms of leverage points for taking sustainability actions: from rebuilding social-ecological capital stocks (e.g., land itself or ecosystem), to redirecting social-ecological interaction flows (e.g., land use and ecosystem services), to reforming social-ecological system architectures (e.g., land system governance institutions), and to re-/co-piloting the common journey toward sustainability of intercoupled social-ecological systems—including those pericoupled and telecoupled (e.g., sustainability upscaling of spatially coupled land systems).

Notably, in the fourth realm we have extended the original focus on cultural and ideological efforts like changing worldviews, mindsets, and paradigms (e.g., via sustainability education) within a single system to include making such efforts to also influence its coupled systems, so as to re-/co-pilot the community of coupled destiny toward shared future blueprints. With this extension, the four realms of sustainability action are not only not mutually exclusive, but also complementary and indispensable to one another, leveraging different depths of social transformation across the varying timeframes of sustainability transitions. To some extent, our extended leverage points perspective complements the original, as we conceptualize the leverage points as a progression and scaling process from individual land patches (i.e., sustainability of land or the Earth subsystem) to land-related processes (i.e., sustainability of land functions or ecosystem services), then to social-ecological harmony of a single system (i.e., sustainability of human wellbeing at the local scale), and ultimately to the interactions among multiple systems (i.e., sustainability of human wellbeing at the regional and higher scales). As the road to global sustainability needs a hierarchical approach that balances weak and strong sustainability at different scales [64], only by deliberately integrating these four nested realms of (land-related) sustainability efforts can our generation hope to achieve systemic transformations in time.

Inspiringly, land and landscape scholars have shown growing ambitions in contributing to sustainability, applying increasingly broadening problem entries of land [11]. In this Special Issue particularly, Wang et al. [65] explicitly take a land system perspective on unpacking the social-ecological mechanisms underlying the rise and fall of rural specialty agriculture, where the roles of land have long been simplified as trivial. With a comparative review on smart growth and smart shrinkage, Yang et al. [66] contend that urban growth and shrinkage are spatially coupled, so that sustainable urbanization efforts necessitate the extension from considering a single city to the simultaneous treatment of both growing and shrinking areas. In a review on urbanization-associated farmland research in China, Yang, et al. [67] find the sustainability focus to gradually evolve from the protection of farmland loss and mitigating farmland degradation, to sustaining the farmland's food production potential, to improving the integrated sustainability of agri-farmland systems, and to governing the sustainability dynamics of coupled urban-rural systems. A similar tendency is observed in the review of urban fringe research in China, with the review authors arguing further for situating urban fringe as a land system type of its own kind for co-piloting the coupled urban-urban fringe-rural system toward sustainability [68]. Even more hearteningly, more scholars like Wang, et al. [69] start to appreciate the underrecognized value of the leverage points perspective on advancing sustainability. Looking ahead, integrating the four realms of land-related levers to advance sustainability will become increasingly recognized and practiced—to this end, a transdisciplinary rather than technocratic culture must first be cultivated [70].

4. Conclusions—Call to Action

Advancing sustainability through land-related approaches necessitates a paradigm shift: from fragmented, discipline-specific efforts to integrated, systems-level transformations. Such a shift is urgently needed to serve the contemporary global sustainability movement, and is achievable once society grasps "the big picture." Our paper contributes by illustrating this big picture through first identifying the four analytical entries of land into sustainability problems and then illuminating how the four entries align with the four nested leverage realms for taking holistic sustainability actions. In essence, our study provides a hierarchical roadmap for researchers, policymakers, and practitioners to act collectively and boldly. Researchers must transcend disciplinary silos and embrace transdisciplinary approaches to deepen insights into these leverage points and their interactions, while policymakers should adopt inclusive governance mechanisms that account for long-term transformations. Practitioners, meanwhile, must translate these insights into adaptive, collaborative solutions, scaling successful models and fostering innovation. Particularly for the upscaling of sustainability, top-down actions alone are insufficient; it is crucial to give due attention to the local variations when integrating multi-level land-related approaches.

The stakes are immense, but so are the opportunities. Land is not just a resource—it is the foundation for the World-Earth system's sustainability and a vital arena for taking transformative actions. For our generation to leave a legacy of sustainability for the future generations, we must harness the full potential of land-related approaches without delay. The time to transform our research, policy, and practice is not tomorrow, but today.

Supplementary Materials: The following supporting information can be downloaded at: https://www.mdpi.com/article/10.3390/land14040756/s1.

Author Contributions: Conceptualization, B.-B.Z.; methodology, J.L. and X.W.; software, X.W.; validation, B.-B.Z.; formal analysis, B.-B.Z.; investigation, B.-B.Z.; resources, B.-B.Z.; data curation, X.W. and J.L.; writing—original draft preparation, B.-B.Z.; writing—review and editing, B.-B.Z., J.L. and X.W.; visualization, X.W.; supervision, B.-B.Z.; project administration, B.-B.Z.; funding acquisition, B.-B.Z. All authors have read and agreed to the published version of the manuscript.

Funding: This research was jointly funded by the National Natural Science Foundation of China (No. 42301131) and the China Scholarship Council (No. 202406330022).

Data Availability Statement: The data for content analysis and grounded coding is provided online as Supplementary Materials (Table S1) associated with this paper.

Conflicts of Interest: The authors declare no conflicts of interest.

References

1. WCED. *Our Common Future: Report of World Commission on Environment and Development*; WCED: Cape Town, South Africa, 1987.
2. United Nations. *Transforming Our World: The 2030 Agenda for Sustainable Development*; UN: New York, NY, USA, 2015.
3. United Nations. *The Sustainable Development Goals Report 2024*; UN: New York, NY, USA, 2024.
4. Janssen, M.A.; Anderies, J.M. *Infrastructure for Sustainability*; Arizona State University: Tempe, AZ, USA, 2023.
5. Fischer, J.; Riechers, M. A leverage points perspective on sustainability. *People Nat.* **2019**, *1*, 115–120. [CrossRef]
6. Blaikie, P.; Brookfield, H. *Land Degradation and Society*; Routledge: London, UK, 1987.
7. Prinz, D.; Rauch, F. The Bamenda model. Development of a sustainable land-use system in the highlands of West Cameroon. *Agrofor. Syst.* **1987**, *5*, 463–474. [CrossRef]
8. Thayer, R.L. The experience of sustainable landscapes. *Landsc. J.* **1989**, *8*, 101–110. [CrossRef]
9. Forman, R.T. Ecologically sustainable landscapes: The role of spatial configuration. In *Changing Landscapes: An Ecological Perspective*; Zonneveld, I.S., Forman, R.T.T., Eds.; Springer: New York, NY, USA, 1990; pp. 261–278.
10. UN General Assembly. *Report of the United Nations Conference on the Human Environment, Stockholm, 5–16 June 1972*; UN: New York, NY, USA, 1973.
11. Zhou, B.-B.; Wu, J.; Anderies, J.M. Sustainable landscapes and landscape sustainability: A tale of two concepts. *Landsc. Urban Plan.* **2019**, *189*, 274–284. [CrossRef]
12. Xie, H.; Zhang, Y.; Zeng, X.; He, Y. Sustainable land use and management research: A scientometric review. *Landsc. Ecol.* **2020**, *35*, 2381–2411. [CrossRef]
13. Meyfroidt, P.; de Bremond, A.; Ryan, C.M.; Archer, E.; Aspinall, R.; Chhabra, A.; Camara, G.; Corbera, E.; DeFries, R.; Díaz, S.; et al. Ten facts about land systems for sustainability. *Proc. Natl. Acad. Sci. USA* **2022**, *119*, e2109217118. [CrossRef] [PubMed]
14. Turner II, B.L.; Lambin, E.F.; Reenberg, A. The emergence of land change science for global environmental change and sustainability. *Proc. Natl. Acad. Sci. USA* **2007**, *104*, 20666–20671.

15. Verburg, P.H.; Erb, K.-H.; Mertz, O.; Espindola, G. Land System Science: Between global challenges and local realities. *Curr. Opin. Environ. Sustain.* **2013**, *5*, 433–437. [CrossRef]
16. Wiens, J.A. Is landscape sustainability a useful concept in a changing world? *Landsc. Ecol.* **2013**, *28*, 1047–1052.
17. Wu, J. Landscape sustainability science: Ecosystem services and human well-being in changing landscapes. *Landsc. Ecol.* **2013**, *28*, 999–1023.
18. National Research Council. *Our Common Journey: A Transition Toward Sustainability*; National Academies Press: Washington, DC, USA, 1999.
19. Bernard, H.R.; Wutich, A.; Ryan, G.W. *Analyzing Qualitative Data: Systematic Approaches*, 2nd ed.; SAGE: London, UK, 2017.
20. Fang, X.; Zhou, B.-B.; Tu, X.; Ma, Q.; Wu, J. "What kind of a science is sustainability science?" An evidence-based reexamination. *Sustainability* **2018**, *10*, 1478. [CrossRef]
21. Baker, W.L. A review of models of landscape change. *Landsc. Ecol.* **1989**, *2*, 111–133.
22. Bren d'Amour, C.; Reitsma, F.; Baiocchi, G.; Barthel, S.; Güneralp, B.; Erb, K.-H.; Haberl, H.; Creutzig, F.; Seto, K.C. Future urban land expansion and implications for global croplands. *Proc. Natl. Acad. Sci. USA* **2017**, *114*, 8939–8944. [PubMed]
23. Mao, D.; Wang, Z.; Wu, J.; Wu, B.; Zeng, Y.; Song, K.; Yi, K.; Luo, L. China's wetlands loss to urban expansion. *Land Degrad. Dev.* **2018**, *29*, 2644–2657.
24. Ren, Q.; He, C.; Huang, Q.; Shi, P.; Zhang, D.; Güneralp, B. Impacts of urban expansion on natural habitats in global drylands. *Nat. Sustain.* **2022**, *5*, 869–878.
25. Hu, T.; Peng, J.; Liu, Y.; Wu, J.; Li, W.; Zhou, B.-B. Evidence of green space sparing to ecosystem service improvement in urban regions: A case study of China's Ecological Red Line policy. *J. Clean. Prod.* **2020**, *251*, 119678.
26. He, P.; Gao, J.; Zhang, W.; Rao, S.; Zou, C.; Du, J.; Liu, W. China integrating conservation areas into red lines for stricter and unified management. *Land Use Policy* **2018**, *71*, 245–248.
27. Gardner, T.A.; Von Hase, A.; Brownlie, S.; Ekstrom, J.M.M.; Pilgrim, J.D.; Savy, C.E.; Stephens, R.T.T.; Treweek, J.O.; Ussher, G.T.; Ward, G.; et al. Biodiversity offsets and the challenge of achieving no net loss. *Conserv. Biol.* **2013**, *27*, 1254–1264.
28. Bull, J.W.; Suttle, K.B.; Gordon, A.; Singh, N.J.; Milner-Gulland, E.J. Biodiversity offsets in theory and practice. *Oryx* **2013**, *47*, 369–380.
29. Liu, Y.; Fang, F.; Li, Y. Key issues of land use in China and implications for policy making. *Land Use Policy* **2014**, *40*, 6–12.
30. Hu, X.; Dong, C.; Zhang, Y. Impacts of cropland utilization patterns on the sustainable use efficiency of cropland based on the human–land perspective. *Land* **2024**, *13*, 863. [CrossRef]
31. Jin, Y.; He, R.; Hong, J.; Luo, D.; Xiong, G. Assessing the accessibility and equity of urban green spaces from supply and demand perspectives: A case study of a mountainous city in China. *Land* **2023**, *12*, 1793. [CrossRef]
32. Liu, Y.; Zhou, Y. Territory spatial planning and national governance system in China. *Land Use Policy* **2021**, *102*, 105288.
33. Zhou, B.-B.; Lv, L. Understanding the dynamics of farmland loss in a rapidly urbanizing region: A problem-driven, diagnostic approach to landscape sustainability. *Landsc. Ecol.* **2020**, *35*, 2471–2486. [CrossRef]
34. Xie, Z.; Zhou, B.-B.; Xu, H.; Zhang, L.; Wang, J. An agent-based sustainability perspective on Payment for Ecosystem Services: Analytical framework and empirical application. *Sustainability* **2021**, *13*, 253.
35. Stuhlmacher, M.; Andrade, R.; Turner II, B.L.; Frazier, A.; Li, W. Environmental outcomes of urban land system change: Comparing riparian design approaches in the Phoenix Metropolitan Area. *Land Use Policy* **2020**, *99*, 104615. [CrossRef]
36. Pei, X.; Zheng, X.; Wu, C. How part-time farming affects cultivated land use sustainability: Survey-based assessment in China. *Land* **2024**, *13*, 1242. [CrossRef]
37. German Advisory Council on Global Change (WBGU). *World in Transition: The Research Challenge*; Springer: Berlin/Heidelberg, Germany, 1997.
38. Banson, K.E.; Nguyen, N.C.; Bosch, O.J.H. Using system archetypes to identify drivers and barriers for sustainable agriculture in Africa: A case study in Ghana. *Syst. Res. Behav. Sci.* **2016**, *33*, 79–99. [CrossRef]
39. Levers, C.; Müller, D.; Erb, K.; Haberl, H.; Jepsen, M.R.; Metzger, M.J.; Meyfroidt, P.; Plieninger, T.; Plutzar, C.; Stürck, J.; et al. Archetypical patterns and trajectories of land systems in Europe. *Reg. Environ. Chang.* **2018**, *18*, 715–732.
40. Yang, Y.; Bao, W.; de Sherbinin, A. Mapping fine-resolution nested social-ecological system archetypes to reveal archetypical human-environmental interactions. *Landsc. Urban Plan.* **2023**, *239*, 104863. [CrossRef]
41. Gao, L.; Bryan, B.A. Finding pathways to national-scale land-sector sustainability. *Nature* **2017**, *544*, 217–222. [CrossRef] [PubMed]
42. Zhang, Y.; Fu, B. Impact of China's establishment of ecological civilization pilot zones on carbon dioxide emissions. *J. Environ. Manag.* **2023**, *325*, 116652. [CrossRef]
43. Li, B.; Tu, Y.; Liu, J. Impact of Industrial Transformation and Upgrading Demonstration Zones on carbon emissions intensity in resource-based cities. *J. Environ. Plan. Manag.* **2024**, *68*, 1037–1056. [CrossRef]
44. Ostrom, E. *Governing the Commons: The Evolution of Institutions for Collective Action*; Cambridge University Press: New York, NY, USA, 1990.

45. Pearson, A. Pursuing Sustainability: A Q&A on Connecting Science and Practice, Balancing Conservation with Use. Available online: https://news.harvard.edu/gazette/story/2016/04/pursuing-sustainability/ (accessed on 28 October 2024).
46. Turner, M.G. A landscape perspective on sustainability science. In *Toward a Science of Sustainability*; Levin, S., Clark, W.C., Eds.; Princeton University: Princeton, NJ, USA, 2010; pp. 79–82.
47. Wu, J. A landscape approach for sustainability science. In *Sustainability Science: The Emerging Paradigm and the Urban Environment*; Weinstein, M.P., Turner, R.E., Eds.; Springer: New York, NY, USA; Dordrecht, The Netherlands; Berlin/Heidelberg, Germany; London, UK, 2012; pp. 59–77.
48. Levin, S. Foreword. In *History of Landscape Ecology in the United States*; Barrett, G.W., Barrett, T.L., Wu, J., Eds.; Springer: New York, NY, USA, 2015; pp. 13–30.
49. Geels, F.W. The multi-level perspective on sustainability transitions: Responses to seven criticisms. *Environ. Innov. Soc. Transit.* **2011**, *1*, 24–40. [CrossRef]
50. Liu, J.; Herzberger, A.; Kapsar, K.; Carlson, A.K.; Connor, T. What is telecoupling? In *Telecoupling: Exploring Land-Use Change in a Globalised World*; Friis, C., Nielsen, J.Ø., Eds.; Springer International Publishing: Cham, Switzerland, 2019; pp. 19–48.
51. Tan, L.; Cui, Q.; Chen, L.; Wang, L. An exploratory study on spatial governance toward urban–rural integration: Theoretical analysis with case demonstration. *Land* **2024**, *13*, 2035. [CrossRef]
52. Krugman, P. Increasing returns, monopolistic competition, and international trade. *J. Int. Econ.* **1979**, *9*, 469–479. [CrossRef]
53. Meentemeyer, R.K.; Tang, W.; Dorning, M.A.; Vogler, J.B.; Cunniffe, N.J.; Shoemaker, D.A. FUTURES: Multilevel simulations of emerging urban–rural landscape structure using a stochastic patch-growing algorithm. *Ann. Assoc. Am. Geogr.* **2013**, *103*, 785–807. [CrossRef]
54. zu Ermgassen, E.K.H.J.; Godar, J.; Lathuillière, M.J.; Löfgren, P.; Gardner, T.; Vasconcelos, A.; Meyfroidt, P. The origin, supply chain, and deforestation risk of Brazil's beef exports. *Proc. Natl. Acad. Sci. USA* **2020**, *117*, 31770–31779. [CrossRef]
55. Carvalho, R.; Rausch, L.; Gibbs, H.K.; Bastos Lima, M.G.; Bernasconi, P.; Valdiones, A.P.; Vasconcelos, A.; Silgueiro, V. Illegal deforestation in Mato Grosso: How loopholes in implementing Brazil's forest code endanger the soy sector. *Land* **2024**, *13*, 1828. [CrossRef]
56. Meyfroidt, P.; Rudel, T.K.; Lambin, E.F. Forest transitions, trade, and the global displacement of land use. *Proc. Natl. Acad. Sci. USA* **2010**, *107*, 20917–20922. [CrossRef]
57. Poteete, A.R.; Janssen, M.A.; Ostrom, E. *Working Together: Collective Action, the Commons, and Multiple Methods in Practice*; Princeton University Press: New Jersey, NJ, USA, 2010.
58. Sattler, C.; Schröter, B. Collective action across boundaries: Collaborative network initiatives as boundary organizations to improve ecosystem services governance. *Ecosyst. Serv.* **2022**, *56*, 101452.
59. Oberlack, C.; Boillat, S.; Brönnimann, S.; Gerber, J.-D.; Heinimann, A.; Speranza, C.; Messerli, P.; Rist, S.; Wiesmann, U. Polycentric governance in telecoupled resource systems. *Ecol. Soc.* **2018**, *23*, 16.
60. Wang, B.; Wan, J.; Zhu, Y. River chief system: An institutional analysis to address watershed governance in China. *Water Policy* **2021**, *23*, 1435–1444.
61. Xu, H.; Xu, M.; Chen, W. Formation of relational poverty governance and its impacts: How Chinese local governments implement poverty alleviation programs. *J. Chin. Political Sci.* **2022**, *27*, 295–317.
62. Yeh, A.G.O.; Xu, J. *China's Pan-Pearl River Delta: Regional Cooperation and Development*; Hong Kong University Press: Hong Kong, China, 2010.
63. Abson, D.J.; Fischer, J.; Leventon, J.; Newig, J.; Schomerus, T.; Vilsmaier, U.; von Wehrden, H.; Abernethy, P.; Ives, C.D.; Jager, N.W.; et al. Leverage points for sustainability transformation. *Ambio* **2017**, *46*, 30–39. [PubMed]
64. Gan, X.; Fernandez, I.C.; Guo, J.; Wilson, M.; Zhao, Y.; Zhou, B.-B.; Wu, J. When to use what: Methods for weighting and aggregating sustainability indicators. *Ecol. Indic.* **2017**, *81*, 491–502.
65. Wang, X.; Wang, M.; Gong, L.; Yu, C. Understanding the rise and fall of rural specialty agriculture from social-ecological land system perspective: A longitudinal case study in China. *Land* **2025**, *14*, 254. [CrossRef]
66. Yang, Y.; Dong, Z.; Zhou, B.-B.; Liu, Y. Smart growth and smart shrinkage: A comparative review for advancing urban sustainability. *Land* **2024**, *13*, 660. [CrossRef]
67. Yang, Q.; Pu, L.; Huang, S. Review of urbanization-associated farmland research in China: A sustainability perspective. *Land* **2024**, *13*, 534. [CrossRef]
68. Li, Z.; Zhang, L.; Hu, T.; Wu, Y. History, progress, and prospects of urban fringe research in China: An idiosyncratic synthesis from a spatial perspective. *Land* **2025**, *14*, 248. [CrossRef]

69. Wang, J.; Mao, Z.; Zhang, Z. A leverage points perspective on China's governance of marine protected areas: Current state and ways forward. *Land* **2025**, *14*, 425. [CrossRef]
70. Lang, D.J.; Wiek, A.; Bergmann, M.; Stauffacher, M.; Martens, P.; Moll, P.; Swilling, M.; Thomas, C.J. Transdisciplinary research in sustainability science: Practice, principles, and challenges. *Sustain. Sci.* **2012**, *7*, 25–43.

land

MDPI

Article

Review of Urbanization-Associated Farmland Research in China: A Sustainability Perspective

Qiqi Yang [1,2], Lijie Pu [1,2,3,4,5,*] and Sihua Huang [3,4,5]

1 School of Geography and Ocean Science, Nanjing University, Nanjing 210023, China; yangqiqi@smail.nju.edu.cn
2 The Key Laboratory of the Coastal Zone Exploitation and Protection, Ministry of Natural Resources, Nanjing 210023, China
3 Nanjing Institute of Technology, School of Environmental Engineering, Nanjing 211167, China
4 NJIT Research Center, The Key Laboratory of Carbon Neutrality and Territory Optimization, Ministry of Natural Resources, Nanjing 211167, China
5 International Joint Laboratory of Green & Low Carbon Development, Nanjing 211167, China
* Correspondence: ljpu@nju.edu.cn

Abstract: Farmland loss in drastically urbanizing landscapes has long been a research concern for resource management, landscape planning, and spatial governance, especially in the context of China. In recent years, the issue of urbanization-associated farmland loss (UAFL) seems to be increasingly recognized as relevant to sustainability. To date, however, existing studies have not yet comprehensively addressed the research gap between UAFL and sustainability. Here, we aim to help fill this knowledge gap by considering UAFL research as an example of the broader land/landscape-related literature, in a hope of informing future studies to better advance sustainability through land-related approaches. Specifically, we combined bibliometric analyses with code-based content analysis to reveal the knowledge base, thematic evolution, and historiographic paths of the literature on UAFL across China and the empirical case studies' relevance to sustainability. Our main findings include: (1) the examined literature barely draws insights from sustainability science and sustainability only started to arise as a notable topic at around 2016; (2) over half of the empirical studies show awareness in advancing sustainability and interest in understanding the social-environmental drivers and processes underlying landscape dynamics, yet few demonstrate methodological transdisciplinarity; (3) those sustainability-relevant studies either frame UAFL as depletion of the farmland resource that may threat China's food security and consequently hinder sustainable urbanization or frame UAFL as part of widespread landscape dynamics that affect the environmental outcome(s) or social–environmental tradeoffs of landscape multi-functions; and (4) existing empirical studies are disproportionately focused on 1991–2006, national, regional, and city scales, and some of China's most developed areas. Our findings provide an overview of this specific research avenue on UAFL and, more importantly, point to the imperative for land/landscape scholars to break out of their disciplinary silos, especially in the natural sciences, to generate more actionable sustainability insights.

Keywords: bibliographic review; cropland loss; urban sustainability; land system science; landscape sustainability science; social–environmental system

1. Introduction

Urbanization is a long-term global trend with multi-faceted social–environmental impacts and profound sustainability implications [1–5]. Particularly, the phenomenon of drastic farmland loss along with China's dramatic urbanization since its opening up to the world system in 1978 has long been raising socio-political and scholarly concerns from home and abroad [6–8]. In fact, the issue of urbanization-associated farmland loss (UAFL) has been widely researched across the world, especially in places that have undergone rapid urbanization [9–11]. The case of China is perhaps the most prominent, given the pressure to feed China's world-leading population and potential impacts on the global food market [12]. Aside from the significance for China *per se*, studies of the Chinese case may provide insights for other places facing or to face the UAFL challenge [13]. In addition to the recognized practical significance of researching UAFL in China, Zhou and Lv [14] made an argument for the scientific significance of deciphering UAFL as a prototype of general sustainability issues—i.e., what/where to sustain versus what/where to develop—in advancing landscape sustainability [15–17] and sustainability science [18–20]. In this regard, one may wonder what sustainability insights the growing literature on UAFL has offered, if any. With such a big picture of the existing research landscape, there may be more fruitful discussions on how future UAFL studies can better contribute to land-related sustainability discourses.

To be fair, some two decades ago, founders of modern land-system science had synthesized the global patterns and causes of the few dominating land use/cover changes including farmland change [21–24]. More recently, Li and Li [25] reviewed existing studies about the patterns, causes, and consequences of global farmland abandonment as well as related policies. For farmland research particularly in the context of China, Wang and Li [26] conducted a systematic review of 169 case studies in 123 publications in the English and Chinese languages to synthesize the patterns, drivers, and change trajectories of farmland use intensification/deintensification. Also, there are studies synthesizing a plurality of datasets to reveal historical nationwide patterns of China's farmland change (e.g., [27,28]). These noted synthetic or review studies have made valuable contributions to depicting the general transition dynamics of the whole land system, including farmland change, or summarizing some specific aspects of farmland change such as abandonment and de/intensification. Yet, these studies tend to conceptualize farmland change as purely an issue of natural resource management instead of a lens for revealing the underlying human–environment architecture and dynamics [14,17]. In other words, the traditional natural resource perspective limits the potential of farmland research in providing broader sustainability implications [29,30].

In the above context, it is imperative to bridge the gap between UAFL and sustainability. Timely attempts should be made to synthesize the sustainability insights embedded in the substantial UAFL literature. This research aims to bridge the gap by conducting such a synthetic review of the existing literature about the UAFL issue that explicitly takes a sustainability perspective [31,32]—namely, examining farmland change in relation to urbanization, with farmland as part of what is desired to be sustained versus urban land as part of what is intended to be developed. Prior to the synthesis, an overarching question must be addressed first, i.e., to what extent has the UAFL issue been studied as a prototype of the more general sustainability issues? Following from this, it might be possible to synthesize what has been known about UAFL's empirical patterns, key social–environmental processes, and common sustainability consequences. Here, we present a quantitative–qualitative review [16,30] of the existing literature on UAFL in the context of China as well as a preliminary meta-analysis [25,26] of the empirical studies therein. On top of existing review methodologies, our study integrates the comprehensiveness of bibliometric reviews [16], the idiosyncratic depth of traditional expert-based reviews [33], and the empirical insights of meta-analysis reviews [34]. Furthermore, our focus is on China, given that UAFL has been most extensively researched in China and that land-related sustainability studies are gaining increasing momentum. The analyses contained in our paper will help clarify how future UAFL research can better advance land-related approaches toward local, regional, and global social–environmental sustainability. The remainder of this paper is organized as follows: Section 2 documents data and methodological details; Section 3 reports the respective findings of the review and case synthesis; Section 4 discusses implications for advancing land-related approaches to sustainability, with the literature on China's UALF considered as a case of broader land-related research; and Section 5 presents the concluding remarks.

2. Materials and Methods

2.1. Literature Search and Data Collection

The data collection for this study followed the Preferred Reporting Items for Systematic Reviews and Meta-Analyses (PRISMA) protocol [35] (Figure 1). Specifically, a literature search was conducted on 23 March 2021 using the Scopus Core Collection database (www.scopus.com). Scopus is more comprehensive than some other popular bibliographic databases, e.g., Web of Science Core Collection [36], and, thus, has been widely used. The first author searched for research articles published before 2021 that contained the terms of "urban*" AND "China" AND ("farmland" OR "cropland" OR "arable land" OR "cultivated land") in their titles, abstracts, or keywords, which resulted in 1496 items. The first author then did a follow-up screening of the titles and abstracts of the papers to exclude any study that did not focus on farmland in the context of urbanizing China, which resulted in 615 "relevant" papers. The bibliographic information of these papers was then exported and saved as our first database in this study for subsequent bibliographic analyses. Furthermore, the authors made their best efforts to download the full texts of the 615 papers (see Table S1 for bibliographic overview), which unfortunately included 9 papers with only abstracts available and even 3 papers missing their abstracts (see Table S2 for details). Next, the first author read the full texts of the 603 papers to select studies for qualitative coding of empirical cases that met the following four criteria: (1) categorized as empirical research; (2) focused on at least one of the three aspects of farmland, including spatiotemporal characteristics of farmland change, drivers of farmland change, and social–environmental consequences of farmland change; (3) employed quantitative analysis of any kind; and (4) researched an area at the city scale or above (n.b. a city in China corresponds to a county in the USA in terms of its administrative level). The selection resulted in 103 eligible papers as our second database for subsequent thematic coding of a range of variables for case synthesis.

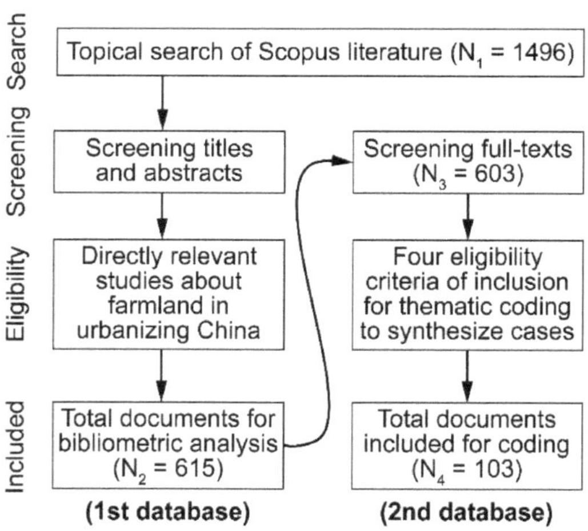

Figure 1. Flow diagram of data collection based on the Preferred Reporting Items for Systematic Reviews and Meta-Analyses (PRISMA) protocol. See text for the four criteria to filter eligible studies for thematic coding and case synthesis.

2.2. Bibliometric Analyses: Topic Mining, Historiography Mapping, and Beyond

To depict the big picture of the progress in researching farmland in the context of urbanizing China, we collected the bibliographic information of the 615 relevant papers for bibliometric analyses. First, we used the Bibliometrix R Package 3.2.1 (R version 4.2.0) [37] to gain an overview of the research landscape, with the 615 papers spanning from 1973 to 2020 (Table S1). Then, we deployed VOSviewer 1.6.13 [38] to mine the prominent research topics from the keywords, titles, and abstracts, respectively, as implemented in Zhou, Wu, and Anderies [16], with topics mined from the keywords interpreted as expressed topics and those from titles and abstracts as latent topics. Furthermore, we harnessed HistCite^{TM} 12.3.17 (by Clarivate, London, UK) [39] to map the historiography of the research field (i.e., the literature's development paths over time) based on the top 25 papers with the most citations from within the 615 sampled papers (i.e., local citations). The full texts of the 25 seminal papers were read carefully and in relation to each other, as so recommended in Zhou, Wu, and Anderies [16]. Lastly, we also utilized HistCite^{TM} to identify the 20 most-cited references of the 615 papers for careful and relational reading of their full texts to understand the knowledge base of the research field. Note that our applications of VOSviewer and HistCite^{TM} adopted default settings.

2.3. *Thematic Coding and Meta-Analysis of Case Studies*

To understand the scope and depth of empirical studies examining farmland change in the context of urbanizing China, with regard to sustainability, we applied thematic coding [32,40] to the 103 empirical studies. First was the proposal of the initial codebook, based on informal coding, discussion, and revisions. Specifically, based on the first and second authors' full-text reading of the 25 top-cited sampled papers and the 20 most-cited references, as noted above, they discussed and proposed an initial set of themes. Based on the pilot coding by the first author and feedback from all the other authors, the formally coded themes were limited to: (1) whether the study was oriented toward improving sustainability; (2) whether the study adopted a social–environmental systems (SES) perspective; (3) the methodology of the study, i.e., qualitative, quantitative, or mixed, as so advocated by the sustainability science literature; (4) study period; (5) geographical scale of the study; and (6) study area. Particularly, the theme of sustainability orientation was included to synthesize whether a study goes beyond UAFL *per se* to further address UAFL's social-environmental impacts, and the theme of SES perspective to synthesize whether a study delves deeper into the social–environmental processes or mechanisms that underlie various types of UAFL. Zhou, Wu, and Anderies [16] distinguish three types of sustainability interpretations, namely, the sustainability of natural capital (i.e., landscape *per se*), sustainability of ecosystem services (i.e., landscape-based functions or services), and sustainability of human wellbeing (i.e., landscape-based social–environmental systems). Here, those focusing on sustaining farmland *per se* were not coded as sustainability-oriented. Moreover, according to the synthesis by Zhou, Wu, and Anderies [32], sustainability science is characterized by the social–environmental systems perspective and the transdisciplinary perspective. Here, we code the transdisciplinary perspective as a qualitative–quantitative mixed methodological orientation, which is less controversial. While coding, the first author also noted down some of the 103 studies' data sources, research methods, and empirical findings on the spatiotemporal characteristics, drivers, and social–environmental consequences of farmland change, although these additional codings are not presented as part of the results.

3. Results

3.1. *Knowledge Base, Thematic Evolution, and the Literature Development Paths*

The knowledge base of the 615 sampled studies on urbanization-associated farmland in China can be inferred from their references, of which we analyzed the top-cited 20 (Table 1). These key references were published mainly in *Land Use Policy* (45%), followed by general journals (i.e., *Science*, *Nature*, and *Proceedings of the National Academy of Sciences of the United States of America*; 20%), with the rest published in journals related to the environment or geography (e.g., *Remote Sensing of Environment*, *Catena*, and *Journal of Geographical Science*; 35%). Full-text reading of these references showed that the majority were relevant empirical studies, with Tan et al. [41] and Deng et al. [42] directly addressing UAFL in China and some others focusing on either farmland [43–47], urbanization [48,49], or land use/cover change in general [50–56]. Among the remaining four that are not so closely related to

UAFL in China, Lambin et al. [21] and Foley et al. [24] provide theoretical insights about land use change trends and transitions, Grimm et al. [57] introduce the ecological basis of urban sustainability in particular in an era of global change, and the ecosystem services evaluation method proposed in Costanza et al. [58] has been widely used in assessing the impacts of land use/cover change. Overall, these references seem to confirm our tentative observation in the Introduction that the existing studies of UAFL in China mostly take a natural resource management perspective, lacking a sustainability perspective to treat farmland loss and urban expansion as two intertwined "landscape signals" of their underlying social–environmental architecture.

Table 1. The 20 top-cited references of the 615 sampled papers on UAFL in China.

Publication	Title	Source
Tan, M.H. et al. (2005) [41]	Urban land expansion and arable land loss in China—a case study of the Beijing–Tianjin–Hebei region	*Land Use Policy*
Lichtenberg & Ding (2008) [46]	Assessing farmland protection policy in China	*Land Use Policy*
Foley et al. (2005) [24]	Global consequences of land use	*Science*
Liu, J.Y. et al. (2005) [44]	Spatial and temporal patterns of China's cropland during 1990–2000: An analysis based on Landsat TM data	*Remote Sensing of Environment*
Liu, J.Y. et al. (2003) [50]	Study on spatial pattern of land-use change in China during 1995–2000	*Science in China Series D-Earth Sciences*
Costanza et al. (1997) [58]	The value of the world's ecosystem services and natural capital	*Nature*
Liu, Y.S. et al. (2014) [56]	Key issues of land use in China and implications for policy making	*Land Use Policy*
Chen (2007) [48]	Rapid urbanization in China: a real challenge to soil protection and food security	*Catena*
Long et al. (2009) [47]	Spatio-temporal dynamic patterns of farmland and rural settlements in the Su–Xi–Chang region: Implications for building a new countryside in coastal China	*Land Use Policy*
Liu, J.Y. et al. (2014) [55]	Spatiotemporal characteristics, patterns, and causes of land-use changes in China since the late 1980s	*Journal of Geographical Sciences*
Long et al. (2012) [54]	Accelerated restructuring in rural China fueled by 'increasing vs. decreasing balance' land-use policy for dealing with hollowed villages	*Land Use Policy*
Liu, J.Y. et al. (2010) [53]	Spatial patterns and driving forces of land use change in China during the early 21st century	*Journal of Geographical Sciences*
Deng, X.Z. et al. (2015) [42]	Impact of urbanization on cultivated land changes in China	*Land Use Policy*
Yang & Li (2000) [43]	Cultivated land and food supply in China	*Land Use Policy*
Seto et al. (2012) [49]	Global forecasts of urban expansion to 2030 and direct impacts on biodiversity and carbon pools	*Proceedings of the National Academy of Sciences of the United States of America*
Long et al. (2007) [52]	Socio-economic driving forces of land-use change in Kunshan, the Yangtze River Delta Economic Area of China	*Journal of Environmental Management*
Deng, X.Z. et al. (2006) [45]	Cultivated land conversion and potential agricultural productivity in China	*Land Use Policy*
Grimm et al. (2008) [57]	Global change and the ecology of cities	*Science*
Lambin et al. (2001) [21]	The causes of land-use and land-cover change: Moving beyond the myths	*Global Environmental Change*
Lin & Ho (2003) [51]	China's land resources and land-use change: Insights from the 1996 land survey	*Land Use Policy*

To provide more direct evidence for the lack of a sustainability perspective in studies about China's UAFL, text mining was conducted to reveal the prominent topics of the 615 sampled papers. The expressed topics that the papers claimed to cover (i.e., *keywords*) fall mainly into four clusters (Figure 2A): (1) the red cluster on urban planning, development, and growth considering farmland for sustainability; (2) the blue cluster on techniques and tools for monitoring land use/cover change; (3) the green cluster on environmental protection and natural resource conservation in the face of economic development and climate change; and (4) the yellow cluster on the ecological, economic, and social impacts of landscape dynamics and related ecosystem services tradeoffs for sustainable development. The latent topics that the papers actually address (in their *titles* and *abstracts*) also cluster into four groups (Figure 2B): (1) the red cluster on farmland and construction land research in the context of rural transition and sustainability; (2) the yellow cluster on conversions between multiple land uses including farmland; (3) the blue cluster on the impacts of urbanization; and (4) the green cluster on place-based and scenario-based studies of urban agglomerations and their impacts such as on ecosystem services value, ecological space, and carbon storage. The two topical landscapes seem to reaffirm the above-noted observation—more specifically, the existing studies relevant to UAFL in China mostly address land use/cover change in general, sustainable landscape urbanization considering multi-dimensional impacts, or farmland use for food security and rural sustainability. Furthermore, the thematic evolution of the topics suggests that sustainability is still a nascent topic within UAFL research in China, which emerged as only mildly noticeable around 2016 (Figure 2C,D).

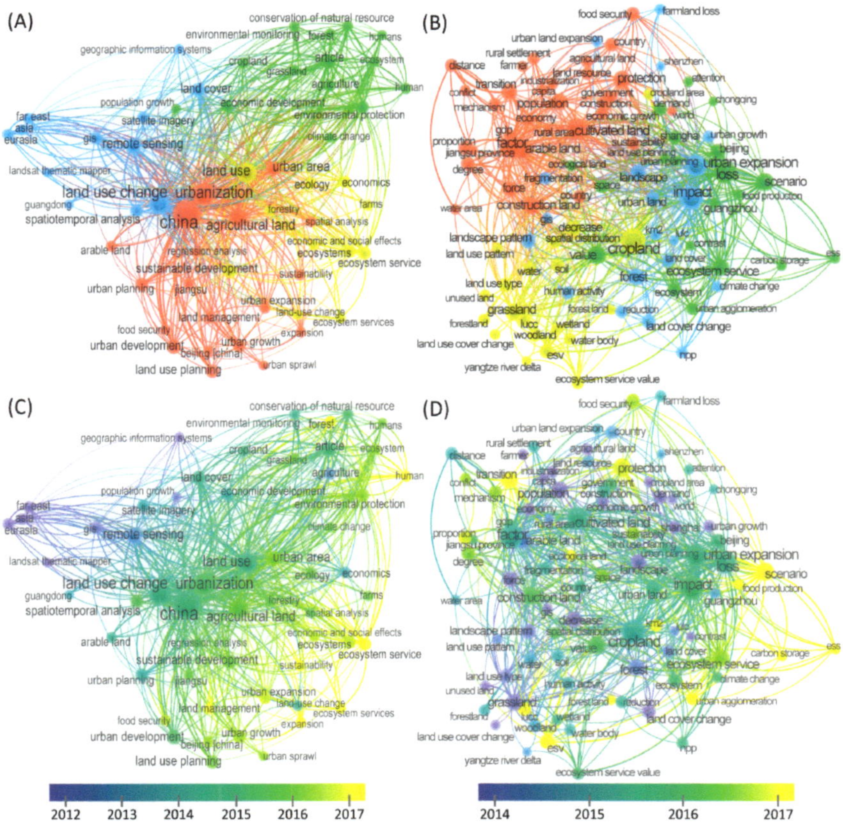

Figure 2. Thematic clustering of the topics identified from the *keywords* (**A**), *titles*, and *abstracts* (**B**), as well as their temporal evolution (**C**,**D**). The node size of a phrase is proportionate to its relative frequency. The analyses adopted VOSviewer's default settings; the association strength method was used for normalizing the strength of the links between nodes.

To provide further and more in-depth insights into how UAFL research in China has evolved, a historiograph of the literature development paths was mapped out based on key citation linkages among the 615 sampled papers. The closely intertwined topology of the historiograph suggests the existence of a cohesive research community studying UAFL in China, which has come into formation especially since around 2005 (Figure 3). The future-oriented simulation study of China's land use/cover change by Verburg et al. [59] was perhaps among the earliest to raise scholarly concerns over China's farmland loss and reduction in production capacity along with urbanization (and desertification and afforestation) as well as the quality of lost farmland. These concerns have been intensively researched by subsequent studies of UAFL in China, e.g., the area and quality of lost farmland, urban uptake of farmland, urbanization impacts on farmland change, and food production capacity reduction due to farmland loss.

The majority of the seminal papers after Verburg, Veldkamp, and Fresco [59] are those empirically examining land use/cover change in China or its different regions often based on either remote sensing or official survey data, and have reported widespread farmland loss and urban expansion [44,47,51,60–68]. Yet, based on official survey data of national land use, Lin and Ho [51] found that China's farmland loss since the 1990s was due mainly to agricultural restructuring and, following that, non-agricultural developments (Tables 7–9 therein). The view that urban expansion is mainly to blame for farmland loss was revisited by Liu et al. [69] based on the same survey data. They further revealed that, of the farmland lost to non-agricultural developments, the primary consumer was China's zeal for developing economic development zones, followed by transportation and rural settlements, while urban expansion (i.e., cities and towns) contributed the least. Relatedly, Tan et al. (2005b) [41] studied the urbanization side in more depth, and investigated the relationship between urban uptake of farmland and the administrative level of an urban region, based on remote sensing data. They found that about 74% of the newly urbanized land across the Beijing–Tianjin–Hebei region was converted from farmland, with the ratio tending to be higher for small cities. In this regard, Deng, Huang, Rozelle, Zhang, and Li [42] employed econometrics to model the potentially different impacts of urbanization modes (city, town, versus village modes) on farmland change, and concluded that the impact of urbanization was marginal and varied with the urbanization mode. Also using econometrics, Jiang et al. [70] directly modeled how the area of farmland converted for urban development was influenced by multi-level socioeconomic and policy factors, and demonstrated the dominant role of local factors and the unexpected role of agricultural investment in driving urban uptake of farmland. Further, Jiang et al. [71] modeled urbanization's impact on farmland use intensity (with the multi-cropping index as a proxy) and showed that farmland scarcity, agricultural investment, and land suitability have positive impacts on farmland intensification while urban uptake of farmland has a small and negative impact (Table 2 therein). Insightful as they are, the above-noted studies focus primarily on the "tree" *per se* instead of the "forest"—UAFL has been studied mainly for better farmland protection or smarter urban growth, lacking interest in UAFL's broader implications for social–environmental sustainability—in line with the preceding bibliographic analyses (Table 1, Figure 2).

There are a few UAFL papers out of the 25 mapped that contribute to sustainability, and these fall into two strands. The more traditional strand of papers frames UAFL as an issue with significance for food security and as a sustainability challenge. With proxies such as net primary productivity for farmland food production potential, Xu et al. [72], Yan et al. [73], and He et al. [74] assessed the impact of reduction in food production potential due to farmland loss caused by urban expansion or land use/cover change in general. A related earlier study by Chen [48] discussed how farmland loss affects soil erosion and food security. The other growing strand of papers frames UAFL as an issue significant for social–environmental sustainability, often in a rural context. Long, Liu, Wu, and Dong [47] linked the land use/cover change in the Su–Xi–Chang area with China's national strategy of "Building a New Countryside" and discussed the implications of farmland and rural settlement changes for rural sustainability, arguing for rural–urban integrated development. Liu, Wang, and Long [65] made a similar case for the farmland and rural settlement changes in southern Jiangsu Province in terms of their effects on rural sustainability. More recently, by using ecosystem services value as a conceptual instrument for assessing social–environmental tradeoffs, Long, Liu, Hou, Li, and Li [67] examined the implications of urbanization-associated land use/cover change for sustainability. Li et al. [75] explicitly studied farmland change as a consequence of rural–urban system dynamics, extending the focus onto sustainability of the coupled rural–urban systems. Subsequently, Liu et al. [76] used the loss of rural land use/cover types (i.e., farmland and rural settlement) as a proxy for studying non-agriculturalization in relation to urbanization and discussed the implications for coupled rural–urban sustainability.

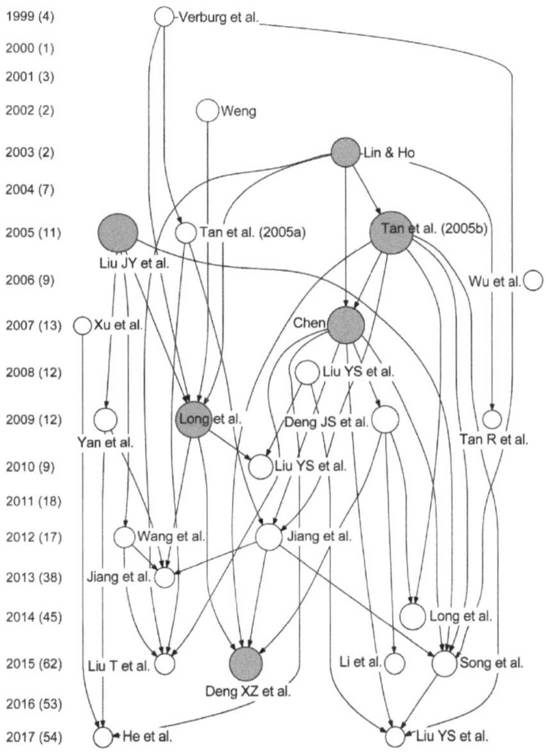

Figure 3. Main paths of the development of UAFL research literature in China, based on citation linkages among the top 25 papers with the most local citations (from within the sampled 615 papers) (see Table S3 for bibliographic details). The six seminal papers shaded gray are among the top 20 cited references within the 615 papers (detailed in Table 1), suggesting their relatively larger influences on subsequent research about UAFL in China. Note that the two seminal papers published in 2005 by Tan and his colleagues is distinguished in the main text by Tan et al. (2005a) [61] and Tan et al. (2005) [41].

Full-text reading of the mapped seminal papers reaffirmed that sustainability studies of UAFL in China are still marginal relative the UAFL literature, yet such research seems to be on its way into the mainstream. In this vein, despite policymaking having often been a central topic in UAFL studies, further efforts are needed for more rigorous policy studies that could delve into the social–environmental architectures beneath the landscape change of UAFL—such as the one by Tan et al. [77] on the governance structures of managing farmland conversion in China, Germany, and the Netherlands, which appear to be less connected with and recognized by the mainstream UAFL literature (Figure 3).

3.2. Meta-Analysis of Empirical Case Studies

Out of the 103 empirical studies about China's UAFL, 50.49% are sustainability-oriented (Figure 4A), with the other half focusing on landscape change *per se*, including farmland, urban land, or land use/cover in general. Note that the 50.49% excludes studies oriented toward farmland protection, which is an administrative task in China. In fact, as revealed above in Figure 3, those sustainability-oriented empirical studies address mainly one or more aspects of UAFL's social–environmental impacts, with only a few exceptions assessing UAFL's overall sustainability impact by integrating multidimensional synergies and tradeoffs (e.g., [67]). There is no single study that assesses empirically UAFL's impact on the sustainability of farmers' or households' human wellbeing. Furthermore, regarding the SES perspective as advanced by sustainability science, 59.22% of the coded studies explicitly adopt this stance. Most of these studies are like the conventional research that investigates the drivers underlying one or more specific land use/cover changes, with only a few exceptions focusing on social–environmental processes and their sustainability implications instead of merely landscape changes (e.g., [76]). In addition, as sustainability science advocates for transdisciplinarity, which translates into mixed methodological approaches to researching sustainability issues (e.g., [78]) including UAFL, the methodological aspect was also coded in the present study, highlighting that only 8.74% of the quantitative case studies were also mixed with qualitative methods. These few exceptions use mostly deductive reasoning to assist their quantitative methods (e.g., [63,75,79,80]), insufficient for taking advantage of the broad ranges of qualitative methods (e.g., participatory observation, focus group, interview) for researching social processes and developing theories.

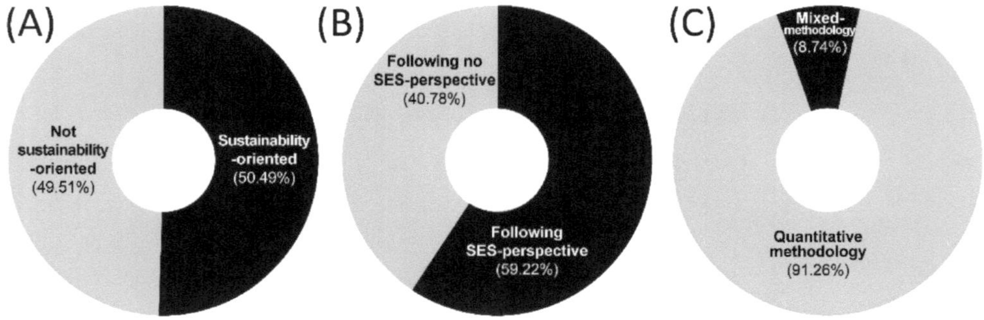

Figure 4. Sustainability features of UAFL case studies in China (N = 103): (**A**) oriented toward addressing social–environmental sustainability impacts or not; (**B**) delving into underlying social–environmental processes or not; and (**C**) adopting qualitative, quantitative, or mixed research methodology.

Additionally, the empirical contexts of the 103 cases were also coded. These empirical studies covered mostly the period from 1991 to 2006, of which the years 1992, 1995, 1997, 2001, 2002 were less covered; China's UAFL prior to 1991 and after 2006 have also been insufficiently studied (Figure 5A). Furthermore, 39.81% of the studies were conducted at the above-city regional scale, 33.01% at the national scale, and 22.33% focus on a city (c.f., a county in the United States), while 4.85% examine the area of a district (c.f., a city in the United States) or even smaller. The dominance of large-scale studies hinders the employment of qualitative methods usually more applicable at the local scale. Lastly, in terms of study area, these studies cover mostly China, including the Three Gorges Reservoir area as well as the relative more developed areas such as Beijing, the Beijing–Tianjin–Hebei region, Jiangsu, Huang–Huai–Hai Plain, Pearl River Delta, Guangzhou, and Nanjing. The spatial concentration of the study areas is insufficient to reveal the full spectrum of the social–environmental complexities related to UAFL.

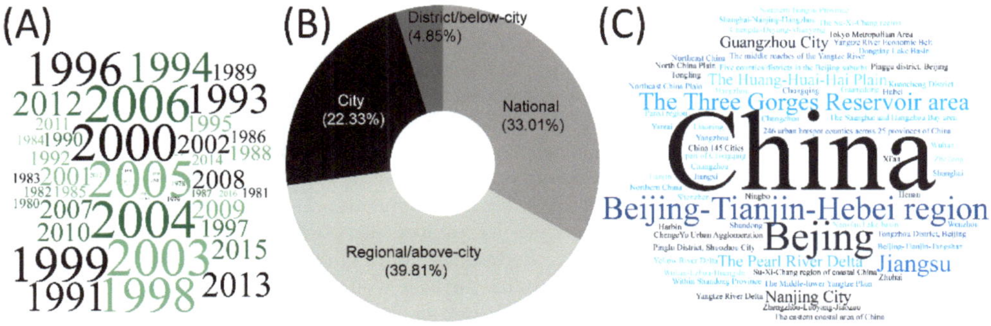

Figure 5. Empirical focuses of UAFL case studies in China (N = 103): (**A**) study period; (**B**) geographic scale; and (**C**) study area. Word size in panels A and C are proportionate to its relative frequency.

4. Discussion

Sustainability is the theme of our era, and our common journey toward sustainability calls for the promotion of sustainability thinking among scholars, policymakers, and other stakeholders alike. Sustainability science characterizes sustainability thinking through the ontological lens of complex social–environmental systems and the transdisciplinary epistemic perspective [16,18,32]. In our research context, sustainability thinking calls for understanding and addressing urbanization-associated farmland loss (UAFL) as an emergent "landscape syndrome" of coupled social–environmental systems based on transdisciplinary approaches. Regretfully, our findings reveal that such sustainability thinking is still marginal in the existing literature about China's UAFL (Table 1, Figures 2 and 3)—though it is on its way into the mainstream (Figures 2 and 3). On the other hand, our findings show that over half of the related empirical studies are indeed sustainability-oriented and explicitly follow the SES perspective that characterizes sustainability science (Figure 4A,B). Here, a contradiction seems to arise. One possibility is that, although there do exist considerable scholarly interests in researching UAFL in China for advancing sustainability, the sustainability insights of such studies remain underappreciated. Alternatively, one may speculate that existing studies have not yet provided sufficient significant sustainability insights. We believe that both aspects have an impact and that they may even be the two sides of the same coin.

The issue of UAFL was originally examined in the field of natural resource management [9,81] and blossomed as part of the land use/cover change research program in the 1990s thanks to the fast development of remote sensing techniques [82]. This research program later developed into a so-called land change science that aims to link pixels to people by going beyond land use/cover change *per se* to understand the underlying drivers and (social–environmental) processes [83,84]. With sustainability science exerting increasing inter-and trans-disciplinary influence since the early 2000s, science-oriented land change research has gradually developed into a use-inspired basic science that is nowadays often called land system science, emphasizing its unique contribution to sustainability [85,86]. Turner II et al. [87] elaborated on this development history of land system science and its link to sustainability. On the other hand, it was not until early 2010s that sustainability science was systematically introduced to China [88,89]. Thus, it is unsurprising that recent studies by Chinese scholars have been more inclined to frame UAFL in relation to farmland sustainability [48,61], rural sustainability [47,65], sustainability of coupled rural–urban systems [75,76], and the sustainability of core–periphery systems [12,14]. In this vein, it is reasonable to expect that future studies of UAFL and other significant land use/cover changes would gain more and more research momentum that explicitly takes a sustainability perspective, thus providing more sustainability insights.

Relatedly, it is also unsurprising that sustainability might have been abused as a buzzword in the existing literature on China's UAFL—thus failing to fulfill the potential of generating actionable sustainability insights. In short, the mainstream and latest developments of sustainability science were selectively introduced to Chinese geographers, ecologists, and other land/landscape-related research communities mainly through another research strand coined as landscape sustainability science [15,90]. Roy Chowdhury and Turner II [91] elaborated on the paralleled and increasingly confluent developments of land system science and landscape sustainability science. In general, however, it is fair to say that the so-called landscape sustainability science is mostly based on landscape ecology (the intersection of geography and ecology [92,93]) and highly science-oriented (or ecology-biased), less useful for integrating with social sciences (especially governance and policy sciences) for generating actionable knowledge (see Figure 5 in [16]). This background can also explain why land/spatial governance has not been well integrated into China's UAFL research (Figure 3) and why mixed methodology is so scarcely seen in the empirical studies (Figure 4C). In contrast to landscape sustainability science, land system science is explicit in its use-inspired, basic research nature; for example, Turner II et al. [94], Verburg et al. [95] and Munroe and Müller [96] explicitly advocate for the idea of land system architecture (and governance). In fact, land system architecture is very similar to the idea and practice of (comprehensive) land consolidation in China [30,34], although the land-related Chinese scholars have not yet widely realized this similarity. In this regard, it is reasonable to propose that for future studies of UAFL to provide more actionable sustainability insights, it is imperative to break down the epistemic biases and interest barriers of any single discipline and advance real inter-and trans-disciplinary studies on UAFL and other landscape changes alike. Only by so doing might we see the hope of deciphering the black box of social and social–environmental architectures beneath (un)sustainable "landscape signals".

This paper was originally designed also to synthesize UAFL's landscape patterns, key social–environmental processes, and common sustainability consequences that were commonly reported in the 103 empirical cases. However, the reported findings are highly context-dependent, thus it is challenging to conduct comparable coding. Our preliminary coding shows that over half of the studies reported net farmland loss and that less than a quarter reported fluctuation, with two cases reporting farmland increase and others reporting no such information. The drivers of farmland change examined or discussed encompass a wide range of factors, such as agricultural restructuring, natural hazards, construction, urbanization, economic growth, population increase, afforestation, and reforestation. In general, these findings are in line with the proposals made by Zhou, Aggarwal, Wu, and Lv [12] that frame UAFL as an emergent and complex "landscape syndrome" [14,16] of coupled core–periphery systems [97], of which the coupled rural–urban system is an archetype. Regarding the common sustainability consequences of UAFL in China, existing studies refer to certain farmland functions such as food production, carbon storage, and land pollution, as well as using ecosystem services value to evaluate multifunctional tradeoffs toward social–environmental sustainability. However, such tradeoff-based sustainability studies of UAFL are still limited. Fortunately, recent publications have increasingly taken this more comprehensive sustainability perspective, as has been widely recognized in sustainability science, see Fan et al. [98], Han et al. [99], Zhang et al. [100], Li et al. [101], and Zou et al. [102] for examples.

Together, our findings show that the existing literature in English on China's UAFL demonstrates strong sustainability awareness yet is still insufficient in providing authentic actionable sustainability insights, due to its long research tradition of a natural resource management perspective on UAFL and the associated methodological dominance of related natural sciences like remote sensing, land change science, and landscape ecology. Looking into the future, it is urgently necessary to conduct more conceptual and theoretical studies that explicitly frame UAFL and other complex landscape syndromes as emergent properties of dynamic social–environmental landscapes/systems. Particularly, governance, the largely missing piece in the existing literature, needs to be explicitly incorporated into the framing of social–environmental systems [103–106]. Future studies can go one step further than Tan, Beckmann, van den Berg, and Qu [77] (Figure 3) to examine the common governance structures of social–environmental systems and those peculiar governance architectures causing specific challenges to sustainability, like UAFL. Furthermore, methodological plurality needs to be embraced by sustainability scholars, especially those social science methods that could engage non-academic stakeholders such as ethnographic observation, field interviews, focus groups, and case studies [40,78]. In this regard, promoting the idea and practice of transdisciplinarity [107–109] among Chinese land-related sustainability scholars remain daunting challenges in the foreseeable future. Lastly, our meta-analysis points to understudied issues that future empirical studies on China's UAFL can further explore, such as periods prior to 1991 and after 2006, below the city scale and at local scales, and in less developed and urbanized areas. Nonetheless, it should be noted that the identified gaps are limited to the literature in English before 2021. More recent studies may well align with the directions we propose above. Also, relevant studies published in other languages such as Chinese may present a more encouraging research landscape than what we have found. Our findings should be interpreted with caveats.

5. Concluding Remarks

Advancing sustainability through land-related approaches has been and will continue to be a fruitful research mission that has its unique theoretical and practical niches. Yet, for such research to generate more actionable sustainability insights, it is time to move beyond the narrow focus on land *per se* as a natural resource toward, on the one hand, seeing land as a landscape "X-ray" for diagnosing SESs, and on the other hand, a land-based governance instrument for intervening in SESs. In this paper, we applied such a sustainability perspective to conducting a bibliometric analysis of 615 sample papers concerning the widespread sustainability-related UAFL issue and a preliminary meta-analysis of the 103 empirical case studies therein. Our study bridges the gap between UAFL and sustainability, which has not been comprehensively addressed in the existing literature. We provide an overview of the existing literature in English involving China's UAFL, clarify the status quo of embedded sustainability thinking, and identify future research directions for better contributing to land-related approaches toward social–environmental sustainability in the context of China.

Our bibliometric analysis shows that the examined studies barely refer to the core sustainability science literature and that it was not until around 2016 that sustainability started to arise as a notable topic. The results also suggest that those sustainability-relevant studies are framed in one of two ways: with farmland loss as a threat to China's food security and consequently a challenge to sustainable urbanization; or with UAFL as part of featured modern landscape dynamics affecting environmental outcome(s) or social–environmental tradeoffs of a multi-functional landscape. The meta-analysis indicates that, though lacking sustainability insights and methodological transdisciplinarity, over half of the empirical studies demonstrate awareness in advancing sustainability and interest in understanding the social–environmental drivers and processes underlying landscape dynamics. The analysis also highlights understudied time periods, areas, and scales that remain potentially fruitful for conducting theory-driven empirical studies, especially those capturing the complexity of urbanization and farmland change. With the examined literature on China's UAFL as representative of the existing land/landscape-related literature, our more general conclusion is that sustainability thinking is still marginal in these traditional fields, yet it has been attracting growing interests and is on its way into the mainstream. We argue that a major barrier to advancing sustainability through land-related approaches is the epistemic and methodological superiority held (implicitly or explicitly, intentionally or unintentionally) by some natural sciences like landscape ecology and perhaps also remote sensing, that have long contributed to researching land/landscape. It is necessary to further utilize the lens and transdisciplinary perspective of social–environmental systems in future studies, for which better understanding the socio-political subsystem and integrating social science methods are imperative. In this vein, the two increasingly integrated fields of land system science and landscape sustainability science still have a long way to go.

Supplementary Materials: The following supporting information can be downloaded at: https://www.mdpi.com/article/10.3390/land13040534/s1, Table S1: Overview of the 615 examined papers on China's UAFL; Table S2: Twelve papers excluded from scanning due to unavailable full texts; and Table S3: Bibliography of the top 25 papers with the most local citations from among the sampled 615 papers.

Author Contributions: Conceptualization, methodology, data curation, formal analysis, visualization, writing—original draft, writing—review and editing, Q.Y.; conceptualization, supervision, project management, funding acquisition, writing—review and editing, L.P.; software, validation, visualization, S.H. All authors have read and agreed to the published version of the manuscript.

Funding: This research was financially supported by the National Natural Science Foundation of China (No. 42171245) and marine science and technology innovation special projects of Jiangsu Province, China (No. JSZRHYKJ202212).

Data Availability Statement: The raw data of the paper is available upon reasonable request to the corresponding author.

Acknowledgments: We thank the guest editor Bing-Bing Zhou for constructive comments during the preparation and revision of the manuscript.

Conflicts of Interest: The authors declare no conflicts of interest.

References

1. Turner, B.L., II; Clark, W.C.; Kates, R.W.; Richards, J.F.; Mathews, J.T.; Meyer, W.B. *The Earth as Transformed by Human Action: Global and Regional Changes in the Biosphere over the Past 300 Years*; Cambridge University Press: New York, NY, USA, 1990.
2. Kundu, D.; Pandey, A.K. World urbanisation: Trends and patterns. In *Developing National Urban Policies: Ways Forward to Green and Smart Cities*; Kundu, D., Sietchiping, R., Kinyanjui, M., Eds.; Springer: Singapore, 2020; pp. 13–49.
3. Kates, R.W.; Parris, T.M. Long-term trends and a sustainability transition. *Proc. Natl. Acad. Sci. USA* **2003**, *100*, 8062–8067. [CrossRef] [PubMed]
4. Gao, J.; O'Neill, B.C. Mapping global urban land for the 21st century with data-driven simulations and Shared Socioeconomic Pathways. *Nat. Commun.* **2020**, *11*, 2302. [CrossRef] [PubMed]
5. Zhong, C.; Guo, H.; Swan, I.; Gao, P.; Yao, Q.; Li, H. Evaluating trends, profits, and risks of global cities in recent urban expansion for advancing sustainable development. *Habitat Int.* **2023**, *138*, 102869. [CrossRef]
6. Seto, K.C.; Kaufmann, R.K.; Woodcock, C.E. Landsat reveals China's farmland reserves, but they're vanishing fast. *Nature* **2000**, *406*, 121. [CrossRef]
7. Brown, L.R. *Who Will Feed China? Wake-Up Call for a Small Planet*; WW Norton & Company: New York, NY, USA, 1995.
8. Li, X.; Yeh, A.G.O. Principal component analysis of stacked multi-temporal images for the monitoring of rapid urban expansion in the Pearl River Delta. *Int. J. Remote Sens.* **1998**, *19*, 1501–1518. [CrossRef]
9. Fischel, W.A. The urbanization of agricultural land: A review of the National Agricultural Lands Study. *Land Econ.* **1982**, *58*, 236–259. [CrossRef]
10. Edgens, J.G.; Staley, S.R. The myth of farmland loss. *Forum Appl. Res. Public Policy* **1999**, *14*, 29–34.
11. Francis, C.A.; Hansen, T.E.; Fox, A.A.; Hesje, P.J.; Nelson, H.E.; Lawseth, A.E.; English, A. Farmland conversion to non-agricultural uses in the US and Canada: Current impacts and concerns for the future. *Int. J. Agric. Sustain.* **2012**, *10*, 8–24. [CrossRef]
12. Zhou, B.-B.; Aggarwal, R.; Wu, J.; Lv, L. Urbanization-associated farmland loss: A macro-micro comparative study in China. *Land Use Policy* **2021**, *101*, 105228. [CrossRef]
13. Bren d'Amour, C.; Reitsma, F.; Baiocchi, G.; Barthel, S.; Güneralp, B.; Erb, K.-H.; Haberl, H.; Creutzig, F.; Seto, K.C. Future urban land expansion and implications for global croplands. *Proc. Natl. Acad. Sci. USA* **2017**, *114*, 8939–8944. [CrossRef]
14. Zhou, B.-B.; Lv, L. Understanding the dynamics of farmland loss in a rapidly urbanizing region: A problem-driven, diagnostic approach to landscape sustainability. *Landsc. Ecol.* **2020**, *35*, 2471–2486. [CrossRef]
15. Wu, J. Landscape sustainability science: Ecosystem services and human well-being in changing landscapes. *Landsc. Ecol.* **2013**, *28*, 999–1023. [CrossRef]
16. Zhou, B.-B.; Wu, J.; Anderies, J.M. Sustainable landscapes and landscape sustainability: A tale of two concepts. *Landsc. Urban Plan.* **2019**, *189*, 274–284. [CrossRef]
17. Mao, D.; Ma, Q.; Zhou, B.-B. Sustainability of human–environment systems through the lens of landscape. *Landsc. Ecol.* **2020**, *35*, 2375–2379. [CrossRef]
18. NRC. *Our Common Journey: A Transition toward Sustainability*; National Academies Press: Washington, DC, USA, 1999.
19. Kates, R.W.; Clark, W.C.; Hall, J.M.; Jaeger, C.; Lowe, I.; McCarthy, J.J.; Schellnhuber, H.J.; Bolin, B.; Dickson, N.M.; Faucheux, S.; et al. Sustainability science. *Science* **2001**, *292*, 641–642. [CrossRef] [PubMed]
20. Kates, R.W.; Parris, T.M.; Leiserowitz, A.A. What is sustainable development? Goals, indicators, values, and practice. *Environ. Sci. Policy Sustain. Dev.* **2005**, *47*, 8–21.

21. Lambin, E.F.; Turner II, B.L.; Geist, H.J.; Agbola, S.B.; Angelsen, A.; Bruce, J.W.; Coomes, O.T.; Dirzo, R.; Fischer, G.; Folke, C. The causes of land-use and land-cover change: Moving beyond the myths. *Glob. Environ. Chang.* **2001**, *11*, 261–269. [CrossRef]
22. Turner, B.L., II. Land-use and land-cover change: Advances in 1.5 decades of sustained international research. *GAIA-Ecol. Perspect. Sci. Soc.* **2001**, *10*, 269–272. [CrossRef]
23. Mustard, J.F.; DeFries, R.S.; Fisher, T.; Moran, E. Land-use and land-cover change pathways and impacts. In *Land Change Science: Observing, Monitoring and Understanding Trajectories of Change on the Earth's Surface*; Gutman, G., Janetos, A.C., Justice, C.O., Moran, E.F., Mustard, J.F., Rindfuss, R.R., Skole, D., Turner, B.L., II, Cochrane, M.A., Eds.; Springer Science & Business Media: Dordrecht, The Netherlands, 2004.
24. Foley, J.A.; DeFries, R.; Asner, G.P.; Barford, C.; Bonan, G.; Carpenter, S.R.; Chapin, F.S.; Coe, M.T.; Daily, G.C.; Gibbs, H.K. Global consequences of land use. *Science* **2005**, *309*, 570–574. [CrossRef]
25. Li, S.; Li, X. Global understanding of farmland abandonment: A review and prospects. *J. Geogr. Sci.* **2017**, *27*, 1123–1150. [CrossRef]
26. Wang, X.; Li, X. China's agricultural land use change and its underlying drivers: A literature review. *J. Geogr. Sci.* **2021**, *31*, 1222–1242. [CrossRef]
27. Miao, L.; Zhu, F.; He, B.; Ferrat, M.; Liu, Q.; Cao, X.; Cui, X. Synthesis of China's land use in the past 300 years. *Glob. Planet. Change* **2013**, *100*, 224–233. [CrossRef]
28. Chou, J.; Dong, W.; Wang, S.; Fu, Y. Quantitative analysis of agricultural land use change in China. *Phys. Chem. Earth Parts A/B/C* **2015**, *87–88*, 3–9. [CrossRef]
29. Meyfroidt, P.; de Bremond, A.; Ryan, C.M.; Archer, E.; Aspinall, R.; Chhabra, A.; Camara, G.; Corbera, E.; DeFries, R.; Díaz, S.; et al. Ten facts about land systems for sustainability. *Proc. Natl. Acad. Sci. USA* **2022**, *119*, e2109217118. [CrossRef] [PubMed]
30. Zang, Y.; Yang, Y.; Liu, Y. Toward serving land consolidation on the table of sustainability: An overview of the research landscape and future directions. *Land Use Policy* **2021**, *109*, 105696. [CrossRef]
31. Kates, R.W. What kind of a science is sustainability science? *Proc. Natl. Acad. Sci. USA* **2011**, *108*, 19449–19450. [CrossRef] [PubMed]
32. Fang, X.; Zhou, B.-B.; Tu, X.; Ma, Q.; Wu, J. "What kind of a science is sustainability science?" An evidence-based reexamination. *Sustainability* **2018**, *10*, 1478.
33. Wu, J.; Hobbs, R. Key issues and research priorities in landscape ecology: An idiosyncratic synthesis. *Landsc. Ecol.* **2002**, *17*, 355–365. [CrossRef]
34. Yin, Q.; Sui, X.; Ye, B.; Zhou, Y.; Li, C.; Zou, M.; Zhou, S. What role does land consolidation play in the multi-dimensional rural revitalization in China? A research synthesis. *Land Use Policy* **2022**, *120*, 106261. [CrossRef]
35. Liberati, A.; Altman, D.; Tetzlaff, J.; Mulrow, C.; Gøtzsche, P.; Ioannidis, J.; Clarke, M.; Devereaux, P.J.; Kleijnen, J.; Moher, D. The PRISMA Statement for reporting systematic reviews and meta-analyses of studies that evaluate health care interventions: Explanation and elaboration. *PLoS Med.* **2009**, *6*, e1000100. [CrossRef]
36. Falagas, M.E.; Pitsouni, E.I.; Malietzis, G.A.; Pappas, G. Comparison of PubMed, Scopus, Web of Science, and Google Scholar: Strengths and weaknesses. *FASEB J.* **2008**, *22*, 338–342. [CrossRef] [PubMed]
37. Aria, M.; Cuccurullo, C. bibliometrix: An R-tool for comprehensive science mapping analysis. *J. Informetr.* **2017**, *11*, 959–975. [CrossRef]
38. van Eck, N.J.; Waltman, L. Software survey: VOSviewer, a computer program for bibliometric mapping. *Scientometrics* **2010**, *84*, 523–538. [CrossRef]
39. Garfield, E.; Paris, S.; Stock, W.G. HistCite^{TM}: A software tool for informetric analysis of citation linkage. *Inf. Wiss. Und Prax.* **2006**, *57*, 391–400.
40. Bernard, H.R.; Wutich, A.; Ryan, G.W. *Analyzing Qualitative Data: Systematic Approaches*, 2nd ed.; SAGE: London, UK, 2017.
41. Tan, M.; Li, X.; Xie, H.; Lu, C. Urban land expansion and arable land loss in China—A case study of Beijing–Tianjin–Hebei region. *Land Use Policy* **2005**, *22*, 187–196. [CrossRef]
42. Deng, X.Z.; Huang, J.; Rozelle, S.; Zhang, J.; Li, Z. Impact of urbanization on cultivated land changes in China. *Land Use Policy* **2015**, *45*, 1–7. [CrossRef]
43. Yang, H.; Li, X. Cultivated land and food supply in China. *Land Use Policy* **2000**, *17*, 73–88. [CrossRef]
44. Liu, J.; Liu, M.; Tian, H.; Zhuang, D.; Zhang, Z.; Zhang, W.; Tang, X.; Deng, X. Spatial and temporal patterns of China's cropland during 1990–2000: An analysis based on Landsat TM data. *Remote Sens. Environ.* **2005**, *98*, 442–456. [CrossRef]
45. Deng, X.; Huang, J.; Rozelle, S.; Uchida, E. Cultivated land conversion and potential agricultural productivity in China. *Land Use Policy* **2006**, *23*, 372–384. [CrossRef]
46. Lichtenberg, E.; Ding, C. Assessing farmland protection policy in China. *Land Use Policy* **2008**, *25*, 59–68. [CrossRef]
47. Long, H.; Liu, Y.; Wu, X.; Dong, G. Spatio-temporal dynamic patterns of farmland and rural settlements in Su–Xi–Chang region: Implications for building a new countryside in coastal China. *Land Use Policy* **2009**, *26*, 322–333. [CrossRef]
48. Chen, J. Rapid urbanization in China: A real challenge to soil protection and food security. *Catena* **2007**, *69*, 1–15. [CrossRef]
49. Seto, K.C.; Güneralp, B.; Hutyra, L.R. Global forecasts of urban expansion to 2030 and direct impacts on biodiversity and carbon pools. *Proc. Natl. Acad. Sci. USA* **2012**, *109*, 16083–16088. [CrossRef] [PubMed]
50. Liu, J.; Liu, M.; Zhuang, D.; Zhang, Z.; Deng, X. Study on spatial pattern of land-use change in China during 1995–2000. *Sci. China D-Earth Sci.* **2003**, *46*, 373–384. [CrossRef]

51. Lin, G.C.S.; Ho, S.P.S. China's land resources and land-use change: Insights from the 1996 land survey. *Land Use Policy* **2003**, *20*, 87–107. [CrossRef]
52. Long, H.; Tang, G.; Li, X.; Heilig, G.K. Socio-economic driving forces of land-use change in Kunshan, the Yangtze River Delta economic area of China. *J. Environ. Manag.* **2007**, *83*, 351–364. [CrossRef] [PubMed]
53. Liu, J.; Zhang, Z.; Xu, X.; Kuang, W.; Zhou, W.; Zhang, S.; Li, R.; Yan, C.; Yu, D.; Wu, S.; et al. Spatial patterns and driving forces of land use change in China during the early 21st century. *J. Geogr. Sci.* **2010**, *20*, 483–494. [CrossRef]
54. Long, H.; Li, Y.; Liu, Y.; Woods, M.; Zou, J. Accelerated restructuring in rural China fueled by 'increasing vs. decreasing balance' land-use policy for dealing with hollowed villages. *Land Use Policy* **2012**, *29*, 11–22. [CrossRef]
55. Liu, J.; Kuang, W.; Zhang, Z.; Xu, X.; Qin, Y.; Ning, J.; Zhou, W.; Zhang, S.; Li, R.; Yan, C. Spatiotemporal characteristics, patterns, and causes of land-use changes in China since the late 1980s. *J. Geogr. Sci.* **2014**, *24*, 195–210. [CrossRef]
56. Liu, Y.; Fang, F.; Li, Y. Key issues of land use in China and implications for policy making. *Land Use Policy* **2014**, *40*, 6–12. [CrossRef]
57. Grimm, N.B.; Faeth, S.H.; Golubiewski, N.E.; Redman, C.L.; Wu, J.; Bai, X.; Briggs, J.M. Global change and the ecology of cities. *Science* **2008**, *319*, 756–760. [CrossRef]
58. Costanza, R.; d'Arge, R.; de Groot, R.; Farber, S.; Grasso, M.; Hannon, B.; Naeem, S.; Limburg, K.; Paruelo, J.; O'Neill, R.V. The value of the world's ecosystem services and natural capital. *Nature* **1997**, *387*, 253–260. [CrossRef]
59. Verburg, P.H.; Veldkamp, A.; Fresco, L.O. Simulation of changes in the spatial pattern of land use in China. *Appl. Geogr.* **1999**, *19*, 211–233. [CrossRef]
60. Weng, Q. Land use change analysis in the Zhujiang Delta of China using satellite remote sensing, GIS and stochastic modelling. *J. Environ. Manag.* **2002**, *64*, 273–284. [CrossRef] [PubMed]
61. Tan, M.; Li, X.; Lu, C. Urban land expansion and arable land loss of the major cities in China in the 1990s. *Sci. China Ser. D Earth Sci.* **2005**, *48*, 1492–1500. [CrossRef]
62. Wu, Q.; Li, H.-q.; Wang, R.-s.; Paulussen, J.; He, Y.; Wang, M.; Wang, B.-h.; Wang, Z. Monitoring and predicting land use change in Beijing using remote sensing and GIS. *Landsc. Urban Plan.* **2006**, *78*, 322–333. [CrossRef]
63. Liu, Y.; Wang, L.; Long, H. Spatio-temporal analysis of land-use conversion in the eastern coastal China during 1996–2005. *J. Geogr. Sci.* **2008**, *18*, 274–282. [CrossRef]
64. Deng, J.S.; Wang, K.; Hong, Y.; Qi, J.G. Spatio-temporal dynamics and evolution of land use change and landscape pattern in response to rapid urbanization. *Landsc. Urban Plan.* **2009**, *92*, 187–198. [CrossRef]
65. Liu, Y.S.; Wang, J.Y.; Long, H.L. Analysis of arable land loss and its impact on rural sustainability in Southern Jiangsu Province of China. *J. Environ. Manag.* **2010**, *91*, 646–653. [CrossRef]
66. Wang, L.; Li, C.; Ying, Q.; Cheng, X.; Wang, X.; Li, X.; Hu, L.; Liang, L.; Yu, L.; Huang, H.; et al. China's urban expansion from 1990 to 2010 determined with satellite remote sensing. *Chin. Sci. Bull.* **2012**, *57*, 2802–2812. [CrossRef]
67. Long, H.; Liu, Y.; Hou, X.; Li, T.; Li, Y. Effects of land use transitions due to rapid urbanization on ecosystem services: Implications for urban planning in the new developing area of China. *Habitat Int.* **2014**, *44*, 536–544. [CrossRef]
68. Song, W.; Pijanowski, B.C.; Tayyebi, A. Urban expansion and its consumption of high-quality farmland in Beijing, China. *Ecol. Indic.* **2015**, *54*, 60–70. [CrossRef]
69. Liu, T.; Liu, H.; Qi, Y. Construction land expansion and cultivated land protection in urbanizing China: Insights from national land surveys, 1996–2006. *Habitat Int.* **2015**, *46*, 13–22. [CrossRef]
70. Jiang, L.; Deng, X.; Seto, K.C. Multi-level modeling of urban expansion and cultivated land conversion for urban hotspot counties in China. *Landsc. Urban Plan.* **2012**, *108*, 131–139. [CrossRef]
71. Jiang, L.; Deng, X.; Seto, K.C. The impact of urban expansion on agricultural land use intensity in China. *Land Use Policy* **2013**, *35*, 33–39. [CrossRef]
72. Xu, C.; Liu, M.; An, S.; Chen, J.M.; Yan, P. Assessing the impact of urbanization on regional net primary productivity in Jiangyin County, China. *J. Environ. Manag.* **2007**, *85*, 597–606. [CrossRef]
73. Yan, H.; Liu, J.; Huang, H.Q.; Tao, B.; Cao, M. Assessing the consequence of land use change on agricultural productivity in China. *Glob. Planet. Chang.* **2009**, *67*, 13–19. [CrossRef]
74. He, C.; Liu, Z.; Xu, M.; Ma, Q.; Dou, Y. Urban expansion brought stress to food security in China: Evidence from decreased cropland net primary productivity. *Sci. Total Environ.* **2017**, *576*, 660–670. [CrossRef]
75. Li, Y.; Li, Y.; Westlund, H.; Liu, Y. Urban–rural transformation in relation to cultivated land conversion in China: Implications for optimizing land use and balanced regional development. *Land Use Policy* **2015**, *47*, 218–224. [CrossRef]
76. Liu, Y.; Yang, Y.; Li, Y.; Li, J. Conversion from rural settlements and arable land under rapid urbanization in Beijing during 1985–2010. *J. Rural Stud.* **2017**, *51*, 141–150. [CrossRef]
77. Tan, R.; Beckmann, V.; van den Berg, L.; Qu, F. Governing farmland conversion: Comparing China with the Netherlands and Germany. *Land Use Policy* **2009**, *26*, 961–974. [CrossRef]
78. Poteete, A.R.; Janssen, M.A.; Ostrom, E. *Working Together: Collective Action, the Commons, and Multiple Methods in Practice*; Princeton University Press: Princeton, NJ, USA, 2010.
79. Long, H.; Ge, D.; Zhang, Y.; Tu, S.; Qu, Y.; Ma, L. Changing man-land interrelations in China's farming area under urbanization and its implications for food security. *J. Environ. Manag.* **2018**, *209*, 440–451. [CrossRef] [PubMed]
80. Ma, L.; Long, H.; Tu, S.; Zhang, Y.; Zheng, Y. Farmland transition in China and its policy implications. *Land Use Policy* **2020**, *92*, 104470. [CrossRef]

81. Platt, R.H. The farmland conversion debate: NALS and Beyond. *Prof. Geogr.* **1985**, *37*, 433–442. [CrossRef]
82. Turner II, B.L.; Skole, D.; Sanderson, S.; Fischer, G.; Fresco, L.; Leemans, R. *Land-Use and Land-Cover Change: Science/Research Plan*; International Geosphere-Biosphere Programme: Stockholm, Sweden, 1995.
83. Gutman, G.; Janetos, A.C.; Justice, C.O.; Moran, E.F.; Mustard, J.F.; Rindfuss, R.R.; Skole, D.; Turner II, B.L.; Cochrane, M.A. *Land Change Science: Observing, Monitoring and Understanding Trajectories of Change on the Earth's Surface*; Springer Science & Business Media: Dordrecht, The Netherlands, 2004; Volume 6.
84. Turner II, B.L.; Lambin, E.F.; Reenberg, A. The emergence of land change science for global environmental change and sustainability. *Proc. Natl. Acad. Sci. USA* **2007**, *104*, 20666–20671. [CrossRef] [PubMed]
85. Rounsevell, M.D.; Pedroli, B.; Erb, K.-H.; Gramberger, M.; Busck, A.G.; Haberl, H.; Kristensen, S.; Kuemmerle, T.; Lavorel, S.; Lindner, M. Challenges for land system science. *Land Use Policy* **2012**, *29*, 899–910. [CrossRef]
86. Verburg, P.H.; Erb, K.-H.; Mertz, O.; Espindola, G. Land System Science: Between global challenges and local realities. *Curr. Opin. Environ. Sustain.* **2013**, *5*, 433–437. [CrossRef] [PubMed]
87. Turner II, B.L.; Lambin, E.F.; Verburg, P.H. From land-use/land-cover to land system science. *Ambio* **2021**, *50*, 1291–1294. [CrossRef] [PubMed]
88. Wu, J.; Guo, X.; Yang, J.; Qian, G.; Niu, J.; Liang, C.; Zhang, Q.; Li, A. What is sustainability science? *Chin. J. Appl. Ecol.* **2014**, *25*, 1–11.
89. Zhou, B.-B.; Ma, Q.; Wu, J.; Hu, G.; Mao, D.; Zeng, X.; Guo, J.; Fang, X.; Liu, Y.; Lyu, L. Sustainability science revisited: Recent advances and new opportunities. *Chin. J. Appl. Ecol.* **2019**, *30*, 325–336. (In Chinese with English Abstract)
90. Wu, J. Landscape sustainability science (II): Core questions and key approaches. *Landsc. Ecol.* **2021**, *36*, 2453–2485. [CrossRef]
91. Roy Chowdhury, R.; Turner II, B.L. The parallel trajectories and increasing integration of landscape ecology and land system science. *J. Land Use Sci.* **2019**, *14*, 135–154. [CrossRef]
92. Risser, P.G.; Karr, J.R.; Forman, R.T.T. *Landscape Ecology: Directions and Approaches, a Workshop Held at Allerton Park, Piatt County, Illinois, April 1983*; Illinois Natural History Survey: Champaign, IL, USA, 1984; p. 18.
93. Milovanović, A.; Milovanović Rodić, D.; Maruna, M. Eighty-year review of the evolution of landscape ecology: From a spatial planning perspective. *Landsc. Ecol.* **2020**, *35*, 2141–2161. [CrossRef]
94. Turner II, B.L.; Janetos, A.C.; Verburg, P.H.; Murray, A.T. Land system architecture: Using land systems to adapt and mitigate global environmental change. *Glob. Environ. Change* **2013**, *23*, 395–397. [CrossRef]
95. Verburg, P.H.; Crossman, N.; Ellis, E.C.; Heinimann, A.; Hostert, P.; Mertz, O.; Nagendra, H.; Sikor, T.; Erb, K.-H.; Golubiewski, N. Land system science and sustainable development of the earth system: A global land project perspective. *Anthropocene* **2015**, *12*, 29–41. [CrossRef]
96. Munroe, D.K.; Müller, D. Land-system science to support achieving the sustainable development goals. *J. Land Use Sci.* **2020**, *15*, 477–481. [CrossRef]
97. Krugman, P. *Increasing Returns and Economic Geography*; National Bureau of Economic Research: Cambridge, MA, USA, 1991.
98. Fan, Y.; Jin, X.; Gan, L.; Yang, Q.; Wang, L.; Lyu, L.; Li, Y. Exploring an integrated framework for "dynamic-mechanism-clustering" of multiple cultivated land functions in the Yangtze River Delta region. *Appl. Geogr.* **2023**, *159*, 103061. [CrossRef]
99. Han, B.; Jin, X.; Sun, R.; Li, H.; Liang, X.; Zhou, Y. Understanding land-use sustainability with a systematical framework: An evaluation case of China. *Land Use Policy* **2023**, *132*, 106767. [CrossRef]
100. Zhang, Z.; Wang, X.; Zhang, Y.; Gao, Y.; Liu, Y.; Sun, X.; Zhi, J.; Yin, S. Simulating land use change for sustainable land management in rapid urbanization regions: A case study of the Yangtze River Delta region. *Landsc. Ecol.* **2023**, *38*, 1807–1830. [CrossRef]
101. Li, Y.; Han, X.; Zhou, B.-B.; Lv, L.; Fan, Y. Farmland dynamics and its grain production efficiency and ecological security in China's Major Grain-Producing Regions between 2000 and 2020. *Land* **2023**, *12*, 1404. [CrossRef]
102. Zou, Y.; Meng, J.; Zhu, L.; Han, Z.; Ma, Y. Characterizing land use transition in China by accounting for the conflicts underlying land use structure and function. *J. Environ. Manag.* **2024**, *349*, 119311. [CrossRef] [PubMed]
103. Clark, W.C.; Harley, A.G. Sustainability science: Toward a synthesis. *Annu. Rev. Environ. Resour.* **2020**, *45*, 331–386. [CrossRef]
104. Betsill, M.M.; Benney, T.M.; Gerlak, A.K. (Eds.) *Agency in Earth System Governance*; Cambridge University Press: Cambridge, UK, 2020.
105. Agrawal, A.; Brandhorst, S.; Jain, M.; Liao, C.; Pradhan, N.; Solomon, D. From environmental governance to governance for sustainability. *One Earth* **2022**, *5*, 615–621. [CrossRef]
106. Aggarwal, R.M.; Anderies, J.M. Understanding how governance emerges in social-ecological systems: Insights from archetype analysis. *Ecol. Soc.* **2023**, *28*, 2. [CrossRef]
107. Lang, D.J.; Wiek, A.; Bergmann, M.; Stauffacher, M.; Martens, P.; Moll, P.; Swilling, M.; Thomas, C.J. Transdisciplinary research in sustainability science: Practice, principles, and challenges. *Sustain. Sci.* **2012**, *7*, 25–43. [CrossRef]
108. Weichselgartner, J.; Truffer, B. From knowledge co-production to transdisciplinary research: Lessons from the quest to produce socially robust knowledge. In *Global Sustainability: Cultural Perspectives and Challenges for Transdisciplinary Integrated Research*; Werlen, B., Ed.; Springer: Cham, Switzerland, 2015; pp. 89–106.
109. Scholz, R.W. Transdisciplinarity: Science for and with society in light of the university's roles and functions. *Sustain. Sci.* **2020**, *15*, 1033–1049. [CrossRef]

land

MDPI

Article

Impacts of Cropland Utilization Patterns on the Sustainable Use Efficiency of Cropland Based on the Human–Land Perspective

Xinyu Hu [1,2], Chun Dong [2,*] and Yu Zhang [2,*]

[1] School of Geomatics, Liaoning Technical University, Fuxin 123000, China; 47211043@stu.lntu.edu.cn
[2] Chinese Academy of Surveying and Mapping, Beijing 100039, China
* Correspondence: dongchun@casm.ac.cn (C.D.); zhangyu@casm.ac.cn (Y.Z.)

Abstract: Confronted with China's burgeoning population and finite arable land resources, the enhancement of sustainable arable land efficiency is of paramount importance. This study, grounded in the United Nations Sustainable Development Goals (SDGs), introduces a robust framework for assessing sustainable arable land use. Utilizing the Sustainable Utilization of Arable Land (SUA) indicator system, the DGA–Super-SBM model, the Malmquist–Luenberger production index, and the TO–Fisher–OSM algorithm, we evaluated the efficiency of sustainable utilization of arable land (ESUA) in 52 prefecture-level cities within China's major grain-producing regions of the Yellow and Huaihai Seas. We analyzed the cropland utilization patterns from 2010 to 2020, examining the influence of these patterns on sustainable utilization efficiency. Our findings indicate that between 2010 and 2020, the arable land usage in these regions exhibited minimal transformation, primarily shifting towards construction land and conversely from grassland and water systems. Notably, the ESUA of arable land demonstrated an upward trend, characterized by pronounced spatial clustering, enduring high efficiency in the northern regions, and a significant surge in the southern sectors. The declining ESUA (D-ESUA) trend was general but increased in half of the cities. The change in the center of gravity of ESUA correlated with the north–south movement of the proportion of cultivated land area, the turn-in rate, and the turn-out rate, yet moved in the opposite direction to that of cultivated land density and yield per unit area. Variables such as the replanting index, cropland density, yield per unit area, and cropland turn-in rate significantly affected ESUA. These findings offer a scientific basis and decision-making support for optimizing the utilization pattern of arable land and achieving a rational allocation of arable land resources.

Keywords: human–land perspective; sustainable utilization of arable land resources; the main grain-producing areas of the Yellow Huaihai; high-dimensional indicators; impact analysis; SDG

1. Introduction

Human society is dependent on arable land resources, which directly determine food security and sustainable development [1]. Over recent decades, arable land resources worldwide have experienced severe degradation and decline. The United Nations Food and Agriculture Organization (FAO) estimates that between 1980 and 2018, the world's arable land area shrank by approximately 4% [2]. This decline can be attributed to transitional exploitation and inefficient use of arable land resources, resulting in the deterioration of land quality, ecological degradation, and soil infertility, which limit agricultural production and reduce the ability of humans to address their food needs [3] and run counter to goals such as Zero Hunger and Climate Action, as outlined in the United Nations Sustainable Development Goals (SDGs). China is the largest developing country in the world, accounting for 20% of the world's population; however, only 9% of its land is arable [4]. Protecting China's limited arable land has received significant attention from the Chinese government, who have been establishing a comprehensive set of arable land protection policies and national strategic decisions [5]. However, owing to accelerated urbanization, the expansion of industrial land, and need for ecological and environmental protection, some of the country's arable land is now being used for other purposes [6]. National Bureau of Statistics data suggest that from 2010 to 2020, China's arable land reduced by approximately 7,533,000 ha, which is approximately the size of the Ningxia Hui Autonomous Region; moreover, the natural ecology around arable land has been affected by the excessive use of chemical fertilizers, pesticides, and agricultural films, leading to low production and inefficient land use. According to the 2020 China Ecological and Environmental Status Bulletin, 19.4% of China's arable land is contaminated; the cost to restore this land would be at least USD 116.3 billion [7]. These facts prompt the need to re-examine China's cropland resource utilization patterns and explore effective ways to improve the efficiency of sustainable utilization of arable land (ESUA) in order to achieve the UN Sustainable Development Goals (SDGs) such as "Zero Hunger", "Climate Action", and "Clean Water and Sanitation" of the UN Sustainable Development Goals (SDGs).

Agricultural land resource utilization is a multifaceted concept that encapsulates the complexities of land planning, distribution, usage, and management. Researchers in this domain are tasked with not only quantifying and qualifying arable land but also assessing the distribution and efficiency of its application. The pattern of cultivated land use encompasses the spatial distribution, quantifiable extent, functional roles, and level of utilization of arable land, among other nuanced characteristics [8]. In the scholarly exploration of agricultural land utilization patterns, domestic and international scholars have expounded extensively on the drivers of land utilization [9], the transformations in land use patterns [10], and the responses to shifts in ecological values [11]. These studies have augmented our comprehension of agricultural land use dynamics. Notably, the analysis of the evolving agricultural land use pattern has been bolstered by the deployment of remote sensing data [12] and socio-economic statistics [13] to delve into the process of land use change [14], its efficiency [13], and its temporal and spatial attributes [15]. These investigations span a spectrum of scales, from the macroscopic intercontinental [11] to the micro-county level [16], offering invaluable insights into the nuances of agricultural land use alterations. In aggregate, the academic discourse on agricultural land use changes has largely focused on the quantity of arable land, shifts in land use patterns, and reactions to ecological value changes, yet it has overlooked the repercussions of land use patterns on the sustainable utilization of agricultural land within the broader context of the United Nations Sustainable Development Goals (SDGs).

The concept of sustainable use was first introduced by the German geographer Hans Carl von Carlowitz in his published work Sylvicultura oeconomica, where he introduced the concept of "Nachhaltigkeit", which states that in the use of resources, not only immediate but also long-term benefits and efficiencies need to be considered to protect the natural environment and socio-economic sustainability. The proposal of sustainable utilization has aroused a strong reaction at the international level. At the 1990 International Symposium on Sustainable Utilization Systems held in New Delhi, India, the concept of sustainable land use was put forward based on the concept of Nachhaltigkeit; at the 1993 Seminar on Sustainable Land Use Management in the 21st Century, held in Canada, the International FAO put forward the "Evaluation Outline of Sustainable Land Management", which set out criteria for the evaluation of sustainable utilization. In-depth studies on the sustainable use of land resources have laid a research foundation and provided models on the sustainable use of cropland, and many studies on the sustainable use of cropland have been carried out worldwide, including those on the theoretical connotations of sustainable cropland use [17,18] and models for sustainable cropland use [19–21]. In 1994, the State Council of China adopted the white paper "China's Agenda 21—Population, Environment and Development", and sustainable development began to involve the field of agriculture. As a major component of agriculture, the sustainable development of arable land has received significant attention. Many subsequent studies have focused on the establishment of indicator evaluation systems and evaluation methods for sustainable use and influencing factors, including the ecological–economic–social model, productive–safety–

protection–economic–social acceptability model, pressure–state–response indicator system, and driving force–pressure-state–impact–response model. There are a range of evaluation methods of sustainable cropland use, including hierarchical analysis, the gray prediction model, coefficient of variation method, multi-factor integrated evaluation method, multiple linear regression model [22], coupled coordination model, structural equation model [23], and ecological footprint model [24], among others. However, most existing research methods and evaluation systems are limited to the statistics and prediction of single system indicators of arable land resources; they cannot meet the comprehensive and accurate evaluation of the sustainable utilization of arable (SUA) land and production efficiency. Moreover, there is a lack of research on the grading of arable land ESUA and the characteristics of the time-space double scale, which is not conducive to the analysis of the SUA evaluation as well as the evolution of the spatial and temporal processes which are even more detrimental to the estimation of the development trend of the sustainable development of arable land.

Considering this context, the present study introduces a comprehensive evaluation framework for the sustainable utilization of arable land, grounded in the United Nations Sustainable Development Goals (SDGs) framework. This framework originates from the interplay between human–land relationships and the balance of inputs and outputs. It constructs an SUA indicator evaluation system that is aligned with the sustainable objectives of production inputs, whilst also encapsulating the economic, social, and ecological expectations, as well as the non-expectations, associated with arable land utilization. Furthermore, the study establishes a sophisticated high-dimensional indicator system evaluation framework that integrates the coupled dimensionally reduced genetic algorithm with the super-efficiency SBM model (DGA–Super-SBM), the Malmquist–Luenberger production index, and the time-sequence optimized optimal segmentation algorithm (TO–Fisher–OSM). This framework is employed to conduct an ESUA evaluation of 52 prefecture-level cities within the major grain-producing regions of the Yellow and Huaihai Seas (YHHRB) in China. Analysis of the shifts in cropland utilization patterns from 2010 to 2020 is conducted, with a keen eye on the essence of cropland utilization patterns. Four key indicators are selected to probe the influence of cropland utilization patterns on the sustainable efficiency of cropland utilization: variations in cropland area, shifts in cropland function, alterations in cropland configuration, and modifications in cropland utilization efficiency. These indicators serve to inform policy makers and provide a robust decision-making tool for the formulation of relevant policies and the sustainable management of cropland resources.

Compared with previous studies, the novelties of this manuscript are centered on three distinct aspects. ① Perspective Innovation. The paper introduces a novel evaluation system for the Sustainable Utilization of Arable land (SUA), drawing upon the United Nations Sustainable Development Goals (SDGs) system. This system is uniquely constructed from the dual vantage points of human–land relationships and the input–output dynamics. Such an approach is unprecedented in the extant scholarly literature, offering a fresh perspective on the assessment of arable land sustainability. ② Methodological Innovation. Leveraging the power of genetic algorithms to optimize a super-efficient Source-Based Model (SBM), this study transcends the limitations of conventional SBM evaluation, which often struggles with high-dimensional parameter spaces. This innovative methodology enables the comprehensive evaluation and analysis of complex, high-dimensional indicator systems. It thus facilitates a profound understanding of the status quo of arable land's sustainable development and promotes the efficient, intensive, and economic utilization of regional arable land. ③ Theoretical Innovation. The paper introduces the concept of cropland utilization patterns, expanding the discourse to encompass four dimensions that influence the ESUA of cropland: area, distribution, function, and utilization degree. This holistic analysis provides a novel theoretical foundation for the sustainable management of cropland resources, offering insights into how these resources can be preserved and utilized for the benefit of future generations.

2. Data and Methods

The comprehensive SUA evaluation system was constructed based on the structure of the economic–social–ecological system (Figure 1). Then, the spatiotemporal evolution of ESUA in the YHHRB was analyzed using DGA–Super-SBM and Malmquist–Luenberger production indices. Finally, ESUA was classified and graded by using TO–Fisher–OSM.

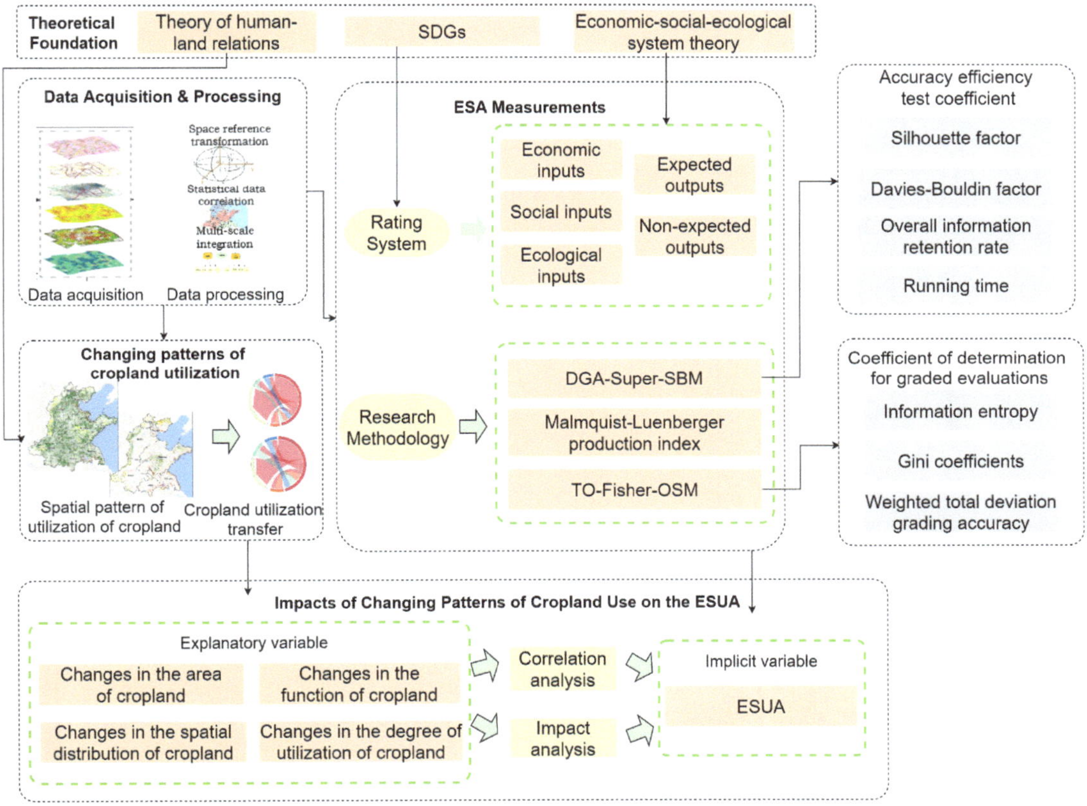

Figure 1. Technology roadmap.

2.1. Conceptual Definition of Cropland Use Patterns

Arable land stands at the heart of agricultural endeavors, shaping a landscape that is decidedly anthropogenic [25]. Subject to immediate human dominion and manipulation, it has evolved in step with the advancement of human civilization. The conceptualization of this vital resource has evolved through the annals of history, its definition broadening to encompass enhanced productivity and innovative land management techniques [26]. Physically, arable land is characterized as a precious terrestrial asset that humanity has reclaimed and put to use to satisfy the demands of life and progress, facilitated by current technological capabilities. It is soil that is hospitable to the growth of crops, endowed with fertility, distinct qualitative attributes, and a favorable soil structure [27].

The utilization pattern of cultivated land encompasses a nuanced and multifaceted construct, spanning the spatial distribution, quantitative dimensions, functionalities, and the degree of utilization of agricultural land, and is a fundamental component of the agricultural production milieu [28]. The spatial distribution of arable land manifests the geographical positioning and extension of this vital resource, serving as a critical determinant that influences the efficiency of arable land utilization and agricultural productivity [29].

The extant area of arable land is intrinsically linked to the scale of food production and the trajectory of agricultural development [30]. The functional evolution of arable land mirrors the diversification of its applications, incorporating multifaceted roles such as food provision, ecological conservation, and recreational tourism [31]. The level of arable land utilization underscores the efficiency and technological sophistication of human utilization, serving as a paramount indicator of the modernization level within agricultural production [32]. Variations in the utilization patterns of cropland encapsulate the alterations in the quantity, location, function, and degree of utilization of cropland across different regions and over temporal intervals. These shifts are contingent upon a multitude of factors, including policy orientation, market demand, technological advancements, and environmental considerations [33]. Consequently, a thorough examination of changes in cropland utilization patterns necessitates an intricate analysis from diverse perspectives. Firstly, the alteration in the acreage of arable land constitutes a pivotal dimension for scrutinizing shifts in the utilization pattern of arable land. An upsurge or decline in arable land acreage reflects the intensity of human engagement with land resources and the scaling of agricultural production. Secondly, the transformation in the function of arable land elucidates the trend toward multifunctionality and the realignment of the agricultural industrial structure. Conversely, modifications in the layout of arable land signal the recalibration of its spatial distribution and the differentiation of agricultural production across regions. Lastly, modifications in the degree of arable land utilization underscore the progression of agricultural production technologies and the enhancement of land resource management methodologies.

2.2. Study Area

The study area was centered on the YHHRB and located between 112°29′ E and the eastern coastline of China and between 31°21′ N and 40°26′ N. This area covers 380,000 km^2 across the middle and lower reaches of the Yellow, Huaihe, and Haihe river basins, including parts of greater Beijing, Tianjin, Hebei, Lu, Henan, northern Jiangsu, and northern Anhui, with a total of 52 prefecture-level municipalities (Figure 2). The area is flat and fertile and covers approximately 32 million ha of arable land, accounting for ~15% of the total arable land in China. The region is rich in water resources, with a number of rivers and lakes providing a solid basis for agricultural production. The area has a warm temperate monsoon climate, four distinct seasons, and average annual sunshine of 1700–2200 h. Due to its special geographic location and the influence of seasonal and climatic factors, the region has fewer hours of sunshine than other parts of China, resulting in a longer crop growth cycle, with most crops maturing twice a year, and some parts of the northern part of the country maturing three times in two years. The main crops include wheat, rice, corn, sorghum, beans, and oilseed, of which wheat and rice are the main food crops. The region is one of the most important food production bases in the country, and plays an important role in the country's economic development and food security.

We chose the YHHRB as a case study for several reasons. First, at the national policy level, achieving food security and sustainability are important concerns, and grain production in the YHHRB is nationally important. Second, natural conditions in the YHHRB are unique, with abundant water resources providing excellent conditions for crop growth. The region has a large amount of high-quality land resources, providing a rich material basis for food production. Cultivated land production conditions are typical of both the North China Plain region and Yangtze River Basin [34], and the study of SUA evaluation in the region can help to provide a reference for cultivated land protection strategies and policies in other key agricultural regions. Third, rapid urbanization in the region has led to a rising demand for land resources. A large amount of arable land has been utilized for production and living purposes, such as industry and urban construction [35], which threatens food security.

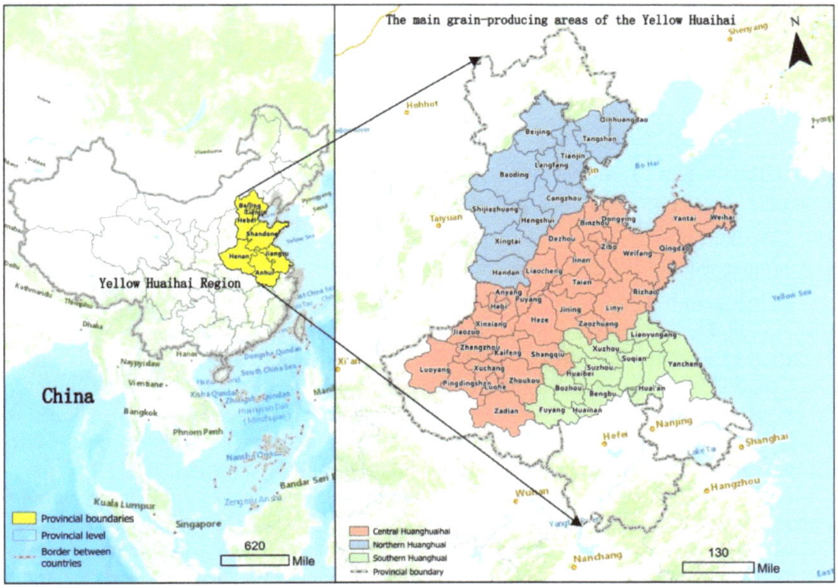

Figure 2. Location map of the project area. Note: This map is based on a standard map [review number GS(2020)4619], retrieved from the Ministry of Natural Resources' standard map service website. The base map is unmodified.

2.3. Research Methods

(1) Calculation of sustainable use efficiency of arable land under high-dimensional indicators using DGA–Super-SBM

Super-SBM is a nonparametric efficiency evaluation method originally proposed by Tone [36]. The model was developed based on the traditional non-expectation SBM model to address input and output slack and directly addresses the problems of excess inputs and insufficient outputs in efficiency assessment. Therefore, Super-SBM is particularly suitable for evaluating ESUA. Recently, Super-SBM has been widely used in the assessment of agro-ecological efficiency and agricultural land use efficiency [37,38]. And in contrast to the conventional CCR model paired with the non-expected SBM framework, the research demonstrates that the super-efficient SBM model exhibits greater precision and trustworthiness. It transcends the constraints encountered by the non-expected SBM model in terms of the scope of its calculated outcomes, thus being able to appraise decision-making units operating at high efficiency levels. However, given that arable land is an intricate and multifaceted ecosystem, a multidimensional indicator system must be considered to assess the sustainable utilization of this resource. Although the super-efficient SBM model has significant utility in evaluating sustainability, the limitations of its modeling parameters can introduce distortions when computing within the context of a high-dimensional indicator system within the super-efficient SBM model. Genetic algorithm-based dimensionality reduction involves the mapping of high-dimensional data into a lower-dimensional space. Unlike conventional dimensionality reduction techniques, such as principal component analysis (PCA) [39], genetic algorithms have the capacity to select a superior projection matrix, thereby more effectively preserving the structural integrity of the original data. Consequently, this study employs an SBM model coupled with a genetic algorithm for dimensionality reduction to evaluate the sustainable efficiency of arable land resource use.

The detailed calculation process of the Super-SBM model is described in the literature. In brief, the Super-SBM model contains non-desired outputs as follows:

$$\min\rho = \frac{\frac{1}{m}\sum_{i=1}^{m} s_i^- / x_{in}}{1+\frac{1}{s_{1+s_2}}\left(\sum_{r=1}^{s_1} s_r^g / y_{rn}^g + \sum_{l=1}^{s_2} s_l^b / y_{ln}^b\right)} \tag{1}$$

where ρ is the target efficiency value, s^- is the amount of slack in the input schedule, s^g is the amount of slack in the desired output, and s^b is the amount of slack in the undesired output.

(2) Calculation of Dynamic sustainable efficiency of cropland (D-ESUA) through the Malmquist–Luenberger Production Index

As the technical efficiency of DGA–Super-SBM assessment is based on 1 year of data, it is suitable for dealing with static changes in ESUA [40]. In analyzing the SUA state, the initial state of utilization changes from period to period, and the conditions under which the ESUA is analyzed are different in each case. For this reason, when calculating ESUA, scholars have proposed evaluation methods that can be compared over time [41]. The dynamic Malmquist production index is based on a distance function to analyze the efficiency changes and technical progress of decision-making units [42–44].

As the traditional Malmquist index ignores the constraints of environment and resources, combined with the actual situation of the YHHRB, in this study, we adopted the Malmquist–Luenberger production index based on the directional distance function of SBM, which is able to analyze the dynamic change of the efficiency of cultivated land use, and to identify how to improve the desired output and reduce pollutant emissions based on the integrated cultivated land and external environmental factors. The Malmquist–Luenberger production index formula is as follows:

$$ML_t^{t+1} = \left[\frac{1+\vec{D}_0^t(x^t,y^t,z^t;y^t,-z^t)}{1+\vec{D}_0^t(x^{t+1},y^{t+1},z^{t+1};y^{t+1},-z^{t+1})} \times \frac{1+\vec{D}_0^{t+1}(x^t,y^t,z^t;y^t,-z^t)}{1+\vec{D}_0^{t+1}(x^{t+1},y^{t+1},z^{t+1};y^{t+1},-z^{t+1})}\right]^{1/2} \tag{2}$$

where ML_t^{t+1} is the index of change in production, $\vec{D}_0^t(x^t,y^t,z^t;y^t,-z^t)$ and $\vec{D}_0^t$ $(x^{t+1},y^{t+1},z^{t+1};y^{t+1},-z^{t+1})$ are the technology in period t as the reference technology (distance functions for decision units in periods t and $t+1$), and $\vec{D}_0^{t+1}(x^t,y^t,z^t;y^t,-z^t)$ and $\vec{D}_0^{t+1}(x^{t+1},y^{t+1},z^{t+1};y^{t+1},-z^{t+1})$ are the technology of period $t+1$ as the reference technology (distance function of the decision unit in period t and period $t+1$).

(3) Classification of ESUAs using TO–Fisher–OSM

Fisher's optimal partitioning method is an ordered clustering algorithm first proposed by the British statistician Ronald Fisher in 1958; the purpose of the algorithm is to find an optimal solution that makes the solution of each sub-problem optimal, so as to obtain the optimal solution of the original problem. Fisher's optimal partitioning method has been widely applied in the field of agriculture, and is used to assess the efficiency of agricultural production and land use, including the efficiency of agricultural resource utilization [45], crop growth detection [46], cropland potential assessment [47], and land quality analysis [48], among others. The traditional optimal segmentation algorithm can only obtain a local optimal solution rather than a global optimal solution, owing to the limitation of the local optimal solution, and is only applicable to short time-series and small-scale data; the computation amount and time increase exponentially for long time-series and large-scale data [49]. Therefore, in practical application, the traditional optimal segmentation method cannot meet the hierarchical division of ESUA in the study area. The manipulation of extensive long-term time-series data remains an underdeveloped area of research. Nonetheless, the processing of time-series data is integral to the examination of the sustainable dynamics of arable land. Consequently, this paper integrates Principal Component Analysis (PCA) with Fisher's Optimal Segmentation Method (FOSM) to create the TO–Fisher–OSM model, thereby enhancing its proficiency in handling time-series data. Compared with the commonly used natural breakpoint method, it can effectively segment the series in time and space through mathematical models and algorithms, and can adaptively select appropriate segmentation algorithms according to the characteristics of different series [50]. TO–Fisher–OSM can select the components that best cover the characteristics of the original data as principal components to complete the dimensionality reduction of the time series and serve as categorical variables, and can use as few principal components as possible to express the variability of the data, which not only reduces the number of categorical variables, but also better captures the correlation between the variables so as to improve the accuracy of the model. Its principle is as follows:

$$\begin{cases} L[P(n,2)] = \min_{2\leq j\leq n} D(1,j-1) + D(j,n), \\ L[P(n,k)] = \min_{k\leq j\leq n} \{L[P(n,k)] + D(j,n)\} \end{cases} \tag{3}$$

where n is the sample size, k is the number of splits, b(n, k) is the sample size split into k classes, D(i, j) is the sum of the squares of the deviations, L(i, j) the loss function, and $P(n,k)$ is the minimum of $L[P(n,k)]$, namely $L[P(n,k)] = \min_{1=i_1<i_2<i_g<\cdots<i_k<i_{k+1}=n+1} L[b(n,k)]$.

2.4. SUA Evaluation System Construction

Drawing upon the United Nations Sustainable Development Goals (SDGs) as a framework, this paper references the National Sustainable Agricultural Development Plan (2015~2030) and builds upon the findings of prior research [23,51]. Additionally, considering the regional peculiarities of the YHHRB, and adhering to the principles of systematicity, representativeness, and utility, a comprehensive evaluation system for Sustainable Agricultural (SUA) has been constructed, encompassing economic, social, and ecological dimensions. In total, 23 indicators, including film inputs, machinery inputs, and light inputs, were selected as input indicators; six indicators (arable land value per capita, number of employees in agriculture, forestry, animal husbandry, and fisheries, and sewage treatment rate) were selected as desired output indicators; three indicators (carbon emissions, surface pollution in agriculture, and the intensity of the grey-water footprint of the plantation industry) were selected as non-desired output indicators. The accessibility and operability of the index data were considered, and the comprehensive evaluation system for the sustainable utilization of YHHRB cropland was finally established as shown in Table 1.

Table 1. YHHRB comprehensive evaluation system for sustainable utilization of cultivated land.

System of Indicators	Initial Indicators	Secondary Indicators	Tertiary Indicators	Unit
SUA evaluation system	Economic inputs	Film inputs	Agricultural plastic film use per unit of GDP	t/CNY 1 million
		Fertilizer inputs	Fertilizer application per unit of GDP	t/CNY 1 million
		Pesticide inputs	Pesticide use per unit of GDP	t/CNY 1 million
		Government inputs	Percentage of expenditure on agriculture	%
		Electricity inputs	Rural electricity consumption	KWh
	Social inputs	Labor inputs	Size of rural population	%
			Ease of farming	km
			Cultivated land area per capita	ha/per
			Rotation and fallow status	%
		Mechanized inputs	Degree of agricultural mechanization	kilowatt
		Irrigation inputs	Effective irrigated area	Kha
			Distance to water source	km

Table 1. *Cont.*

System of Indicators	Initial Indicators	Secondary Indicators	Tertiary Indicators	Unit
SUA evaluation system	Ecological inputs	Land inputs	Cropland fragmentation	Blocks/Kha
			Percentage of high-grade arable land	%
			Share of arable land area	%
			Cropland type as a percentage	%
			Percentage of basic farmland area	%
		Resource inputs	Percentage of area graded by slope	%
			NDVI	
			Soil-water harmony	m^3/ha
			Ratio of cropland area to forest and grassland area	
		Light input	Length of exposure	Hour
	Expected outputs	Economic benefits	Value of arable land per capita	CNY/per
			Disposable income per rural household	ten thousand CNY
			Value of ecological services	ten thousand CNY
		Social benefits	Coefficient of income disparity between urban and rural areas	
			Production per unit area of arable land	t/ha
		Ecological benefit	Percentage of forest cover	%
			Sewage treatment rate	%
	Unexpected outputs	Gaseous waste	Carbon footprint	mt
		Solid waste	Agricultural surface pollution	t
		Liquid waste	Graywater footprint intensity of cultivation	m^3/kg

2.5. Data Sources

This study used panel data from 52 prefecture-level cities in the YHHRB from 2010 to 2020. The resources used to compile panel data for the 52 prefecture-level cities from 2011 to 2021 included the China Urban Statistical Yearbook, China Regional Economic Statistical Yearbook, Hebei Statistical Yearbook, Jiangsu Statistical Yearbook, and Third National Land Survey databases; some missing data were supplemented by interpolation or exponential smoothing. Owing to the different positive and negative directions and unit magnitudes of the data for each indicator, it was necessary to standardize the original data using the positive indicator standardization Formula (4) and negative indicator standardization Formula (5) before performing data calculation:

$$b_{ij} = \frac{a_{ij} - a_j^{min}}{a_j^{max} - a_j^{min}} \tag{4}$$

$$b_{ij} = \frac{a_j^{max} - a_{ij}}{a_j^{max} - a_j^{min}} \tag{5}$$

where b_{ij} is the normalized value, a_{ij} is the ith value of the j metrics, and a_j^{max} and a_j^{min} are the maximum and minimum values of the j indicators, respectively.

3. Results and Analysis

3.1. Changing Patterns of Cropland Use

The utilization patterns of arable land in the YHHRB over the period from 2010 to 2020 are delineated in Table 2 and Figure 3. Scrutiny of the temporal aspect reveals that between 2010 and 2015, the allocation of arable land exhibited minimal fluctuation, with a staggering 96% of the arable land within the municipal jurisdiction remaining unaltered. The prevalent outward movement of arable land was toward conversion into construction land, whereas the principal inward movement stemmed from the transformation of grassland. Notably, the extent of arable land ceding from the region between 2015 and 2020 was marginally reduced by 1.1%, registering a general trend of stability. The quantum of land transitioning to construction purposes diminished by a significant 35.41%, whereas there was a discernible upsurge of 15.7% in the conversion of land to shrubbery. The reversion of grassland and forested areas to arable land collectively accounted for 71.2% of the total area subjected to conversion. These findings suggest that throughout the trajectory of arable land conversion, there is a palpable diminution in the expanse of non-agricultural territories, such as construction land and shrub land, concurrent with a gradual expansion of the agricultural landmass.

Table 2. Arable land use transfer matrix, 2010–2020.

2010–2020	Grassland	Cropland	Dioecious	Wasteland	Building Land	Woodland	Wetlands	Water
grassland	33,594.14	4122.92	213.50	38.78	325.30	2331.10	0.01	35.41
cropland	2838.78	1,004,189.64	1.49	2.56	24,753.50	4107.92	0.01	3351.35
dioecious	272.50	3.28	1241.82	0.00	0.02	243.19	0.00	0.00
wasteland	17.62	137.31	0.00	2507.44	1051.03	0.00	0.00	1293.35
building land	0.18	20.75	0.00	1.16	268,226.19	0.04	0.00	1333.50
woodland	15.85	3499.95	344.19	0.00	142.99	125,633.98	0.00	0.86
wetlands	0.00	0.00	0.00	0.00	0.00	0.00	0.00	0.00
water	5.75	3798.17	0.00	355.58	2205.81	9.77	0.00	37,635.35

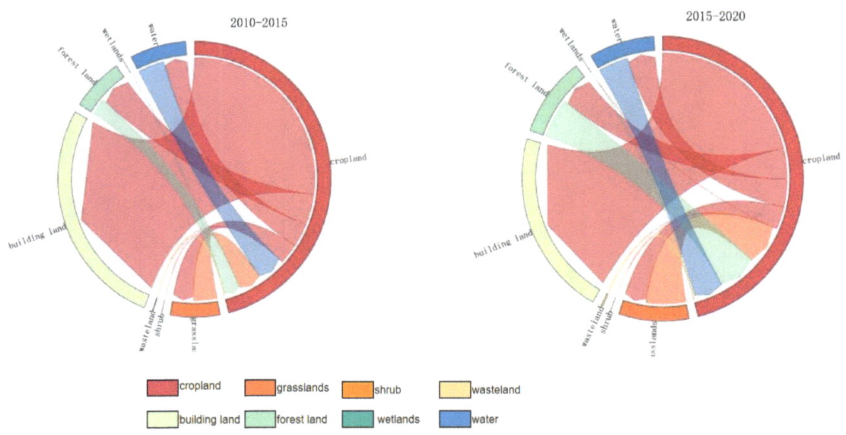

Figure 3. Arable land use transfer chord, 2000–2020.

Spatially, the interconversion of arable land and other land types within the YHHRB exhibits pronounced spatial heterogeneity (Figure 4). The primary sources of arable land conversion in these areas are grassland, aquatic systems, and forested land. The reclamation of grassland is particularly prevalent in the western reaches of Hebei Province, the central and southern districts of Shandong Province, and the western portion of Henan Province, with grassland contributing to 35.59% of the converted arable land. Aquatic systems are the source of 32.79% of the converted arable land, primarily located in the eastern and southern parts of Jiangsu Province, the central region of Tianjin, and the southeastern portion of Anhui Province. Forested land accounts for 30.21% of the converted arable land, primarily distributed across Beijing, Henan Province, and Hebei Province. The transfer of grassland and aquatic systems exhibits a pattern of spatial clustering, whereas the distribution of forested land is more diffuse. Arable land is primarily acquired from construction land and forested land, with the area of arable land transferred to construction land representing 70.61% of the total converted area, primarily occurring in Anhui Province, Hebei Province, and the Municipality of Tianjin. The area transferred to forested land constitutes 11.72% of the converted area, primarily found in the Municipality of Beijing, Jiangsu Province, and Shandong Province.

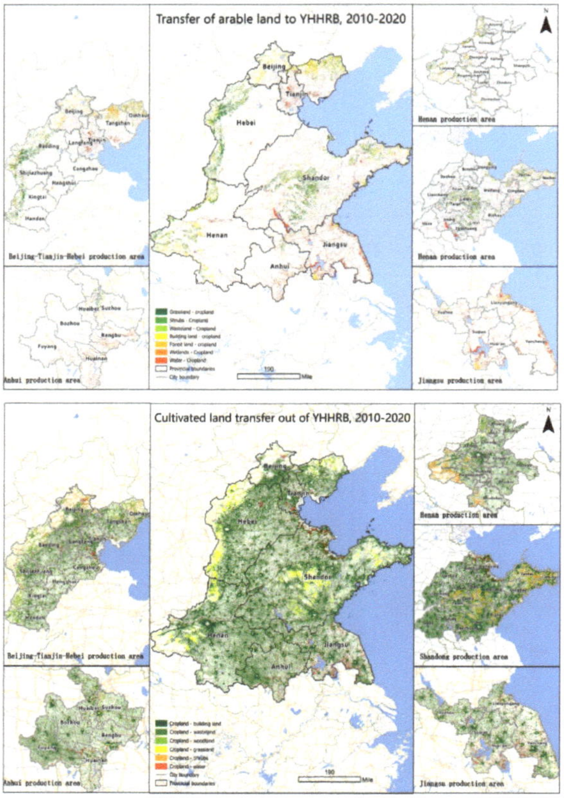

Figure 4. Spatial pattern of changes in cropland use, 2010–2020.

3.2. *Characterization of the Spatial and Temporal Evolution of the Efficiency of the Sustainable Use of Arable Land*

3.2.1. Model Optimization Evaluation

In this study, the Super-SBM, DGA–Super-SBM, and projection-seeking model for evaluating high-dimensional indicators were compared in terms of model performance by being used to evaluate the SUA land resources in a high-dimensional indicator system. In evaluating model performances, we adopted five indicators: the silhouette coefficient, Davies–Bouldin index, overall information retention rate, running time, and visualization test. The silhouette coefficient measures the tightness within the cluster to which each sample belongs and the separation from other clusters. Higher silhouette factors indicate a better distribution of samples and a more accurate delineation of clusters. The Davies–

Bouldin index measures the similarity between coarseness and the closeness within clusters. A lower Davies–Bouldin index indicates a better delineation of clusters. The combined assessment of these two metrics was used to reflect the modeling performances of the three models. The overall information retention rate measures the amount of information retained in the downscaled dataset relative to the original dataset, and the expectation is that the downscaled dataset will retain sufficient information while reducing dimensionality to maintain model accuracy and evaluative power. Running time can directly affect model efficiency and practicality in practical applications. The calculation results of the above four indicators are shown in Table 3. Finally, the three models were visualized and examined by scatter plots, and the distribution and clustering of the dimensionality-reduced data and original data were observed to determine whether the dimensionality-reduced data could effectively express the structure of the original dataset (Figure 5). We found that DGA–Super-SBM was the most suitable model for evaluating the SUA of land resources.

Table 3. Model performance evaluation.

Norm / Model	Silhouette Factor	Davies–Bouldin Index	Overall Information Retention Rate	Running Time
Super-SBM	0.5779	0.4229	—	2s’26
DGA–Super-SBM	0.7604	0.3023	94.96%	0s’91
Projective tracing algorithm	0.5816	0.4819	11.44%	3s’81

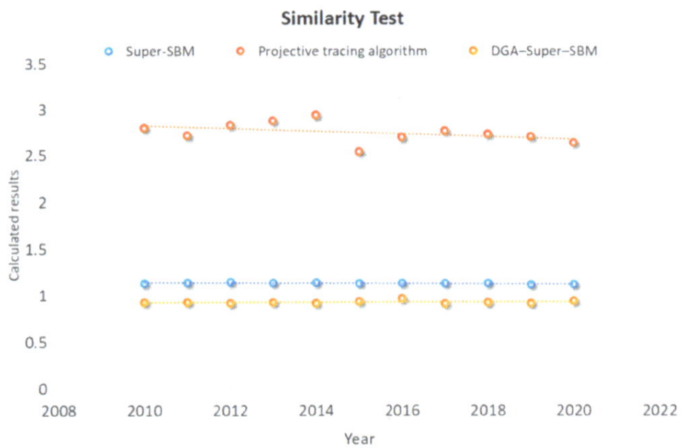

Figure 5. Similarity tests. Note: In the figure, the trend and the amount of information of DGA–Super-SBM and Super-SBM are basically kept the same, while the difference of Projective tracing algorithm is larger.

3.2.2. Characterization of the Time Evolution of Sustainable Use Efficiency of Arable Land

In this study, based on the national strategy of sustainable development of food production, the YHHRB was divided into seven regions according to administrative districts (Table 4). The average value of ESUA of arable land from 2010 to 2020 was calculated on the basis of administrative regions (Figure 6).

Table 4. Yellow–Huai–Hai River Basin (YHHRB) regional divisions.

Administrative Divisions	Province	Prefecture-Level City
Northern YHHRB	Beijing	
	Tianjin	
	Hebei	Shijiazhuang, Qinhuangdao, Tangshan, Langfang, Baoding, Cangzhou, Hengshui, Xingtai, Handan
Central YHHRB	Shandong	Jinan, Dezhou, Binzhou, Dongying, Zibo, Weifang, Yantai, Weihai, Liaocheng, Qingdao, Heze, Jining, Zaozhuang, Linyi, Rizhao, Tai'an
	Henan	Zhengzhou, Anyang, Puyang, Hebi, Xinxiang, Jiaozuo, Luoyang, Kaifeng, Shangqiu, Xuchang, Pingdingshan, Luohe, Zhoukou, Zhumadian
Southern YHHRB	Anhui	Huabei, Bozhou, Cebu, Fuyang, Bengbu, Huainan
	Jiangsu	Xuzhou, Lianyungang, Suqian, Huai'an, Yancheng

Figure 6. ESUA distribution in the Yellow–Huai–Hai River Basin.

From Figure 6, from 2010 to 2020, 44.23% of cities in the YHHRB had arable land with ESUA of >1. Among them, Beijing had the largest value (1.3727), followed by Cangzhou City (1.1347); the lowest value is for Hebi City (0.4717). Overall, the average ESUA value for the YHHRB from 2010 to 2020 was 0.9012, and the number of more efficient and efficient cities increased from 15.38% in 2010 to 23.07% in 2020; the number of cities with medium efficiency or above decreased from 67.31% in 2010 to 71.15% in 2020, indicating that the region as a whole increased from low efficiency to medium efficiency. Figure 5 also shows regional differences in ESUA. At the provincial level, 11.54% of cities in Hebei Province had an ESUA of >1, 7.69% of cities in Henan Province had an ESUA of >1, 5.77% of cities in Jiangsu Province had an ESUA of >1, 13.46% of cities in Shandong Province had an ESUA of >1, and 1.92% of cities in Anhui Province had an ESUA of >1. Except for Beijing and Tianjin, the two municipalities with the highest efficiency, the ESUA of the remaining provinces was in the order of Hebei > Jiangsu > Shandong> Henan > Anhui.

Figure 7 shows temporal changes in ESUA. ESUA for the whole YHHRB showed an increasing and then decreasing trend, with ESUA peaking in 2016 and decreasing from 2016 to 2020. ESUA in the northern YHHRB consistently maintained a high efficiency level during the study period, with a mean value of 1.136. The gap between ESUAs in the central and northern regions gradually narrowed, with a mean value of 0.866; in the southern region, ESUA had a clear upward trend, with a mean value is 0.878. The results show that ESUA is more efficient in the northern part of the YHHRB than in the central and southern parts. In the YHHRB, land management policies as well as high-tech agricultural development strategies are relatively developed in the northern region, where local governments utilize advanced high-tech agricultural applications and environmental protection to spend more resources on improving ESUA [52]. The central region has rich land resources and has experienced sustainable and efficient development of arable land; however, owing to poor agricultural management, a large amount of reclaimed arable land has had a decline in stability and quality, leading to a decline in the ESUA [14]. Although ESUA in the southern region was lower than that in the northern and central provinces in 2010, its favorable geographical location and land resources, coupled with a push to develop green agriculture and increase investment in agricultural science and technology, have allowed the ESUA to gradually improve, contributing to a steady increase in the efficiency of cultivated land in the YHHRB in terms of sustainability.

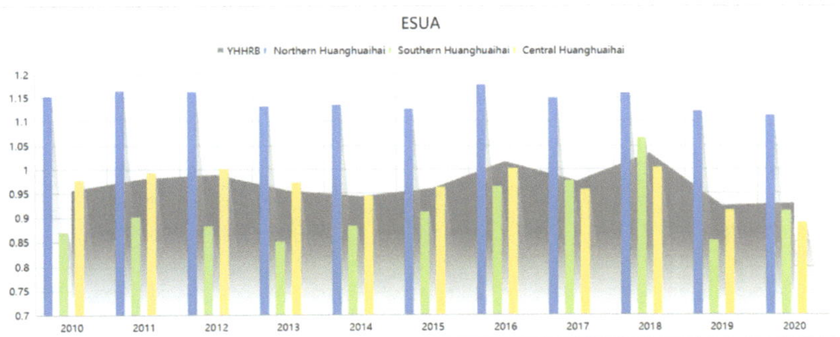

Figure 7. Temporal changes in average ESUA in different regions of the YHHRB.

3.2.3. ESUA Spatial Difference Characterization

We named the evaluation efficiency of the sustainable use of arable land from low to high as the low efficiency zone, lower efficiency zone, medium efficiency zone, higher efficiency zone, and high efficiency zone, and from Figure 6, we can see that there are 30 cities, accounting for 57.7%, in the high efficiency zone and higher efficiency zone, and 10 cities, accounting for 19.23%, in the medium efficiency zone, so it can be seen that the Huanghuaihai region as a whole is in a higher sustainable use efficiency level. From Figure 6, ESUA in the YHHRB increases from west to east, and spatial agglomeration is obvious. The overall high ESUA in the northeast and low ESUA in the southwest are mainly due to differences in the (1) soil and water resources, (2) the level of agricultural science and technology, and (3) land occupation and environmental pollution. First, the northeastern YHHRB is mostly plains, with fertile soil, sufficient water sources, and flat terrain; in contrast, the southwestern YHHRB is mostly mountainous and hilly, with overall poor land quality, shallow soils, low soil fertility, and severe soil erosion, all of which restrict arable land use [53]. Second, the northeastern region of the YHHRB, which has a more advanced economy and level of development, has a much higher level of agricultural mechanization and agricultural science and technology input than the southwestern region; larger government investment in agriculture is indispensable to rapidly improving ESUA [54]. Finally, the southwestern region of the YHHRB suffers from poor land development and environmental pollution caused by urban expansion and industrial development, which has greatly limited the ESUA of local arable land. Supporting the sustainable development of arable land requires increased land governance and environmental protection [55].

3.2.4. Dynamic Sustainable Efficiency of Cropland

Figure 8 shows changes in 2-yearly ESUA data for each province in the YHHRB from 2010 to 2020. We observed a negative overall change in D-ESUA (0.9977) over the research period, with D-ESUA decreasing by 2.3% per year. The average sustainable efficiency of cropland resources from 2010 to 2020 showed a fluctuating trend, indicating that SUA in the YHHRB is still in the exploratory development stage. The number of provinces showing a positive offset (>1) in D-ESUA during the study period accounted for 50% of the entire YHHRB, among which Beijing and Tianjin had the highest average annual growth rates (4.89 and 5.56%, respectively). While the other provinces did not show an offset in efficiency, they also showed an increasing trend yearly. In summary, according to D-ESUA analysis, the YHHRB is still in the exploratory stage in terms of sustainable development of arable land, but it is expected to experience rapid development in the near future.

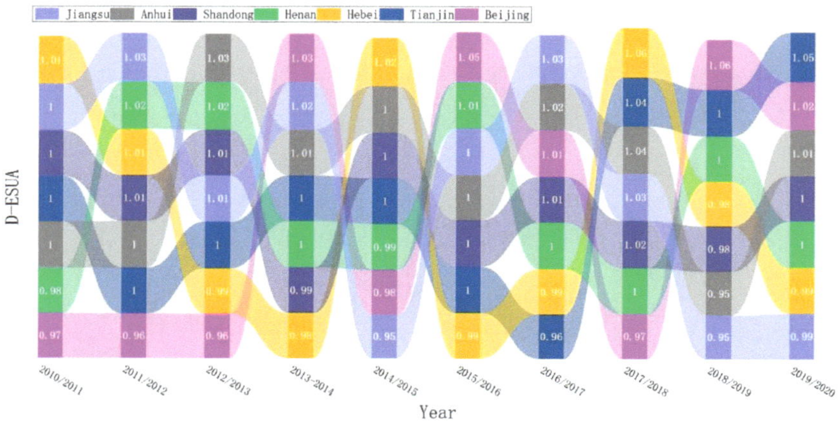

Figure 8. Temporal changes in dynamic sustainable efficiency of cropland for different provinces of the YHHRB.

Figure 9 shows average D-ESUA values for different regions over time. The D-ESUA of the whole YHHRB fluctuated between 2010/2011 and 2017/2018; between 2017/2018 and 2019/2020, it decreased and then increased. D-ESUA values were highest in the southern region from 2010/2011 to 2013/14 and in 2016/2017; they were highest in the northern region from 2014/2015 to 2019/2020 (with the exception of 2016/17). During the study period, D-ESUA in the northern region increased from low to high efficiency, probably owing to local government policies on arable land protection and the rapid development of new agricultural technologies. D-ESUA resources in the central and southern regions were slow to improve, experiencing a fluctuating rising trend, probably because of the transition of land cultivation due to local climate and crop maturity, which led to a decline in land quality and affected the sustainable development efficiency of arable land.

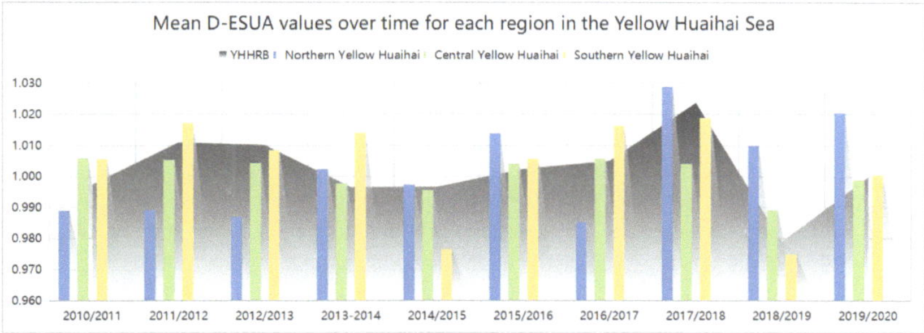

Figure 9. Temporal changes in mean dynamic sustainable efficiency of cropland (D-ESUA) for each region of the Yellow–Huai–Hai River Basin (YHHRB).

3.2.5. Grading of the Efficiency of the Sustainable Use of Arable Land

We used the natural breakpoint, optimal segmentation, and TO–Fisher–OSM methods to classify the SUA of land, to explore the most suitable classification model for SUA in the YHHRB. The natural breakpoint method is automatically generated by ArcGIS; the results of the optimal segmentation method and TO–Fisher–OSM are shown in the following Tables 5 and 6, respectively. According to the actual needs of arable land classification, we set the number of classification to five.

Table 5. Grade classification based on the optimal partitioning method.

Number of Categories	Error Function	Optimal Segmentation Results
2	0.3788	1–15,16–54
3	0.2296	1–3, 4–15, 16–54
4	0.1329	1–3, 4–15, 16–50, 51–54
5	**0.0724**	**1, 2–6, 7–15, 16–50, 51–54**
6	0.049	1, 2–5, 6–11, 12–15, 16–50, 51–54
7	0.0278	1, 2–5, 6–11, 12–15, 16–43, 44–52, 53–54
8	0.0189	1, 2–5, 6–11, 12–15, 16–39, 40–48, 49–52, 53–54
9	0.0128	1, 2–3, 4–6, 7–11, 12–15, 16–39, 40–48, 49–52, 53–54
10	0.0095	1, 2–3, 4–6, 7–11, 12–15, 16–24, 25–41, 42–49, 50–52, 53–54

Note: The color-emphasized fonts in the table are the optimal solutions of the Optimal Segmentation and TO-Fisher-OSM algorithms chosen by the authors.

Table 6. Grade classification based on the TO–Fisher–OSM method.

Number of Categories	Error Function	TO–Fisher–OSM Results
2	0.6249	1–15, 16–54
3	0.3066	1–15, 16–52, 53–54
4	0.1513	1–12, 13–23, 24–52, 53–54
5	**0.0931**	**1–3, 4–12, 13–23, 24–52, 53–54**
6	0.0552	1–3, 4–12, 13–19, 20–38, 39–52, 53–54
7	0.039	1–3, 4–12, 13–17, 18–23, 24–45, 46–52, 53–54
8	0.0293	1–3, 4–12, 13–17, 18–23, 24–36, 37–48, 49–52, 53–54
9	0.0226	1–2, 3–9, 10–12, 13–17, 18–23, 24–36, 37–48, 49–52, 53–54
10	0.0168	1–2, 3–9, 10–12, 13–15, 16–19, 20–23, 24–36, 37–48, 49–52, 53–54

Note: The color-emphasized fonts in the table are the optimal solutions of the Optimal Segmentation and TO-Fisher-OSM algorithms chosen by the authors.

Maintaining the main statistical characteristics of data is an important principle in data classification. There are three commonly used grading evaluation models: information entropy, the Gini coefficient, and weighted total deviation grading accuracy evaluation model. Information entropy measures a dataset's existence, focusing on measuring information uncertainty; the smaller the information entropy, the more reasonable the division. The Gini coefficient is used to extract the Gini impurity to measure the uncertainty or impurity of the rank classification; the smaller the Gini coefficient, the more reasonable the classification. The weighted total deviation grading accuracy model considers the consistency within and between grades and focuses on the evaluation of the grading interval; the closer the grading accuracy model is to 1, the better the model performance. The three evaluation models only consider one or two factors that affect the merits of grading, and do not have universality for all grading results; therefore, they have greater limitations in practical application.

To evaluate the grading results more comprehensively, we comprehensively considered these three evaluation models as the attributes of multi-attribute decision making, that is, as evaluation indices for determining the reasonableness of grading based on the natural breakpoint method, optimal segmentation method, and TO–Fisher–OSM method (Table 7, Figure 10). There were large differences in grading among the three grading methods, with TO–Fisher–OSM offering the best grading accuracy and reasonableness. From Figure 10, the natural breakpoint approach provided good grading visually and a more uniform distribution at all levels; however, the demarcation points did not completely show the data distribution features. The optimal segmentation method emphasized data distribution features; however, there was some data aggregation, and the data distribution at all levels was unequal, obscuring the spatial rank inequalities in the YHHRB. TO–Fisher–OSM could fully highlight the data distribution characteristics while maintaining the overall sense of balance, quantity, and rank of SUA in the YHHRB. Moreover, the visualization effect was better than that of the other methods.

Table 7. Determination of grading evaluation results.

	Information Entropy	Gini Coefficient	Weighted Total Deviation Grading Accuracy
Natural breakpoint method	0.6448	0.1082	0.1805
Optimal segmentation method	0.4945	0.1176	0.1829
TO–Fisher–OSM	0.3987	0.0660	0.4247

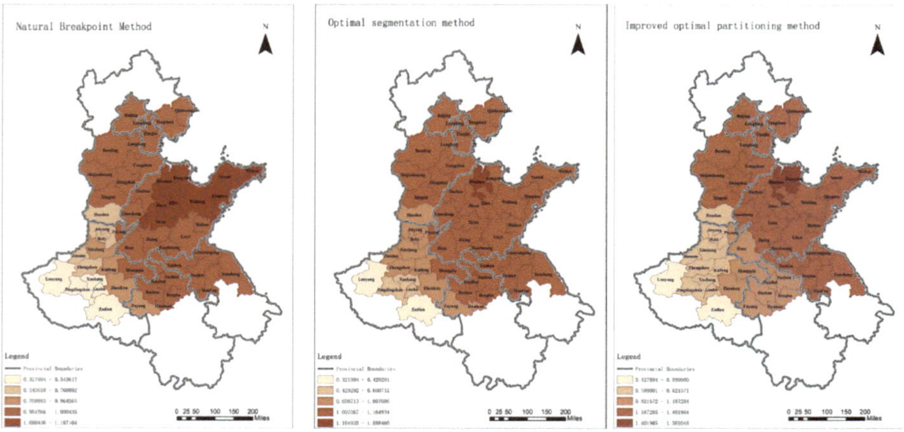

Figure 10. Thematic diagrams of ESUA grade classification.

3.3. Analysis of the Correlation between Changes in Cropland Use Patterns and Sustainable Use Efficiency of Cropland

This investigation isolates variables from four distinct dimensions: arable land area, arable land function, arable land configuration, and arable land utilization intensity. These dimensions are examined to discern the repercussions of alterations in the structure of arable land use on the efficiency of its sustainable utilization. The core indicators encompass

① Arable Land Area Ratio, which quantifies the retention of arable land within the YHHRB by comparing the regional arable land area to the total regional area; ② cultivated land transfer rate, signifying the proportion of land area that transitions from alternative land use categories to arable land; ③ cropland turnover rate, indicating the proportion of land area that converts from arable land to other land use types; ④ Arable Land Shape Index, which evaluates the degree of clustering or dispersal of arable land by assessing the deviation of the regional arable land form from a circular or square shape of equivalent area; ⑤ Arable Land Density, which assesses the clustering of arable land by calculating the number of arable land patches per unit of area; ⑥ Yield per Unit Area, representing the productivity of arable land, or the quantity of foodstuffs yielded per unit of arable land area; ⑦ Arable Land Replanting Index, reflecting the extent of arable land utilization within a given area, computed by the ratio of the sown area of arable land to the total arable land area. Tabular descriptive statistics for each of these indicators are provided in Table 8.

Table 8. Descriptive statistics of variables.

Targets	2010		2015		2020	
	Average Value	Standard Deviation	Average Value	Standard Deviation	Average Value	Standard Deviation
Arable Land Area Ratio	49.003	13.734	48.835	13.668	48.095	14.383
Cultivated land transfer rate	2.697	1.397	3.082	1.534	4.445	2.833
Cropland turnover rate	4.029	2.172	4.133	2.066	5.200	4.239
Arable Land Shape Index	47.397	20.822	51.002	23.116	51.246	22.509
Arable Land Density	1.925	0.840	2.176	0.865	2.102	0.777
Yield per Unit Area	7.626	1.802	7.905	2.539	8.140	2.822
Arable Land Replanting Index	1.259	0.290	1.243	0.463	1.542	0.866

Utilizing the center of gravity model along with statistical data, the coordinates of the center of gravity for cropland use pattern and cropland sustainable use efficiency indicators within the YHHRB between 2010 and 2020 were determined. Subsequent to this, an analysis of the trajectory of the center of gravity was conducted, based on which the spatial–temporal coupling relationship and associated characteristics were investigated, aiming to furnish a theoretical foundation for subsequent research. As depicted in Figure 11, the center of gravity for the sustainable use efficiency of arable land in the YHHRB experienced a northeastward shift from 2010 to 2020. The center of gravity for each indicator of the cropland use pattern exhibited the following movements: the center of gravity for the cropland area share ratio heading northwest from 2010 to 2020; the center of gravity for the cultivated land transfer rate heading northeast from 2010 to 2020; the center of gravity for the cropland turnover rate initially moving eastward from 2010 to 2015, followed by a northeastward shift from 2015 to 2020; the center of gravity for the cropland shape index migrated northwestward during 2010–2015 and southeastward from 2015 to 2020; the center of gravity for cropland density trended southeastward from 2010 to 2015 and southwestward from 2015 to 2020; the center of gravity for yield per unit area shifted southwestward from 2010 to 2015 and northwestward from 2015 to 2020; and the center of gravity for the cropland replanting index experienced a southward and north-westward shift from 2010 to 2015, followed by a further north-westward movement from 2015 to 2020.

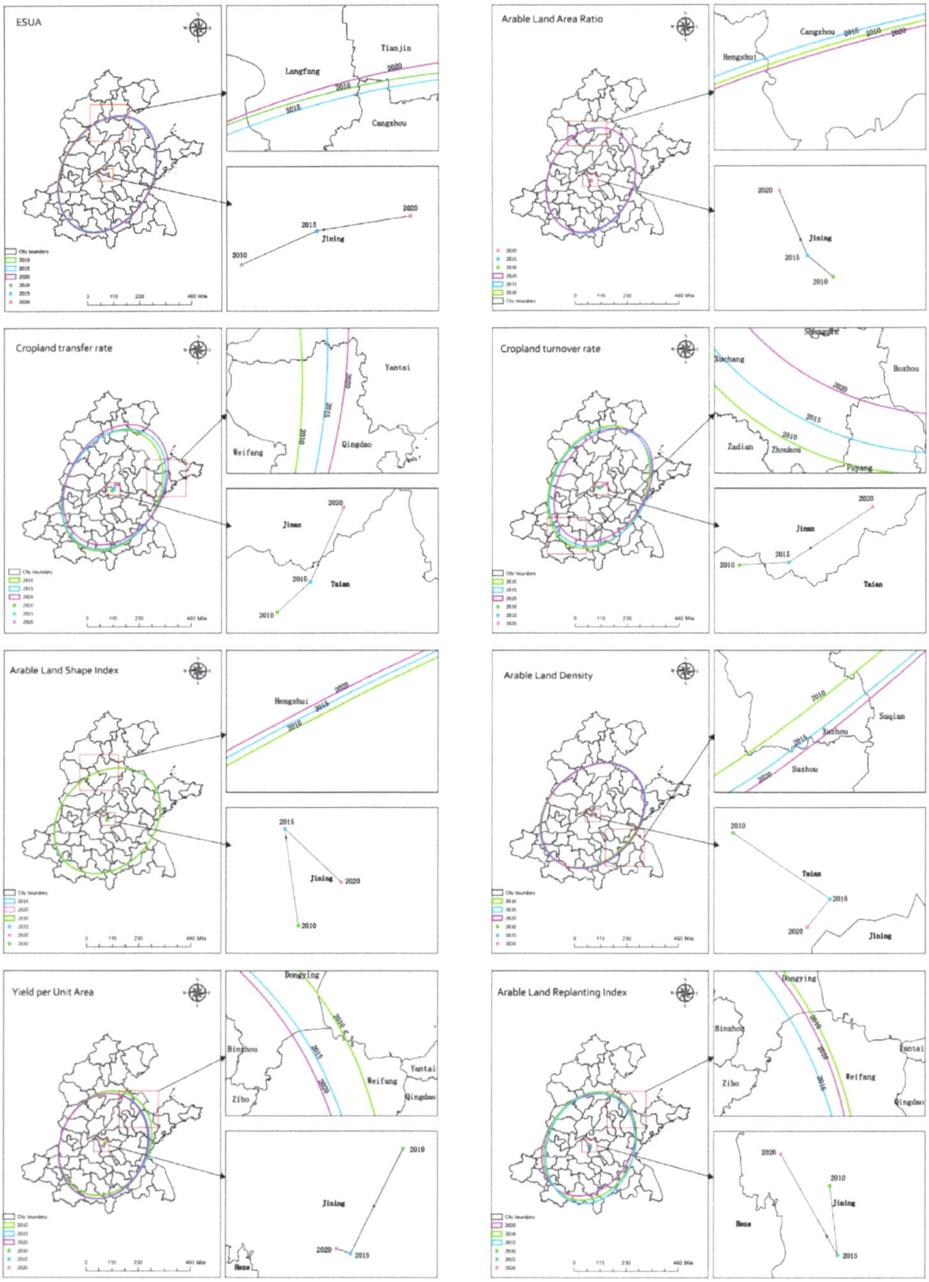

Figure 11. Migration of the center of gravity of the variables of sustainable use efficiency of arable land and arable land use pattern.

The trajectory of the center of gravity for the sustainable use efficiency of arable land aligns in the north–south direction with that of the arable land area share ratio, the arable land conversion rate, and the arable land abandonment rate, while it diverges from the direction of the arable land density and the yield per unit area in the north–south direction.

Similarly, in the east–west direction, the center of gravity for sustainable use efficiency is aligned with the movement of the arable land area share ratio and the arable land abandonment rate, whereas it exhibits a different trend from the arable land density, the yield per unit area, and the replanting index. During the period from 2010 to 2015, the center of gravity for sustainable use efficiency and the arable land shape index moved in the same direction in the north–south direction, whereas in the subsequent period from 2015 to 2020, they exhibited a similar movement in the east–west direction. In contrast, the relationship between the sustainable use efficiency of arable land and the arable land density was inverse in the north–south direction from 2010 to 2020, and they moved in the same direction in the east–west direction from 2010 to 2015. These observations suggest that there are certain correlations between the cultivated land transfer rate, conversion rate, yield per unit area, and replanting index and the efficiency of sustainable arable land use. However, the specific influence of the arable land area ratio, shape index, and density on sustainable arable land use efficiency is not yet fully understood. To delve deeper into the specific impacts of these variables on sustainable arable land use efficiency, further empirical research is warranted.

3.4. Empirical Study on Analyzing the Impact of Changes in Cropland Use Patterns on the Sustainable Use Efficiency of Cropland

In the course of assessing the stability of the selected indicators, we employed the Augmented Dickey–Fuller (ADF) unit root test to gauge the stationarity of the efficiency of sustainable arable land use, the proportion of arable land area, the rate of conversion of arable land into other uses, the rate of abandonment of arable land, the shape index of arable land, the density of arable land patches, the yield per unit area, and the replanting index. The findings reveal that all these indicators exhibit stationary characteristics, thereby refuting the conjecture of the presence of a unit root. To ascertain whether these indicators are interconnected in a long-term equilibrium, we conducted the Engle–Granger cointegration test. Initially, we hypothesized the absence of a cointegration relationship, but the *p*-value of the test result was below the 0.05 significance level, thereby warranting the rejection of the null hypothesis and affirming the presence of a cointegration relationship among these indicators. Consequently, we are justified in conducting a comprehensive analysis of these indicators using a panel data model.

Regarding the selection of a panel data model, researchers commonly confront three primary alternatives: the pooled ordinary least squares (OLS) model, the fixed effects (FE) model, and the random effects (RE) model. To discern the most fitting model, this study engaged in a comparative assessment utilizing the F-test, the Breusch–Pagan (BP) test, and the Hausman test. The outcomes disclosed *p*-values significantly below 0.05, affirming that the FE model is the most suitable choice. Additionally, the study included time as a dummy variable in the OLS regression model, and it was discerned that both the time term and the ESUA of arable land were statistically significant, indicating that temporal dynamics must be considered in the analysis. The regression outcomes for the fixed effects model that omitted the time effect were not significant, hence necessitating the establishment of a model with a time fixed effect.

Building upon the preceding analyses, we employed a time fixed effect (FE) model to examine the repercussions of alterations in cropland use patterns on the sustainable use efficiency of cropland. The regression analysis of the efficiency of sustainable arable land use within the YHHRB, alongside each variable, revealed *p*-values for the F-statistic values that were less than 0.05, corroborating the statistical significance of the regression outcomes (refer to Table 9). Among the variables, the replanting index and the cropland density demonstrated significant positive impacts on the sustainable use index of cropland, with regression coefficients of 0.138 and 0.067, respectively. This implies that for every incremental unit rise in the replanting index, the sustainable use efficiency of cropland escalates by 13.8%, and for every incremental unit increase in cropland density, it increases by 6.7%. These findings suggest that both the cropland replanting index and density have been on the rise from 2010 to 2020, which collectively contribute to the rational utilization of land resources and the enhancement of agricultural productivity, thereby bolstering the sustainable use efficiency of cropland. In contrast, the output per unit area and the cropland transfer rate exhibited significant negative impacts, with regression coefficients of −0.02 and −0.01, respectively. This suggests that for every incremental unit increase in the output per unit area, the sustainable use efficiency of cropland decreases by 2%, and for every incremental unit rise in the cropland transfer rate, it decreases by 1%. These outcomes indicate that the escalation in unit area production and the augmentation of cropland transfer rates exert an intensifying pressure on land resources, thereby hindering the advancement of cropland ESUA. Variables such as the changes in arable land area per unit, the cultivated land transfer rate, and the shape index of arable land manifested negative effects, albeit to a lesser extent and not statistically significant. This suggests that the fluctuations in these metrics do not significantly influence the sustainable use efficiency of arable land. This might be attributable to the relatively minor magnitude of their changes or the influence of other factors, rendering their roles in improving the sustainable use efficiency of arable land less conspicuous. Consequently, these variables are not elaborated upon further in this study.

Table 9. Time fixed effects model regression results.

Variant	Ratio	t-Value	Standard Error
Intercept	1.038 ***	7.444	0.139
Arable Land Area Ratio	−0.003	−1.646	0.002
Cultivated land transfer rate	−0.02 *	−1.649	0.012
Cropland turnover rate	0.138 **	2.394	0.058
Arable Land Shape Index	−0.01 **	−2.235	0.004
Arable Land Density	−0.002	−0.593	0.003
Yield per Unit Area	−0.001	−1.342	0.001
Arable Land Replanting Index	0.067 ***	3.017	0.022

Note: *, ** and *** denote significant at the 10%, 5% and 1% statistical levels, respectively.

4. Discussion

Analysis of the quantitative disparities in the spatial distribution of arable land transfers within the study area reveals distinct patterns. The influx of arable land into the region is notably concentrated in the western mountainous and central hilly zones, whereas in the plain regions, conversion of arable land to other land uses, particularly for construction, is prevalent. This trend reflects a marked preference for "gradual encroachment to offset ruggedness", "territorial expansion to counteract elevation", and "unified occupation to nullify absence", suggesting that human influence on arable land usage is pronounced, especially in flat areas. Notably, in the plains, the encroachment of arable land by construction expansion is particularly significant [56]. Furthermore, the shift in arable land use demonstrates an adaptive response to topographical features and resource allocation, with a significant portion of arable land in Anhui and Jiangsu provinces being reallocated from water bodies. This phenomenon underscores the rapid pace of population growth and urbanization, coupled with escalating demands for land resources. To satisfy the needs of food production and economic advancement, some water-covered areas may be reclaimed for agricultural use [57]. Currently, the Jiangsu provincial government is actively promoting the policy of "converting ploughs to lakes and grasses", signaling a heightened emphasis on ecological environmental conservation and the implementation of measures to curtail the overexploitation and irrational utilization of land resources [58].

During the research period, ESUA decreased in the YHHRB. This may have been due to rapid urban development, which led to environmental pollution and agricultural structure imbalances. This developmental mindset has led to a decline in arable land quality and ecological and environmental crises [59]. Some studies have shown that the overall ESUA is higher in the YHHRB compared with other regions. However, the ESUA trends differed among the provinces in the region (Figure 12), reflecting the different arable land conditions and local development paths. Surprisingly, the ESUAs of Henan (Yu), Shandong (Lu), and Jiangsu (Su), which form the major grain-producing area in the country ('Yulu-Suzhou'), were not high. Yulu-Suzhou is located in the Central Plains and is the country's main food production region; however, the quality of arable land has declined owing to agricultural overuse and environmental pollution. Simultaneously, owing to excessive abstraction of groundwater, the overall lack of surface water has decreased the soil water content and fertility, leading to a decrease in grain production. In 2007, with the release of the National Medium and Long-Term Urbanization Plan (2007–2020), the Yulu-Suzhou region began the process of accelerated urbanization and economic/industrial development while attracting many laborers and rapid development of secondary and tertiary industries. This resulted in serious environmental pollution and a decline in the quality and quantity of arable land [60].

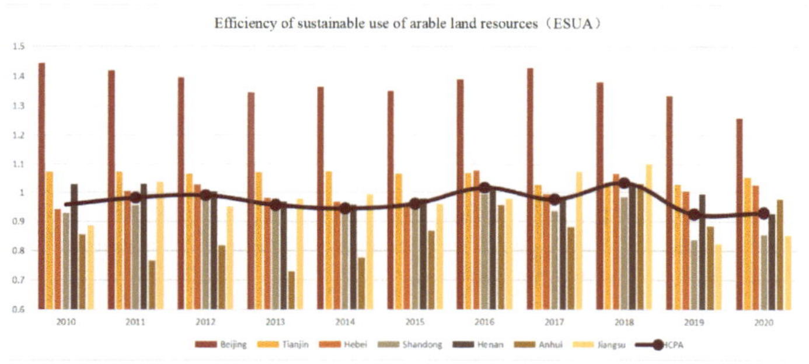

Figure 12. ESUA trends by province in the YHHRB.

The cities with the highest outputs are mainly concentrated in the provinces of Yu, Lu, and Su, which contain cities with the lowest efficiency. ESUAs are determined by desired outputs; however, inputs and undesirable outputs are also important influencing factors. Therefore, we considered not only the relationship between inputs and desired outputs but also the relationship between desired and undesired outputs, and the relationship between inputs and undesired outputs. Accordingly, we suggest two possible causes of low ESUA: (1) more non-desired outputs than desired outputs and (2) increased non-desired outputs due to input overload. The first results in low ESUA of arable land; for the second, we found that economic inputs, social inputs, ecological inputs, and non-desired outputs showed an increasing fluctuating trend in the main grain producing areas of the YHHRB during the study period; the overall growth rate was lower than the non-desired outputs, although the desired outputs also increased (Figure 13). Although the desired outputs also increased, the overall growth rate was lower than the non-desired outputs. During 2019~2020, ecological inputs, social inputs, and non-desired outputs showed a decreasing trend, while economic inputs showed an increasing trend. Therefore, the relationship between input and output factors requires further observation in future works.

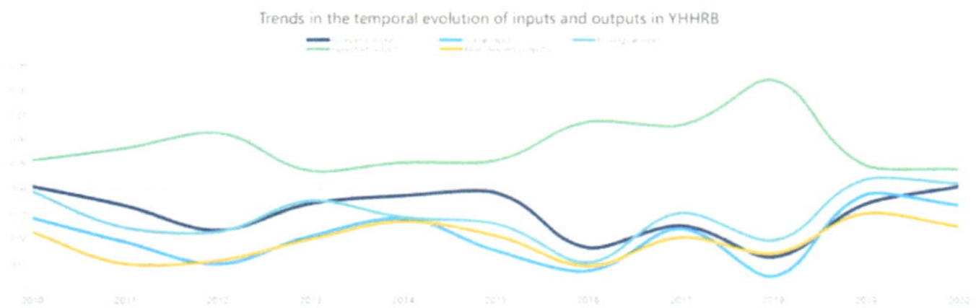

Figure 13. Temporal evolution of input–output in the YHHRB.

5. Conclusions

In this scholarly endeavor, we have developed a comprehensive evaluation index system for Sustainable Utilization of Arable Land (SUA). By integrating the Data Envelopment Analysis–Super-SBM (DGA–Super-SBM) model, the Malmquist–Luenberger production index, and the TO–Fisher–OSM algorithm, we have established a computational framework capable of handling long-time-series, high-dimensional index systems. This framework was subsequently applied to assess the Environmental Sustainable Assessments (ESUAs) for 52 prefecture-level municipalities within China's Huanghuai-Haihai Sea grain-producing regions. Our analysis delved into the trajectory of cropland use changes and the impact of these variables on ESUA. The findings of our study are as follows.

Between 2010 and 2020, the utilization of arable land within the YHHRB underwent minimal alteration, with approximately 90% of the land maintaining its agricultural status. The predominant direction of arable land conversion was to construction purposes, with the primary sources of such transfers being grasslands and aquatic systems. A spatial analysis of these patterns revealed that the conversion of arable land to construction land comprised 70.61% of the total area transferred, principally occurring in Anhui Province, Hebei Province, and the Municipality of Tianjin. Conversely, the reallocation of land from grasslands to arable land represented 35.59% of the transferred area, clustered predominantly in the western reaches of Hebei Province, the central and southern districts of Shandong Province, and the western portion of Henan Province. Additionally, the transformation of arable land from water systems to agricultural use was primarily concentrated in the eastern and southern sectors of Jiangsu Province, the central area of Tianjin Municipality, and the southeastern portion of Anhui Province.

Between the years 2010 and 2020, the Environmental Sustainable Assessment (ESUA) within the YHHRB exhibited a fluctuating trend, reaching its zenith in 2016. Geographically, ESUA progressively augmented from the west to the east, manifesting distinct spatial clustering, with a preponderance of high efficiency in the northeastern quadrant and lower levels in the southwestern reaches. The northern regions maintained their proficiency in ESUA, while the disparity between the central and northern regions lessened, and there was a marked uptick in the southern areas. The Malmquist–Luenberger production index disclosed a general decline in declining ESUA (D-ESUA), with an average annual diminution of 2.3%. However, D-ESUA increased in half of the municipalities, suggesting a transient enhancement in arable land utilization efficiency. The most substantial upsurge in D-ESUA was observed in the northern regions, at 3.2%, with more modest, fluctuating increments in the central southern parts of the country. Collectively, the sustainability of cropland resources in the Yellow and Huaihai regions is commendable.

Utilizing the center of gravity model, it was discovered that the alterations in the ESUA of arable land aligned with the north–south migration of the share of arable land area, as well as the rates of arable land conversion in and out. Conversely, these changes diverged from the movement of arable land density and the yield per unit area. In terms of the east–west progression, the gravitation of sustainable cropland use efficiency aligned with the rates of arable land conversion in and out, whereas it diverged from the trends in arable land area share, yield per unit area, and the replanting index of cropland. A time fixed effects model analysis revealed that the replanting index and cropland density exerted significant positive influences on ESUA. Conversely, the yield per unit area and the cropland turn-in rate demonstrated significant negative impacts. The alterations in cropland unit area, cropland turnout rate, and cropland shape index demonstrated negative effects, albeit minor and insignificant in statistical terms.

Under the environment of accelerating the construction of a strong agricultural country and comprehensively realizing the revitalization of the countryside, SUA is an important driving force to promote green modernization of agriculture and rural areas. It is important to study the evaluation criteria and spatiotemporal evolution of SUA to clarify the direction of sustainable development of arable land, drive the development of peripheral cities by high-efficiency cities, and drive peripheral cities by the main food-producing areas to ultimately achieve strategic sustainable development of arable land across the whole country. Owing to limitation of the model parameters, DGA–Super-SBM can only evaluate static sustainable utilization index systems; however, in reality, arable land changes dynamically. Therefore, it is necessary to identify dynamic indicators for the dynamic monitoring of sustainable utilization. How to evaluate high-dimensional dynamic indicator systems is a research direction for the future.

Author Contributions: Conceptualization, C.D.; methodology, Y.Z. and X.H.; validation, formal analysis, X.H.; data collation, X.H.; writing—original manuscript preparation, X.H. All authors have read and agreed to the published version of the manuscript.

Funding: This study was funded by the Project on Basic Research Operations of Central Research Institutes: Research on the Construction of Natural Resources Survey and Monitoring Technology System and Some Key Technologies (No. AR2203).

Data Availability Statement: The data presented in this study are available on request from the corresponding author due to the confidentiality of the data.

Acknowledgments: This study was made possible by the contributions and support of many parties. Please accept our apologies for any omissions. We would like to thank the institution that funded our research, the Project on Basic Research Operations of Central Research Institutes: Research on the Construction of Natural Resources Survey and Monitoring Technology System and Some Key Technologies (No. AR2203), and the editors and reviewers of Ecological Indicators for their professional comments on this manuscript. Without your contributions, our research would not have been possible. All authors agree to the content of the acknowledgements.

Conflicts of Interest: The authors declare no conflicts of interest.

References

1. Bai, Z.; Caspari, T.; Gonzalez, M.R.; Batjes, N.H.; Mader, P.; Bunemann, E.K.; de Goede, R.; Brussaard, L.; Xu, M.; Ferreira, C.S.S.; et al. Effects of agricultural management practices on soil quality: A review of long-term experiments for Europe and China. *Agric. Ecosyst. Environ.* **2018**, *265*, 1–7. [CrossRef]
2. FAO. *The 10 Elements of Agroecology: Guiding the Transition to Sustainable Food and Agricultural Systems*; Food and Agriculture Organization of the United Nations: Rome, Italy, 2018. Available online: http://www.fao.org/3/i9037en/i9037en.pdf (accessed on 10 May 2024).
3. Rajbanshi, J.; Das, S.; Paul, R. Quantification of the effects of conservation practices on surface runoff and soil erosion in croplands and their trade-off: A meta-analysis. *Sci. Total Environ.* **2023**, *864*, 161015. [CrossRef] [PubMed]
4. Long, H.; Ge, D.; Zhang, Y.; Tu, S.; Qu, Y.; Ma, L. Changing man-land interrelations in China's farming area under urbanization and its implications for food security. *J. Environ. Manag.* **2018**, *209*, 440–451. [CrossRef]
5. Wang, W.; Cao, Y.; Su, R.; Qiu, M.; Song, L.; Zhou, W. Cultivated Land Protection Policy in China: Background Effect and Future Trends. *Chin. J. Agric. Resour. Reg. Plan.* **2020**, *41*, 40–51. [CrossRef]
6. Wang, C.; Sun, X.; Wang, M.; Wang, J.; Ding, Q. Chinese Cropland Quality and Its Temporal and Spatial Changes due to Urbanization in 2000–2015. *J. Resour. Ecol.* **2019**, *10*, 174–183. [CrossRef]
7. Dai, A. Study on soil environmental protection and control strategy. *Huanjingyufazhan* **2020**, *32*, 36–37. [CrossRef]
8. Yu, Z.; Chen, C.; Cui, Z. Impacts of changes in cultivated land use patterns on total factor productivity of grain in the three northeastern provinces, China. *J. Agric. Resour. Environ.* **2024**, *41*, 1–14. [CrossRef]
9. Liu, D.; Liu, F.; Niu, W.; Hao, Z.; Sheng, K.; Zhang, B.; Wang, Z. Spatial-temporal Changes and Driving Factors of Cultivated Land Intensive Use in Inner Mongolia Autonomous Region from 1985 to 2018. *Bull. Soil Water Conserv.* **2022**, *42*, 365–372.
10. Zhao, Y.; Tan, Y. Spatial-temporal Changes of Cultivated Land Use at Provincial Level Since Second National Land Survey in China. *Bull. Soil Water Conserv.* **2020**, *40*, 204–212. [CrossRef]
11. Luo, H.; Pan, L.; Hu, X.; Liu, Z. Study on grain and ecological effects of arable land use change in major grain-producing are as in China. *Acta Agric. Zhejiangensis* **2023**, *35*, 226–237. [CrossRef]
12. Dong, Q. Analysis and Research on Land Use Change of Cultivated Land Extraction from Remote Sensing Images of Changchun City Based on ArcGIS. Master's Thesis, Jilin Agricultural University, Jilin, China, 2022.
13. Wang, Y.-R.; Ji, M.; Guo, F.-B.; Li, Z. Study on Spatio-temporal Variation and Influencing Factors of Farmland Utilization Ecological Efficiency in Shandong Province. *Acta Agric. Jiangxi* **2023**, *35*, 130–137. [CrossRef]
14. Dong, G.; Liu, Z.; Niu, Y.; Jiang, W. Identification of Land Use Conflicts in Shandong Province from an Ecological Security Perspective. *Land* **2022**, *11*, 2196. [CrossRef]
15. Zhao, D.; Xu, C. Structural Model and Empirical Analysis of Bond Pricing in Government Implicit Guarantee. *J. Tongji Univ. Nat. Sci.* **2020**, *48*, 1506–1514. [CrossRef]

16. Gao, R.; Zhang, Z.; Li, W.; Li, M.; Niu, Y.; Zhang, R.; Zhang, G.; Yuan, L.; Zhang, C. Effects of Cultivated Land on Wind Erosion of Annual Land Use Change at Lankao County, Henan Province During 2018–2019. *Bull. Soil Water Conserv.* **2021**, *41*, 112–117+124. [CrossRef]
17. Weiland, U.; Kindler, A.; Banzhaf, E.; Ebert, A.; Reyes-Paecke, S. Indicators for sustainable land use management in Santiago de Chile. *Ecol. Indic.* **2011**, *11*, 1074–1083. [CrossRef]
18. Hurni, H. Assessing sustainable land management (SLM). *Agric. Ecosyst. Environ.* **2000**, *81*, 83–92. [CrossRef]
19. Alipbeki, O.; Alipbekova, C.; Sterenharz, A.; Toleubekova, Z.; Makenova, S.; Aliyev, M.; Mineyev, N. Analysis of Land-Use Change in Shortandy District in Terms of Sustainable Development. *Land* **2020**, *9*, 147. [CrossRef]
20. Uisso, A.M.; Tanrivermis, H. Driving factors and assessment of changes in the use of arable land in Tanzania. *Land Use Policy* **2021**, *104*, 105359. [CrossRef]
21. Zerriffi, H.; Reyes, R.; Maloney, A. Pathways to sustainable land use and food systems in Canada. *Sustain. Sci.* **2023**, *18*, 389–406. [CrossRef] [PubMed]
22. Zhang, W.; Cheng, Q.; Cha, X. Study on the Influencing Factors of Cultivated Land "Non-grain" from the Perspective of Sustainable Development. *Food Sci. Technol. Econ.* **2022**, *47*, 9–14. [CrossRef]
23. Cui, Y.; Long, Y.-L. Study on Evaluation of Sustainable Utilization of Cultivated Land Based on Structural Equation Model. *Acta Agric. Jiangxi* **2020**, *32*, 126–135. [CrossRef]
24. Liu, H.-J. Study on Sustainable Utilization of Cultivated Land in Hunan Province Based on Ecological Footprint Model. *J. Anhui Agric. Sci.* **2022**, *50*, 53–56+93. [CrossRef]
25. Potapov, P.; Turubanova, S.; Hansen, M.C.; Tyukavina, A.; Zalles, V.; Khan, A.; Song, X.-P.; Pickens, A.; Shen, Q.; Cortez, J. Global maps of cropland extent and change show accelerated cropland expansion in the twenty-first century. *Nat. Food* **2022**, *3*, 19–28. [CrossRef] [PubMed]
26. Zabel, F.; Delzeit, R.; Schneider, J.M.; Seppelt, R.; Mauser, W.; Vaclavik, T. Global impacts of future cropland expansion and intensification on agricultural markets and biodiversity. *Nat. Commun.* **2019**, *10*, 2844. [CrossRef] [PubMed]
27. Daqui, Z. *Environmental Geology*; Higher Education Press: Beijing, China, 2000.
28. Liang, X.; Jin, X.; Dou, Y.; Zhang, X.; Li, H.; Wang, S.; Meng, F.; Tan, S.; Zhou, Y. Mapping sustainability-oriented China's cropland use stability. *Comput. Electron. Agric.* **2024**, *219*, 108823. [CrossRef]
29. Wang, X.; Hao, J.-Q.; Dai, Z.-Z.; Haider, S.; Chang, S.; Zhu, Z.-Y.; Duan, J.-J.; Ren, G.-X. Spatial-temporal characteristics of cropland distribution and its landscape fragmentation in China. *Farming Syst.* **2024**, *2*, 100078. [CrossRef]
30. Folberth, C.; Khabarov, N.; Balkovič, J.; Skalský, R.; Visconti, P.; Ciais, P.; Janssens, I.A.; Peñuelas, J.; Obersteiner, M. The global cropland-sparing potential of high-yield farming. *Nat. Sustain.* **2020**, *3*, 281–289. [CrossRef]
31. Sünnemann, M.; Beugnon, R.; Breitkreuz, C.; Buscot, F.; Cesarz, S.; Jones, A.; Lehmann, A.; Lochner, A.; Orgiazzi, A.; Reitz, T.; et al. Climate change and cropland management compromise soil integrity and multifunctionality. *Commun. Earth Environ.* **2023**, *4*, 394. [CrossRef]
32. Zhang, Z.; Du, J.; Shen, Z.; El Asraoui, H.; Song, M. Effects of modern agricultural demonstration zones on cropland utilization efficiency: An empirical study based on county pilot. *J. Environ. Manag.* **2024**, *349*, 119530. [CrossRef] [PubMed]
33. Winkler, K.; Fuchs, R.; Rounsevell, M.; Herold, M. Global land use changes are four times greater than previously estimated. *Nat. Commun.* **2021**, *12*, 2501. [CrossRef] [PubMed]
34. Zheng, Y.; Long, H.; Chen, K. Spatio-temporal patterns and driving mechanism of farmland fragmentation in the Huang-Huai-Hai Plain. *J. Geogr. Sci.* **2022**, *32*, 1020–1038. [CrossRef]
35. Ge, D.; Long, H.; Li, Y.; Zhang, Y.; Tu, S. The Spatio-Temporal Pattern of Multifunctional Transformation of China's Grain Production System in the Process of Urbanization: The Case of Huang-Huai-Hai Plain. *Econ. Geogr.* **2018**, *38*, 147–156+182. [CrossRef]
36. Tone, K. A slacks-based measure of super-efficiency in data envelopment analysis. *Eur. J. Oper. Res.* **2002**, *143*, 32–41. [CrossRef]
37. Chi, M.J.; Guo, Q.Y.; Mi, L.C.; Wang, G.F.; Song, W.M. Spatial Distribution of Agricultural Eco-Efficiency and Agriculture High-Quality Development in China. *Land* **2022**, *11*, 722. [CrossRef]
38. Wei, J.X.; Lei, Y.L.; Yao, H.J.; Ge, J.P.; Wu, S.M.; Liu, L.N. Estimation and influencing factors of agricultural water efficiency in the Yellow River basin, China. *J. Clean. Prod.* **2021**, *308*, 127249. [CrossRef]
39. Lu, Z.-F.; Deng, W.M.; Zhu, W.W. Dynamic Evaluation of Ecological Environment Governance Performance of Jiangsu Province Based on PC-SBVndexano. *East China Econ. Manag.* **2019**, *33*, 32–38. [CrossRef]
40. Pishgar-Komleh, S.H.; Cechura, L.; Kuzmenko, E. Investigating the dynamic eco-efficiency in agriculture sector of the European Union countries. *Environ. Sci. Pollut. Res.* **2021**, *28*, 48942–48954. [CrossRef] [PubMed]
41. Caves, D.W.; Christensen, L.R.; Diewert, W.E. The Economic-Theory of Index Numbers and the Measurement of Input, Output, and Productivity. *Econometrica* **1982**, *50*, 1393–1414. [CrossRef]
42. Aparicio, J.; Santin, D. A note on measuring group performance over time with pseudo-panels. *Eur. J. Oper. Res.* **2018**, *267*, 227–235. [CrossRef]
43. Pastor, J.T.; Lovell, C.A.K.; Aparicio, J. Defining a new graph inefficiency measure for the proportional directional distance function and introducing a new Malmquist productivity index. *Eur. J. Oper. Res.* **2020**, *281*, 222–230. [CrossRef]
44. Pan, W.-T.; Zhuang, M.-E.; Zhou, Y.-Y.; Yang, J.-J. Research on sustainable development and efficiency of China's E-Agriculture based on a data envelopment analysis-Malmquist model. *Technol. Forecast. Soc. Chang.* **2021**, *162*, 120298. [CrossRef]

45. Gosme, M.; Suffert, F.; Jeuffroy, M.H. Intensive versus low-input cropping systems: What is the optimal partitioning of agricultural area in order to reduce pesticide use while maintaining productivity? *Agric. Syst.* **2010**, *103*, 110–116. [CrossRef]
46. Mobe, N.T.; Dzikiti, S.; Dube, T.; Mazvimavi, D.; Ntshidi, Z. Modelling water utilization patterns in apple orchards with varying canopy sizes and different growth stages in semi-arid environments. *Sci. Hortic.* **2021**, *283*, 110051. [CrossRef]
47. Qin, N. Prediction of cultivated land dem and based on Fisher arithmetic and Markov Chamn Model with Weights—A case study of Liao Cheng city. *Sci. Surv. Mapp.* **2011**, *36*, 29–31. [CrossRef]
48. Wang, J.X.; Wang, X.R.; Zhou, S.L.; Wu, S.H.; Zhu, Y.; Lu, C.F. Optimization of Sample Points for Monitoring Arable Land Quality by Simulated Annealing while Considering Spatial Variations. *Int. J. Environ. Res. Public Health* **2016**, *13*, 980. [CrossRef] [PubMed]
49. Song, C.; Wu, B.; Zhao, J.; Xu, Z. An integrated output space partition and optimal control method of multiple-model for nonlinear systems. *Comput. Chem. Eng.* **2018**, *113*, 32–43. [CrossRef]
50. Dinakaran, K.; Rajalakshmi, D.; Valarmathie, P. Efficient pattern matching for uncertain time series data with optimal sampling and dimensionality reduction. *Microprocess. Microsyst.* **2020**, *75*, 103057. [CrossRef]
51. Lyu, X.; Sun, X.; Peng, W.; Niu, S. Spatial-Temporal Differentiation of Sustainable Intensification of Cultivated Land Use in Shenyang City Based on Emergy Analysis. *China Land Sci.* **2022**, *36*, 79–89. [CrossRef]
52. Gong, Y.; Li, J.; Li, Y. Spatiotemporal characteristics and driving mechanisms of arable land in the Beijing-Tianjin-Hebei region during 1990–2015. *Socio-Econ. Plan. Sci.* **2020**, *70*, 100720. [CrossRef]
53. Huang, H.; Wen, L.; Kong, X.; Chen, W.; Sun, X. The Impact of Spatial Pattern Evolution of Cultivated Land on Cultivated Land Suitability in China and Its Policy Implication. *China Land Sci.* **2021**, *35*, 61–70. [CrossRef]
54. Jiao, X.-Q.; Mongol, N.; Zhang, F.-S. The transformation of agriculture in China: Looking back and looking forward. *J. Integr. Agric.* **2018**, *17*, 755–764. [CrossRef]
55. Sun, Y.; Wang, N. Sustainable urban development of the pi-shaped Curve Area in the Yellow River basin under ecological constraints: A study based on the improved ecological footprint model. *J. Clean. Prod.* **2022**, *337*, 130452. [CrossRef]
56. Kuang, W.; Zhang, S.; Du, G. Remotely sensed mapping and analysis of spatio-temporal patterns of land use change across China in 2015–2020. *Acta Geogr. Sin.* **2022**, *77*, 1056–1071. [CrossRef]
57. Lu, D.; Wang, Z.; Su, K.; Zhou, Y.; Li, X.; Lin, A. Understanding the impact of cultivated land-use changes on China's grain production potential and policy implications: A perspective of non-agriculturalization, non-grainization, and marginalization. *J. Clean. Prod.* **2024**, *436*, 140647. [CrossRef]
58. Gao, C.; Luo, Y.; Li, P. Does the Regional Ecological Security Pattern Benefit Eco-Environmental Protection? A Case Study of Yangtze River Delta. *Ecosyst. Health Sustain.* **2023**, *9*, 0017. [CrossRef]
59. Hou, Y.; Chen, Y.; Li, Z.; Wang, Y. Changes in Land Use Pattern and Structure under the Rapid Urbanization of the Tarim River Basin. *Land* **2023**, *12*, 693. [CrossRef]
60. Yin, G.; Lin, Z.; Jiang, X.; Yan, H.; Wang, X. Spatiotemporal differentiations of arable land use intensity—A comparative study of two typical grain producing regions in northern and southern China. *J. Clean. Prod.* **2019**, *208*, 1159–1170. [CrossRef]

land

MDPI

Article

Assessing the Accessibility and Equity of Urban Green Spaces from Supply and Demand Perspectives: A Case Study of a Mountainous City in China

Yawen Jin [1], Rongxiao He [1,*], Jingke Hong [2], Dan Luo [2] and Guoling Xiong [1]

1 School of Tropical Agriculture and Forestry, Hainan University, Haikou 570228, China; jinyawen@hainanu.edu.cn (Y.J.); xgarling@hainanu.edu.cn (G.X.)
2 School of Construction Management and Real Estate, Chongqing University, Chongqing 400045, China; hongjingke@cqu.edu.cn (J.H.); luodan@cqu.edu.cn (D.L.)
* Correspondence: rx.he@hainanu.edu.cn

Abstract: Urban green space accessibility is an essential consideration in determining environmental liveability and the well-being of individuals, and the spatial inequity of urban green space supply and demand has become a research focus. However, few studies have conducted a multidimensional and comprehensive assessment of the influences on accessibility from the different perspectives of supply and demand. To address this, our study centred on the mountainous Chongqing region and established a comprehensive research framework examining the spatial pattern of accessibility of urban green spaces and its correlation with physical geographical elements and socioeconomic factors. We reveal the spatial distribution characteristics of urban green space accessibility by using Gaussian-based two-step floating catchment area and network analysis methods and further observe the spatial clustering features utilising hotspot analysis. The ordinary least squares (OLS) model and the spatial lag model were used to evaluate the physical geographical and socioeconomic disparities. Our findings reveal explicit blind spots in urban green space accessibility, primarily within the 30 min travel threshold in the city's marginal area. A discernible supply–demand imbalance existed in the urban core, constituted by implicit blind spots. Furthermore, we identified that the relationship between urban green space accessibility and elevation under different methods is not always consistently significant over space because spatial heterogeneity may exist. Most concerningly, the study found inequities in urban green space accessibility, particularly impacting vulnerable demographics such as the elderly and lower-income groups. These results can inform urban planners and policymakers about the blind spots of urban green space accessibility and sufficiently consider the physical and socioeconomic heterogeneity of the space to determine where and how to implement inclusive urban greening policies or planning schemes. It is also of great significance in increasing awareness of vulnerable groups and preventing environmental inequality.

Keywords: urban green spaces; accessibility; Gaussian-based two-step floating catchment; network analysis; supply and demand; vulnerable groups

1. Introduction

An urban green space (UGS) is recognised as an urban element that closely mirrors nature, often associated with recreation, socialisation, aesthetics, cultural heritage and ecological functions. It is a crucial component of urban design [1,2]. The relevance of UGSs for urban ecosystem services, urban sustainability and human happiness has been reaffirmed by an extensive amount of research demonstrating their beneficial effects on the environment and society [3–6]. UGSs serve key ecological functions, such as the filtration of air [7], temperature regulation [8] and flood control [9]. Given these substantial benefits, it is crucial that all city dwellers have easy access to UGSs [10].

Assessing the availability of UGSs may provide an in-depth awareness of resource distribution. Recent scholarship highlights the challenges posed by urbanisation, including the shrinking or insufficient supply of UGSs [11,12], uneven spatial distribution [13] and environmental injustices related to access [3,14]. The aforementioned phenomena have sparked heightened scholarly and governmental attention towards the assessment of existing patterns of UGS distribution and their spatial variations [1,15,16]. In the meantime, with the acceleration of global urbanisation, high urban population densities place additional pressures on the upkeep and expansion of UGSs, particularly when it comes to quality and quantity [2]. Therefore, a substantial, well-distributed supply of UGSs has become a crucial indicator of the liveability of a city. However, it is crucial to acknowledge that UGSs may not consistently fulfil the diverse requirements of people. To address this issue, the accessibility and distribution of UGSs need to be understood in relation to population density.

As a crucial indicator in spatial equity studies, accessibility is widely considered a useful tool for evaluating the number and equity of UGS, and its assessment methods have been continuously optimised in recent years. As a crucial indicator in spatial equity studies, accessibility is widely considered a useful tool for evaluating the number and equity of UGS, and its assessment methods have been continuously optimised in recent years. Traditionally, the 'container approach' calculates the per capita UGS by dividing the overall green space by people residing inside a designated border [17,18]. However, this simplistic method overlooks the geographic spread of UGSs and people's mobility patterns, such as the walking distance and cross-border effects [19,20]. Research has sought to integrate accessibility into UGS allocation evaluations, recognising that the uneven geographical distribution of supply and demand creates variations in accessibility. Moreover, various approaches and models have been devised to assess UGS accessibility by using geographic information systems (GIS) [21–23]. These include the buffer zone approach, the weighted distance method, network analysis and the two-step floating catchment area (2SFCA) model. Among these methods, the network analysis method based on topology and operations research may more accurately reflect the spatial distribution of the service area of UGSs because it considers actual road accessibility. In addition, the 2SFCA method is considered an excellent accessibility method for determining potential spatial differences between supply and demand [10]. It was first proposed by Radke and Mu in 2000 and subsequently revised by Luo and Wang in 2003 [24], followed by Luo and Qi's improvement of the method to produce enhanced 2SFCA (E2SFCA) [25]. Using a catchment, it explicitly takes into account resource supplies, population demands and their interactions. However, it still proposes uniform accessibility within each travel time zone. To detect deficit areas more efficiently, Dai introduced the Gaussian function into the 2SFCA and discount access to the catchment area [26], thus making it possible to evaluate the accessibility of the UGS. Therefore, in this study, the Gaussian-based 2SFCA was selected to measure accessibility, and multiple modes of transportation were considered to provide a more realistic estimate of accessibility.

However, most academic research places emphasis on analysing the disparities in the accessibility of UGSs across different locations but often fails to thoroughly investigate the factors that contribute to these diverse outcomes. The varying characteristics of physical geographical elements can directly or indirectly influence human activities and consequently shape social development [27,28]. In addition, it is crucial to assess the correlation between socioeconomic factors and the accessibility of UGSs since social development affects living conditions and is the basis for all social activities. Research on UGS demand has increasingly pivoted from a homogenous population to a relationship-centric focus, considering the correlation between green spaces and demographic characteristics. This shift occurred in the context of a broad movement towards people-centred urban construction that emphasises the needs of specific groups [29,30]. Much of this research focused on developed countries and examined the association between UGSs, socioeconomic status [31], religious attributes [32] and age [33]. Consequently, we must analyse UGS distribution and accessibility in relation to physical geographical elements and socioeconomic factors.

Inspired by the current challenges of UGS accessibility indicators and the potential supply–demand imbalance in Chongqing [34,35], this study has three objectives. First, it introduces a more comprehensive framework for evaluating UGS accessibility in terms of supply and demand. Second, it identifies key locations for optimising the distribution of UGS by comparing the results of different methods and identifying explicit and implicit blind spots. Finally, it also discerns the intrinsic relationships between UGS accessibility, physical geographical elements and socioeconomic factors, which assists urban and regional planners in understanding the disparities between them in order to formulate intervention plans.

The ensuing portions of the study are organised in the following manner: Section 2 provides a comprehensive overview of the data gathering and research method used in this research. Section 3 presents and analyses the results of UGS accessibility calculations for different travel modes from the supply and demand perspectives. This section additionally encompasses the statistical analysis of quantitative data derived from spatial analysis. The last two sections, respectively, are where the discussion and the conclusions are presented.

2. Materials and Methods

2.1. Study Area

We selected Chongqing, a typical mountainous city with a complex urban environment, as our study site. Moreover, Chongqing is a rapidly urbanising megacity with multiple urban centres, while it has a significantly different environment from that of plain cities, and exemplifies the intensive human–environment relationship in China [34,36] The city is intersected by four north–south mountains (the Jinyun, Zhongliang, Tongluo and Mingyue Mountains) and two west–east rivers (the Jialing and Yangtze Rivers). The city's complex topography, which is divided by rivers and valleys with a substantial topographical height difference, constrains the city layout, resulting in a dispersed pattern [37,38]. Moreover, the city's complex transportation systems and limited availability of construction land present challenges to rational UGS planning [39]. Despite these challenges, the point-like three-dimensional layout of green space resources in mountainous cities offers opportunities for an efficient green space supply. The primary area of interest for our study is the Chongqing buildup area, situated predominantly within the outer ring road and consisting of nine administrative districts: Yuzhong, Jiangbei, Yubei, Shapingba, Jiulongpo, Nanan, Dadukou, Beibei and Banan. These districts comprise 128 subdistricts, collectively spanning 3140.28 km^2 (Figure 1). On the basis of local knowledge, we have delineated the area encompassed by the inner ring as the core area, the area within the outer ring as the suburbs and the area outside as the urban marginal area. Our study may provide insights that are applicable to other mountainous cities, particularly those in developing countries.

The UGS referred to in the text is a collated green space with a certain scale (>3000 m^2) and service function, with the main emphasis on the dimension of its recreational function and more concerned with the close relationship between the green space and the daily life of the residents. In total, 724 UGSs were distinguished within (n = 724) the study area, including parklands and gardens, greenbelts along roads (>5000 m^2), greenbelts along rivers, greening squares, green spaces around residences and green spaces around institutions. Water bodies and small green spaces, such as roadside greenery, were not considered.

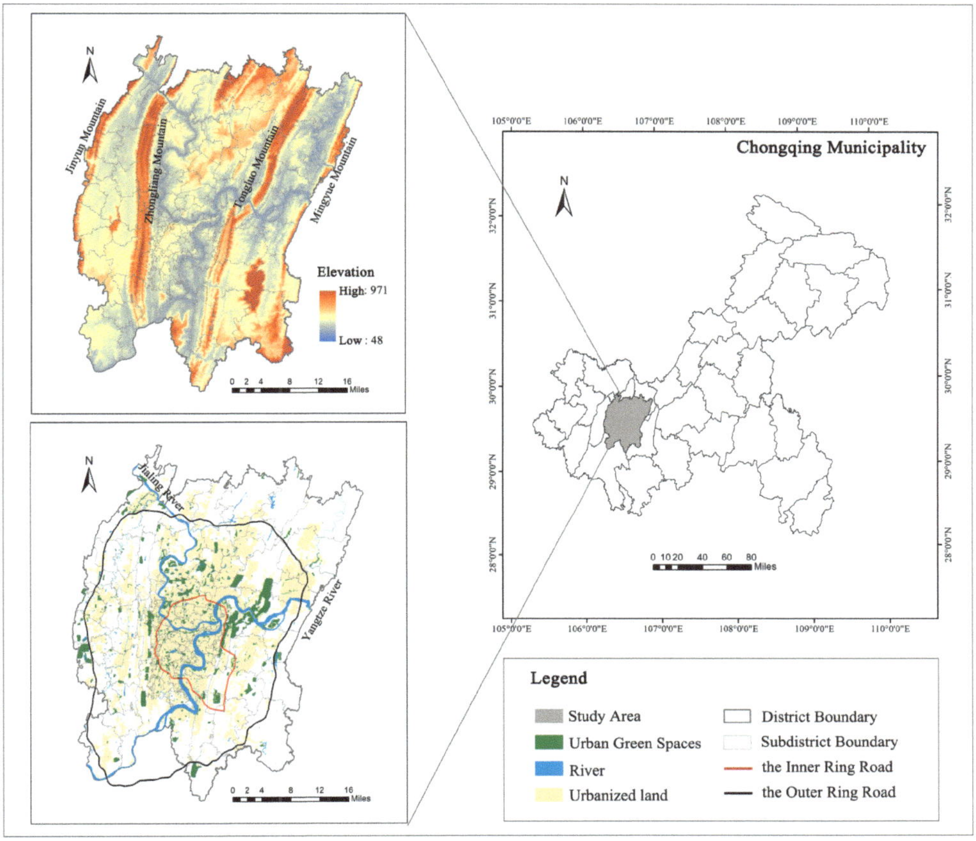

Figure 1. Location of the study area (Chongqing, China).

2.2. Data Collection

The data for this research were compiled from various sources. The first set of data, pertaining to UGSs, was obtained from open street maps. These data were combined with GlobeLand30—a 30-metre resolution land cover dataset of Chongqing for the year 2020—and corroborated and supplemented using geographic data generated by Google Remote Sensing images. To ensure the consistency and accuracy of our study of UGSs information, we referred to the land use survey data (1:10,000 scale) provided by the Land and Resources Bureau. The second set of data, providing details on the residents within each 'subdistrict', was sourced from the Sixth National Population Census of the People's Republic of China 2010, a resource analogous to the US census tracts. The third dataset was obtained from the 'Baidu Sitemap Generator' in 2021 and covers the subdistrict network information. Finally, we obtained data about residential communities from Lianjia's website (http://www.lianjia.com/ (accessed on 29 July 2021)). Lianjia, one of China's largest housing intermediaries, provides real estate and rental services and offers detailed information about residential communities, such as location, coordinates and prices. We collected data for 5935 residential areas, including name, household number, longitude, latitude and average price attributes.

We categorised the selected variables into two dimensions: physical geographic elements and socioeconomic factors. The first dimension includes factors like elevation and the proportion of UGSs. The second encompasses road network density, residents' income and the percentage of the population that is at least 65 years old in each subdistrict. Given the unavailability of public income data for Chinese cities, we used home sale prices as a proxy for residents' income levels [40,41]. Assuming that a higher quality of life is sought by higher-income individuals who can afford commercial housing, while those with lower income opt for affordable housing, provided a basis for our estimations.

2.3. Urban Green Space Accessibility Method

2.3.1. Network Analysis Method

The core purpose of using the network analysis method is to calculate the service area of UGSs to visually determine the level of service at the supply level of UGSs in the study area because it is generally believed that people within the service area are more likely to reach the UGS and enjoy the service. In the traditional radius method, a service area is depicted as a circle with a radius encompassing a UGS. The network analysis method offers a more precise measure than the radius method because it determines a facility's service area at a specific cost on the basis of a particular travel mode's road network (either walking or driving a private car) [32]. This method considers actual travel routes, distances and travel friction. For this study, we selected residents' travel time to UGSs as the defining range of service areas instead of the radius. On the basis of prior research, we set 30 min as the maximum acceptable travel time for residents. We assumed a 5 km/h walking pace [42]. The driving speeds on the highway, the expressway, the main road, the secondary main road and the branch road were set at 100, 80, 60, 40 and 20 km/h, respectively.

2.3.2. Gaussian-Based 2SFCA Method

To further analyse the interaction between the supply of UGS and population demand at different locations, the Gaussian-based 2SFCA approach is employed to assess the UGS accessibility [34]. This method avoids the dichotomy problem that can lead to errors with the 2SFCA approach [26,43]. The formulas used are outlined below.

The initial step involves conducting a search for all population sites i that fall within a specified travel time threshold (d_0) from each UGS location j. This process establishes the catchment area for the UGS location j. Individuals at location i will be assigned weights based on a Gaussian function, denoted as G. The projected number of prospective users for UGS at position j may be determined by summing the weighted individuals living within its catchment region. The UGS to population ratio (R_j) is expressed as follows:

$$R_j = \frac{S_j}{\sum_{i\in\{d_{ij}\leq d_0\}}^{k} D_i \times G(d_{ij})}, \tag{1}$$

$$D_i = K_0 \times H_i, \tag{2}$$

where (D_i) denotes the population at site i whose centroid lies inside the catchment (i.e., $d_{ij} \leq d_0$) of UGS j, and (H_i) indicates the overall number of households in the residential area. (K_0) is the average population per household. According to the average population per household of urban residents in the Chongqing Statistical Yearbook in 2020, the value is 2.98; d_{ij} is the distance people have to walk or drive to travel from population location i to UGS site j; UGS capacity (size in square metres) is denoted by S_j; and G is the friction-of-distance listed below:

$$G(d_{ij}, d_0) = \begin{cases} \frac{e^{-}(1/2) \times (d_{ij}/d_0)^2 - e^{-}(1/2)}{1 - e^{-\frac{1}{2}}}, & d_{ij} \leqslant d_0 \\ 0, & d_{ij} > d_0 \end{cases}, \tag{3}$$

Next, the search is conducted for each population site i to identify all UGSs l that fall within the specified time threshold (d_0) from i. The Gaussian function G is utilised to discount each R. The summation of discounted R values throughout the region of catchment i is used to ascertain the level of accessibility at population site i, using the following equation:

$$A_i = \sum\nolimits_{l \in \{d_{ij} \leqslant d_0\}} R_l \times G(d_{ij}), \tag{4}$$

where R_j denotes all the UGSs serving the population at position i, and all of the other symbols and notations remain consistent with Equation (1). After special weighting, the accessibility value (A_i) can be interpreted as the per capita green area.

The catchment size (d_0) is essential in determining whether a UGS is accessible. In this study, six thresholds ranging from 10 to 30 min with a 5 min increment were tested. The reason this increment was chosen was to examine the variation in accessibility resulting from different thresholds, which is consistent with previous studies [15,44]. Meanwhile, the method considers the road routes based on the actual terrain and the resistance of the residents to travelling in the travelling process. In addition, the accessibility of UGSs in each community was determined using the Gaussian-based 2SFCA approach. Then, the accessibility of each subdistrict was calculated by averaging the UGS accessibility of the communities.

2.4. Hot Spot Analysis Using Getis Ord Gi* Statistic

We used a hot spot analysis (Getis-Ord Gi*) tool to identify the geospatial clustering of hot (high) and cold (low) spots of UGS accessibility and to explore the spatial layout of UGS accessibility under different methods and modes of transportation [45]. One of the benefits of utilising this tool is its capability to ascertain localised connections among observations and their neighbouring entities, hence facilitating the visualisation of spatial interaction patterns through the identification of small clusters or minor outliers [46,47]. The following is the formula used to obtain the Getis-Ord Gi* local statistics [48]:

$$G_i^* = \frac{\sum_{j=1}^{n} \omega_{i,j} x_j - \overline{X} \sum_{j=1}^{n} \omega_{i,j}}{S \sqrt{\frac{n \sum_{j=1}^{n} \omega_{i.j}^2 - \left(\sum_{j=1}^{n} \omega_{i,j} \right)^2}{n-1}}}, \tag{5}$$

where the variable x_j denotes the attribute value associated with item j. The variable $w_{i,j}$ represents the spatial weight between features i and j. Additionally, the variable n corresponds to the overall amount of traits.

$$\overline{X} = \frac{\sum_{j=1}^{n} x_j}{n}, \tag{6}$$

And

$$S = \sqrt{\frac{\sum_{j=1}^{n} x_j^2}{n} - \overline{X}^2}. \tag{7}$$

For each feature in the dataset, the Gi* statistic returns a z-score as its output. A z-score that is significantly positive shows a pronounced concentration of high values, commonly referred to as a 'hot spot'. Conversely, a z-score that is slightly negative signifies a notable concentration of low values, often referred to as a 'cold spot'. This method offers a more comprehensive illustration of the spatial clustering attributes pertaining to the accessibility of UGSs as opposed to solely examining the extreme values of high or low accessibility in isolation.

2.5. Spatial Autoregressive Analyses

To compare the disparities in the calculation of UGS accessibility using the two methods across various influencing factors, the study conducted weight calculations and indicator combinations on the results of two distinct methods. The analysis involved comparing the accessibility results under different travel modes using the entropy weights method. Additionally, we selected variables related to physical geographical elements and socioeconomic factors for statistical analysis. Initially, we constructed an ordinary least squares (OLS) regression model to examine the relationships between the dependent variable (UGS accessibility) and the independent variables (physical geographical elements and socioeconomic factors). For our analysis, we included indicators such as residents' income, elevation, road network density, the proportion of the elderly and the proportion of UGSs. These indicators were selected based on their significance in defining geosocial conditions and specific groups, as emphasised in prior research [49,50]. Given that population density is already incorporated in the formula for the Gaussian-based 2SFCA method, it was not included in the regression analysis. Subsequently, in order to address the possibility of global autocorrelation in the data, the use of Moran's I statistic was employed to evaluate the existence of global autocorrelation in both variables that are dependent and the mistakes of the initial OLS models. Spatial regression was conducted when significant spatial autocorrelation was detected in the study area using Moran's I-statistic. The spatial regression model resolved spatial dependencies that the linear regression analysis was incapable of handling [51]. The utilisation of Lagrange multiplier testing indicated that the spatial lag model (SLM) was the most optimal selection for this study. The selection of the inverse of the distance matrix was considered to be the optimal approach for representing spatial relationships in the data. The equation for the SLM model is as follows:

$$y = \rho W y + X\beta + e \tag{8}$$

where W represents the matrix of spatial weights. The parameter ρ characterises the range of spatial autocorrelation between -1 and 1. Additionally, $\rho W y$ quantifies the spatial dependence present in y. The matrix X represents observations on the covariates, with dimensions $n \times k$. β is the $k \times 1$ vector of regression coefficients, and e is an $n \times 1$ vector of errors that have been independently and uniformly distributed.

3. Results

3.1. Spatial Distribution of Urban Green Space Accessibility

3.1.1. Accessibility Evaluation Based on Supply

The spatial distribution of the accessibility of UGSs is shown in Figure 2, and the coverage of the accessibility of UGSs is illustrated in Table 1, according to the different modes of transportation. Areas not covered by the accessibility range within a certain threshold time are identified as explicit blind spots, meaning 'no supply' zones. The findings indicate that there is a decline in the accessibility of UGSs as one moves from the core area towards the urban marginal area. The coverage of accessibility <15 min within the core area is 69.86%; the same coverages in the suburbs and urban marginal area are 12.17% and 2.68%, respectively. As shown in Table 2, a 14.97% coverage requires more than 30 min to reach a UGS within the inner ring road. These uncovered areas are mainly situated southeast, possibly due to the hindrance posed by the Tongluo Mountain Range that traverses this area. Meanwhile, a 30 min walking range covers 29.09% of the area in the suburbs. Although the built-up area of Chongqing is dominated by good accessibility under the driving mode, 28.81% of the area requires 30 min at least to arrive at a UGS, mainly located in areas outside the outer ring road. Under 30 min of walking and driving, the explicit blind spots accounted for 93.06% and 73.90% of the urban marginal area, respectively.

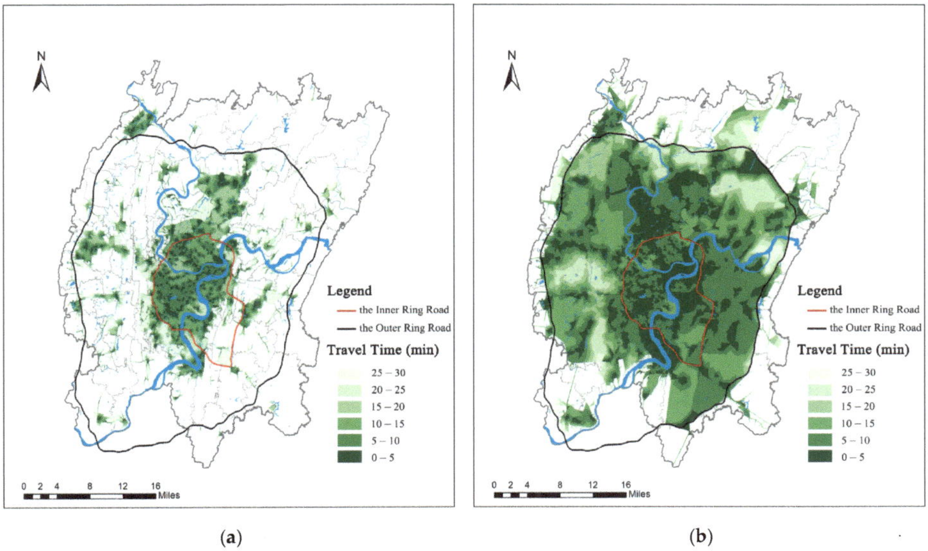

Figure 2. Spatial distribution of service area to urban green spaces under walking and driving modes. (**a**) Walking mode; (**b**) Driving mode.

Table 1. Coverage of accessibility of urban green spaces.

Travel Modes	Time Consumption (min)	The Core Area		The Suburbs		The Urban Marginal Area		Total Coverage	
		Area (km^2)	(%)	Area (km^2)	(%)	Area (km^2)	(%)	Area (km^2)	(%)
Walking	≤5	49.10	16.60	46.92	2.57	5.01	0.49	101.04	3.22
	5–10	97.35	32.92	82.90	4.53	11.23	1.10	191.48	6.10
	10–15	59.61	20.16	92.60	5.07	11.10	1.09	163.31	5.20
	15–20	19.36	6.55	97.85	5.35	13.07	1.29	130.28	4.15
	20–25	14.85	5.02	75.66	4.14	11.09	1.09	101.61	3.24
	25–30	11.19	3.78	135.85	7.43	20.03	1.97	167.07	5.32
Driving	≤5	152.31	51.50	363.14	19.86	28.57	2.81	544.02	17.32
	5–10	114.65	38.77	600.63	32.86	46.39	4.56	761.63	24.25
	10–15	28.78	9.73	382.88	20.94	39.94	3.93	451.61	14.38
	15–20	0	0	155.29	8.49	47.26	4.65	202.55	6.45
	20–25	0	0	144.12	7.88	51.79	5.10	195.91	6.24
	25–30	0	0	28.70	1.57	51.38	5.05	80.08	2.55

Table 2. Regression results of accessibility with physical geographic elements and socioeconomic factors.

Variables	Multiple Regression		Spatial Lag Regression	
	Accessibility Based on Supply	Accessibility Based on Supply–Demand	Accessibility Based on Supply	Accessibility Based on Supply–Demand
Proportion of the elderly_ln	−0.220 **	0.173	−0.243 **	0.212
Income_ln	0.049 **	0.484 ***	0.053 ***	0.484 ***
Density of road network_ln	0.825 ***	−0.736 ***	0.596 ***	−0.623 **
Proportion of green area_ln	0.094 ***	0.200 ***	0.071 ***	0.191 ***
Elevation_ln	−0.477 **	2.357 ***	−0.373	2.231 ***
Constant	−4.326 ***	−22.353 ***	−1.844	−18.20 ***
Lagged Coeff (Rho)	—	—	0.591 ***	0.494 **
R square	0.797	0.453	0.834	0.504
AIC	162.5644	442.6414	149.4884	441.5876
BIC	179.6766	459.7536	172.3047	464.4039

*** $p < 0.01$ and ** $p < 0.05$ indicate significant differences at 1% and 5% test levels, respectively.

3.1.2. Accessibility Evaluation Based on Supply–Demand

The evaluation of accessibility on the basis of supply indicates that the core area exhibits the highest accessibility within a 15 min threshold for various travel modes, encompassing the largest coverage of access to UGSs. However, considering its population density, the core area does not present a significant advantage. The heterogeneity in accessibility observed from these two perspectives may be attributed to the high population density in the core area, leading to a situation where the residents' demand for UGSs exceeds the available supply. Figure 3 depicts the spatial distribution of UGS accessibility for pedestrians and drivers obtained using the Gaussian-based 2SFCA method within a 30 min threshold. At the regional scale, subdistricts with good accessibility decline from the suburbs towards the core area, while subdistricts with poor accessibility are predominantly located in the urban marginal area. Analysing at the subdistrict scale, we found that 16 subdistricts (12.5% of all subdistricts) fail to access UGSs within the threshold time when using the walking mode, mainly in the urban marginal area. Additionally, 67 subdistricts exhibit accessibility scores below 2.42, primarily in the core area. In the driving mode, 11 subdistricts (8.59% of all study subdistricts) have no access to UGSs, and these subdistricts are distributed in the urban marginal area. Comparatively, the accessibility of the core area improves in the driving mode but remains dominated by subdistricts with scores below 2. Overall, the explicitly blind spots with zero accessibility are primarily concentrated in the urban marginal area, where the distribution of UGSs is virtually non-existent.

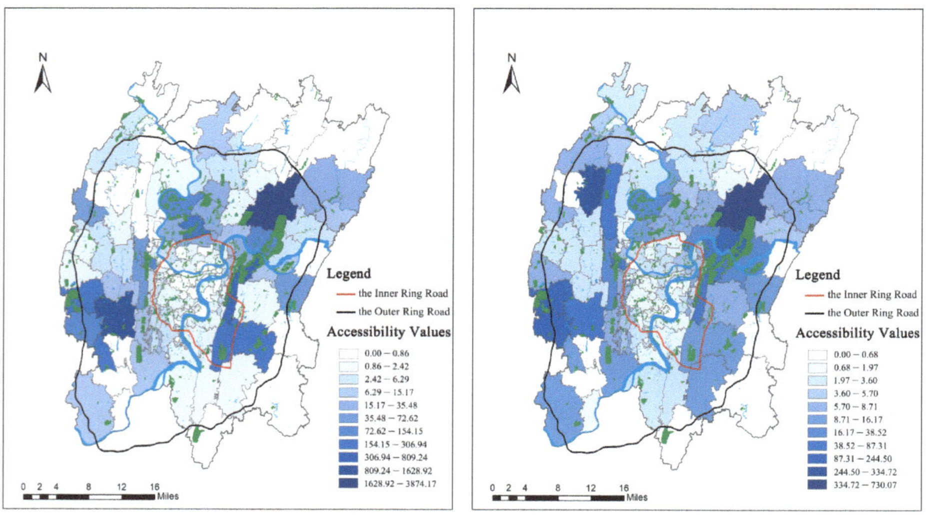

(**a**) (**b**)

Figure 3. Spatial distribution of accessibility of urban green spaces under the Gaussian-based 2SFCA method. (**a**) Walking mode; (**b**) Driving mode.

3.2. Spatial Layout of Urban Green Space Accessibility

On the basis of the 30 min walking and driving modes, the accessibility of UGSs exhibits an unclear agglomeration centre, as depicted in Figures 2 and 3. In order to enhance our comprehension of the spatial layout of accessibility at the subdistrict level, we utilised the hot spot analysis tool referred to as Getis-Ord Gi*. A hot spot is defined as a spatial unit characterised by a high value and surrounded by other spatial units with similarly high values. Conversely, a cold spot represents a spatial unit that features a lower value and neighbouring units that also possess low values. In our analysis, we found a distinct spatial mismatch between the two accessibility aggregation patterns. Figure 4a,b illustrate that the urban marginal area primarily comprises cold spot aggregation areas, while the core area exhibits accessible hot spot aggregations from the supply perspective. Notably, for the accessibility evaluation based on supply and demand, the cold spot aggregation area is primarily concentrated in the core area.

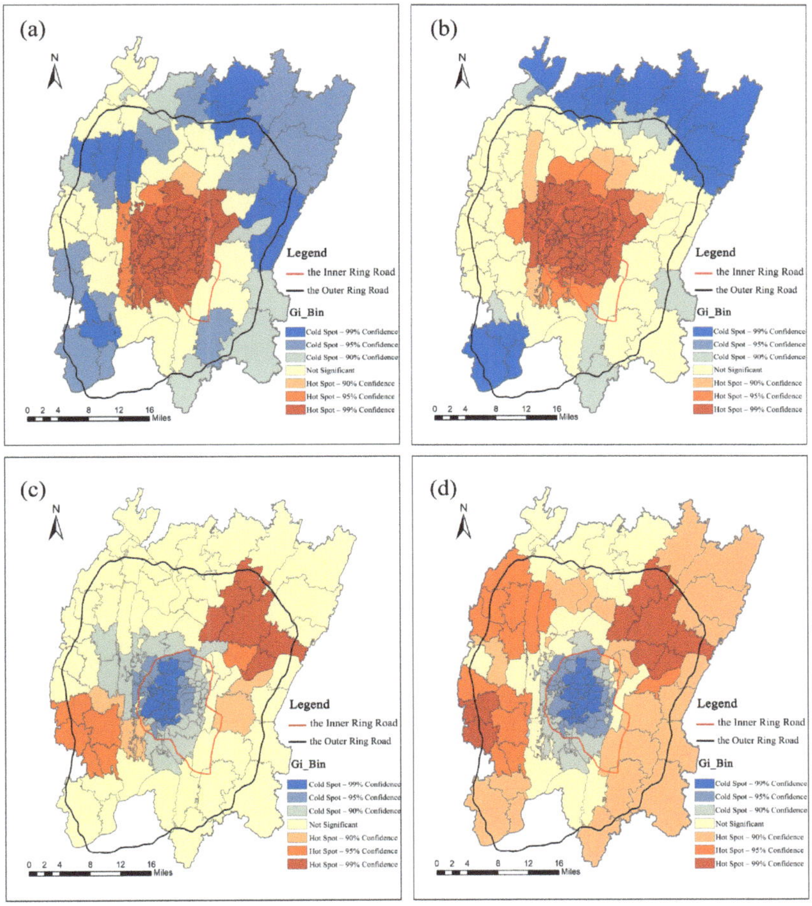

Figure 4. Gi* cluster map for the hot spot analysis of urban green space accessibility under walking and driving modes. (**a**) Accessibility of walking mode under the network analysis method; (**b**) Accessibility of driving mode under the network analysis method; (**c**) Accessibility of walking mode under the Gaussian-based 2SFCA method; (**d**) Accessibility of driving mode under the Gaussian-based 2SFCA method.

3.3. Spatial Correlation between Urban Green Space Accessibility and Related Variables

The findings of the hotspot analysis reveal a spatial mismatch in the accessibility of UGSs as determined by using two distinct methods, with each based on different perspectives. To further investigate the potential influencing factors contributing to variations in accessibility, we examined the physical geographic elements and the socioeconomic factors. As depicted in Figure 5a, the percentage of the population that is aged 65 and older rises from the core area outwards. Figure 5b,d show a similar trend in the way they are distributed, and we find that the core area presents high road density and population density. Moreover, the core area demonstrates a relatively flatter elevation than the surrounding areas. Additionally, the core area has become an implicit blind spot for accessibility partly because of the excessive population density surpassing the capacity of UGSs. Consequently, a significant number of residents lack effective access to UGSs. Figure 5g illustrates the results obtained by normalising and combining these factors, with lower values in the core area compared to other areas.

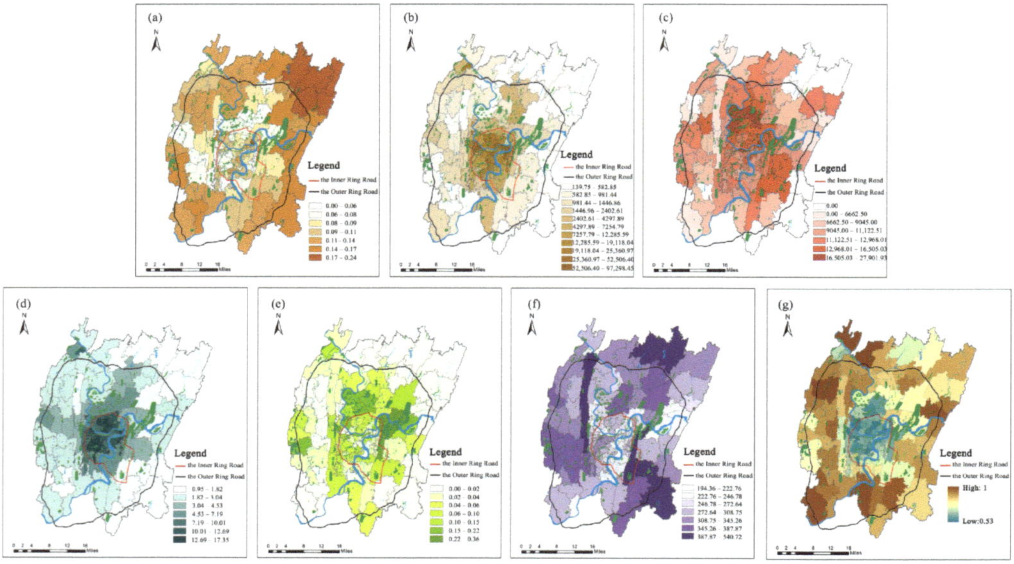

Figure 5. Spatial distribution of physical geographical elements and socioeconomic factors. (**a**) Proportion of the elderly; (**b**) density of population; (**c**) residents' income; (**d**) density of road network; (**e**) proportion of urban green space; (**f**) elevation; (**g**) standardised consolidated values.

This study employs a multiple regression model to explore UGS accessibility and the correlations between various physical geographic elements and socioeconomic factors from diverse viewpoints. According to the OLS analysis, the variance inflation factor was below 2, indicating that the explanatory variables were free of multicollinearity issues. However, the significant Moran's I values obtained from both the network analysis and Gaussian-based 2SFCA methods for the OLS models (Moran's I = 0.090, z score = 2.091, $p < 0.05$; Moran's I = 0.484, z score = 10.509, $p < 0.01$) suggest spatial autocorrelation within the residuals.

Given this spatial autocorrelation, further analyses were conducted using spatial regressions to examine the correlation between UGS accessibility and the selected independent variables using both methods independently. The spatial autoregressive lag model (SARlag) was chosen for this study based on the Lagrange multiplier test, which demonstrated its superiority to the spatial error regression model. Furthermore, the SARlag model outperformed the OLS model, demonstrated by the elevated R^2 and log likelihood values, decreased AIC values and significant spatial lag coefficients (rho), indicating a strong spatial dependence on UGS accessibility. The relationships between UGS accessibility and the index of physical geographic and socioeconomic factors under the two methods showed that the SARlag regression explains about 83.40% and 50.40% of the variations in the changes in UGS accessibility in the 122 subdistricts, respectively.

The regression results (Table 2) show apparent disparities in green space access. For both the OLS and SARlag models, the network analysis method revealed a positive and statistically significant effect of income on UGS accessibility. Notably, a discernible inverse correlation was observed between the proportion of elderly residents and UGS accessibility.

Adding a one unit increase in the natural logarithm of the elderly corresponds to an average of 22.0% and 24.3% decrease in the logarithm of accessibility of UGSs under both models. It may be due to the unique features of the mountainous city that the travel behaviours of the elderly are not fully considered in the urban green space planning process, resulting in inequitable urban green space planning for the elderly. The steepness of the mountain cities exacerbates this inequity. However, no correlation existed between elevation and the accessibility of UGSs from a supply-only aspect.

Alternatively, the Gaussian-based 2SFCA method revealed that higher-income residents had substantially greater access to UGSs ($p < 0.01$), suggesting an income-driven accessibility disparity, which is similar to the results of the network analysis method. Additionally, elevation, a distinguishing geographic factor between mountainous and plain cities, exerted a positive influence on UGS accessibility from a supply and demand perspective. However, contrary to the hypothesis that a higher road network density would enhance accessibility, our findings indicate a negative correlation between these two variables. This unexpected result can be attributed to the overdeveloped street network that imposed restrictions on urban land use and thereby reduced the available UGSs.

4. Discussion

An escalating conflict between supply and demand has arisen in the urban environment due to the increasing constraints and expanding population. Despite studies that have examined the supply and demand of UGS in developed and developing countries [52] and recognised that the inequity of UGS is mainly reflected in the supply and people's demand for UGS, there is a dearth of knowledge that exists regarding the equitable distribution of resources for UGSs among all city residents [19,52]. The contributions of this paper are mainly reflected in the following aspects: Firstly, based on the existing studies, a more sophisticated analytical framework is proposed to analyse the supply and demand of the urban master plan, which integrates the network analysis method, attenuation coefficients, Gaussian functions and travel modes to provide a more realistic evaluation. Meanwhile, the study further enhances the assessment by considering accessibility under both modes. This contribution offers methodological support and a theoretical reference for the evaluation of UGS accessibility in other cities. Secondly, we utilised multidimensional evaluation indicators to identify the heterogeneity in UGS accessibility according to different physical geographical elements and socioeconomic factors, with particular attention to vulnerable populations. Assessment results provide policymakers and planners with decision-making information conducive to achieving equity in the distribution of UGSs.

The marginal area of Chongqing's built-up region has the weakest allocation of UGS resources and suffers from significant service deficits. During the examination of spatial distribution, we noted marked polarisation in the UGS accessibility of supply within a 30 min threshold time across different modes of transportation. High accessibility areas are predominantly distributed in the core area, while explicit blind spots persist in the urban marginal area. This is similar to the findings of previous studies [52,53]. Viewing this through a supply–demand lens, subdistricts lacking accessibility are also primarily situated in the urban fringe, validating the efficacy of the network analysis method in identifying explicit blind spots. For these explicit blind spots devoid of supply, especially the urban marginal areas with inferior UGS resources, future successful green space planning can be realised through spatial excavation by reserving space for greening, effectively allocating UGS and anticipating futural demand.

Assessing the spatial layout, the results underscore the spatial incongruity in the accessibility of two distinct categories of UGSs, with a distinct supply–demand imbalance in the core area. Despite the strengths of the core UGS catchment area, there is a deficiency in meeting all of the residents' daily needs, thus manifesting as an implicit accessibility blind spot. This shortfall is likely attributable to the high population density in the core area that constantly intensifies the pressure on the allocation of UGS and the intensive land development encroaching on the available UGSs. Concurrently, an overly dense road network further restricts and fragments the distribution of UGSs, which discourages the development of UGSs and ultimately hinders efficient resource allocation. Therefore, the mismatch between the supply and demand of UGS needs to be emphasised in the urban greening process. Currently, it may be difficult to build large UGSs due to the fixed and compact land use patterns and land demand restrictions [54,55]. Considering these factors, other greening strategies such as vertical greening and informal green spaces could be implemented in the future to strengthen linkages between nearby green spaces and create more daily recreational opportunities for residents to boost their sense of belonging to a city [56]. Furthermore, the potential for UGS development in the suburbs can be utilised. Population redistribution is facilitated by emphasising the suburbs in order to alleviate land development pressures in the core area. This strategy could enhance the spatial patterns of UGS accessibility in Chongqing and reduce both explicit and implicit blind spots.

According to the findings of the spatial correlation analysis, there is some heterogeneity in the relationship between UGS accessibility and altitude under different methods. Therefore, an effective analysis of the elevation is necessary in order to effectively safeguard access to sufficient green space. Elevation, a crucial physical geographic element that distinguishes mountainous from plain cities, can serve as a foundational parameter for future research on cities in mountainous regions. Chongqing, a city typified by its mountainous and riverine landscape, faces stringent constraints on urban construction land due to its varied topography, including numerous mountains and valleys. Therefore, planning in this city should prominently feature its unique topographical elements. Specifically, in the suburbs—the transition zones between core and marginal regions—where both UGS resources and population density are moderate, planning should accentuate the cities' mountains and rivers. The steep topography of these areas engenders a three-dimensional, high–low fragmented distribution of UGSs, thereby enhancing their service efficiency. Consequently, future urban development should prioritise three-dimensional space development and maximise green spaces amid densification [57].

The high-density road network fails to consistently have a positive influence on accessibility. On the one hand, as a link between UGSs and residents, transportation networks can affect the efficiency of UGSs in providing services to residents. A properly planned road network system can benefit residents in terms of accessibility of UGSs. On the other hand, an increase in road density fails to consistently improve the accessibility of UGS, and an overdeveloped street network instead restricts land use [5,54], thus reducing the amount of available UGS. At the same time, the economic value of land further increases the pressure on UGS. For example, landowners make more profit from urban commercial, industrial and residential land development than from green infrastructure land. To solve the mentioned problems, landscape planners can increase or improve the existing street green space and pocket parks based on rational design and utilisation of road networks or also design UGSs in some redevelopment projects, such as urban village reform, and rebuild abandoned factories to form high-quality living communities using UGS planning and landscape design as well as to strengthen connections with nearby UGS.

Indeed, the widely available housing sales data in China can serve as a valid indicator for different social groups when studying a city because they can reflect many other socioeconomic conditions, such as income levels [45]. Discrepancies in UGS accessibility have been observed pertaining to specific neighbourhood socioeconomic characteristics [58]. Subdistricts with higher social status are favoured in terms of UGS provision, which is in line with previous research. To be specific, our study clearly demonstrates that UGS accessibility is better in areas where high-income populations are located. We argue that higher-income groups are more capable of choosing resource-rich, high-quality living environments, thus suggesting that economically vulnerable families are being deprived of access to UGSs, exposing social inequities. In addition, this study reveals low UGS accessibility in subdistricts with a high level of population aging. The problems of population ageing will exist for some time in China, which may significantly increase the demand for social infrastructure, including the demand for UGS service resources. In this context, inadequate access to UGSs for the elderly may raise issues of environmental injustices that require pre-awareness and effective countermeasures. Therefore, planning and constructing UGSs solely on the basis of supply or demand optimisation, assuming a homogenous population, fails to provide a comprehensive solution. Considering the accessibility of various social groups is an indispensable aspect of green planning [59]. Therefore, residents of disadvantaged communities should be prioritised for UGS additions, optimising the spatial pattern of accessibility to ensure equity [60,61].

Findings from this study have important policy implications and can provide methodological references and theoretical support for planners and policymakers. Chongqing has prioritised a greening strategy where all residents live within a 15 min walk from the nearest UGS. The Liangjiang New Area and the Jiulongpo District are national pilot districts for the 'quarter-hour convenient living circle'. The Jiulongpo District encompasses 18 subdistricts, while the Liangjiang New District spans parts of the Jiangbei, Yubei and Beibei Districts in Chongqing, covering a total of 36 subdistricts. This approach aims to provide residents with UGS services within a 15 min walking radius. However, our study identified a gap, with only 69.68% accessibility coverage within a 15 min walking radius in the core area that has the highest population density. Therefore, striving for complete UGS service coverage should be considered a key priority in UGS planning. Simply increasing the size and improving the functions of UGSs may lead to further spatial inequality. It is crucial to comprehensively consider the nature of the land, the amount and type of UGS, the quality of UGS and the level of supply and demand.

5. Conclusions

Taking the buildup area of Chongqing as an example, this study enhanced the evaluation of UGS accessibility from multiple viewpoints. Regarding spatial distribution, we identified explicit blind spots in the threshold time from a supply perspective, especially in urban marginal areas. These marginal areas were also found to be disadvantaged when assessed from the supply–demand perspective. A comparison of the spatial layouts of the two types of accessibility reveals that although the core area has a generous supply of UGSs, it cannot meet the needs of all residents, rendering it an implicit blind spot in UGS accessibility. However, the suburbs exhibit potential for UGS development. The future expansion and development of UGSs in the suburbs can not only extend their service areas to the periphery but also partially alleviate the land and population pressure in the core area. Moreover, our analysis revealed that there is some heterogeneity in the relationship between accessibility and both elevation and road network density, suggesting that elevation and road network density should be considered critical factors in future studies on mountainous cities. Lastly, we observed that vulnerable groups, such as people aged 65 and over and low-income populations, experience disadvantages in accessing UGSs. Insights into the spatial distribution and demand preferences of these groups will undoubtedly aid in promoting future urban social equity.

In summary, our study suggests that an effective UGS planning policy should emphasise three key aspects: (i) implementing targeted measures to mitigate explicit and implicit blind spots from diverse supply and demand perspectives, (ii) paying attention to the unique physical geography of mountainous cities and considering the development of three-dimensional spaces and (iii) enhancing awareness of vulnerable groups and mitigating social inequality.

Author Contributions: Conceptualisation, Y.J. and J.H.; methodology, Y.J. and G.X.; software, Y.J.; validation, R.H., J.H. and D.L.; data curation, Y.J.; writing—original draft preparation, Y.J.; writing—review and editing, J.H. and R.H.; visualisation, Y.J. and G.X.; supervision, R.H.; project administration, R.H. and J.H.; funding acquisition, R.H. All authors have read and agreed to the published version of the manuscript.

Funding: This research was supported by the Hainan Provincial Natural Science Foundation of China (423MS020) and the China Scholarship Council (202207560001).

Data Availability Statement: The source of all the data used in this study is provided in the manuscript.

Conflicts of Interest: The authors declare no conflict of interest.

References

1. Nasri Roodsari, E.; Hoseini, P. An assessment of the correlation between urban green space supply and socio-economic disparities of Tehran districts—Iran. *Environ. Dev. Sustain.* **2021**, *24*, 12867–12882. [CrossRef]
2. Zhang, Y.; Zhang, T.; Zeng, Y.; Yu, C.; Zheng, S. The rising and heterogeneous demand for urban green space by Chinese urban residents: Evidence from Beijing. *J. Clean. Prod.* **2021**, *313*, 127781. [CrossRef]
3. Chen, Y.; Yue, W.; La Rosa, D. Which communities have better accessibility to green space? An investigation into environmental inequality using big data. *Landsc. Urban Plan.* **2020**, *204*, 103919. [CrossRef]
4. Ma, B.; Zhou, T.; Lei, S.; Wen, Y.; Htun, T.T. Effects of urban green spaces on residents' well-being. *Environ. Dev. Sustain.* **2018**, *21*, 2793–2809. [CrossRef]
5. Fang, X.; Li, J.; Ma, Q. Integrating green infrastructure, ecosystem services and nature-based solutions for urban sustainability: A comprehensive literature review. *Sustain. Cities Soc.* **2023**, *98*, 104843. [CrossRef]
6. Lafortezza, R.; Sanesi, G. Nature-based solutions: Settling the issue of sustainable urbanization. *Environ. Res.* **2019**, *172*, 394–398. [CrossRef]
7. Kroeger, T.; Escobedo, F.J.; Hernandez, J.L.; Varela, S.; Delphin, S.; Fisher, J.R.B.; Waldron, J. Reforestation as a novel abatement and compliance measure for ground-level ozone. *Proc. Natl. Acad. Sci. USA* **2014**, *111*, E4204–E4213. [CrossRef]
8. Rahman, M.A.; Moser, A.; Rötzer, T.; Pauleit, S. Within canopy temperature differences and cooling ability of Tilia cordata trees grown in urban conditions. *Build. Environ.* **2017**, *114*, 118–128. [CrossRef]
9. Zolch, T.; Henze, L.; Keilholz, P.; Pauleit, S. Regulating urban surface runoff through nature-based solutions—An assessment at the micro-scale. *Environ. Res.* **2017**, *157*, 135–144. [CrossRef]
10. Dony, C.C.; Delmelle, E.M.; Delmelle, E.C. Re-conceptualizing accessibility to parks in multi-modal cities: A Variable-width Floating Catchment Area (VFCA) method. *Landsc. Urban Plan.* **2015**, *143*, 90–99. [CrossRef]
11. Chen, Y.; Men, H.; Ke, X. Optimizing urban green space patterns to improve spatial equity using location-allocation model: A case study in Wuhan. *Urban For. Urban Green.* **2023**, *84*, 127922. [CrossRef]
12. Haaland, C.; van den Bosch, C.K. Challenges and strategies for urban green-space planning in cities undergoing densification: A review. *Urban For. Urban Green.* **2015**, *14*, 760–771. [CrossRef]
13. Krekel, C.; Kolbe, J.; Wüstemann, H. The greener, the happier? The effect of urban land use on residential well-being. *Ecol. Econ.* **2016**, *121*, 117–127. [CrossRef]
14. Wüstemann, H.; Kalisch, D.; Kolbe, J. Access to urban green space and environmental inequalities in Germany. *Landsc. Urban Plan.* **2017**, *164*, 124–131. [CrossRef]
15. Liu, B.; Tian, Y.; Guo, M.; Tran, D.; Alwah, A.A.Q.; Xu, D. Evaluating the disparity between supply and demand of park green space using a multi-dimensional spatial equity evaluation framework. *Cities* **2022**, *121*, 103484. [CrossRef]
16. Nesbitt, L.; Meitner, M.J.; Girling, C.; Sheppard, S.R.J.; Lu, Y. Who has access to urban vegetation? A spatial analysis of distributional green equity in 10 US cities. *Landsc. Urban Plan.* **2019**, *181*, 51–79. [CrossRef]
17. Labib, S.M.; Lindley, S.; Huck, J.J. Scale effects in remotely sensed greenspace metrics and how to mitigate them for environmental health exposure assessment. *Comput. Environ. Urban Syst.* **2020**, *82*, 101501. [CrossRef]
18. Rigolon, A.; Browning, M.; Lee, K.; Shin, S. Access to Urban Green Space in Cities of the Global South: A Systematic Literature Review. *Urban Sci.* **2018**, *2*, 67. [CrossRef]
19. Rigolon, A. A complex landscape of inequity in access to urban parks: A literature review. *Landsc. Urban Plan.* **2016**, *153*, 160–169. [CrossRef]
20. Xu, M.; Xin, J.; Su, S.; Weng, M.; Cai, Z. Social inequalities of park accessibility in Shenzhen, China: The role of park quality, transport modes, and hierarchical socioeconomic characteristics. *J. Transp. Geogr.* **2017**, *62*, 38–50. [CrossRef]
21. Saghapour, T.; Moridpour, S.; Thompson, R.G. Public transport accessibility in metropolitan areas: A new approach incorporating population density. *J. Transp. Geogr.* **2016**, *54*, 273–285. [CrossRef]
22. Sun, C.; Lin, T.; Zhao, Q.; Li, X.; Ye, H.; Zhang, G.; Liu, X.; Zhao, Y. Spatial pattern of urban green spaces in a long-term compact urbanization process—A case study in China. *Ecol. Indic.* **2019**, *96*, 111–119. [CrossRef]
23. Ye, C.; Hu, L.; Li, M. Urban green space accessibility changes in a high-density city: A case study of Macau from 2010 to 2015. *J. Transp. Geogr.* **2018**, *66*, 106–115. [CrossRef]
24. Luo, W.; Wang, F. Measures of Spatial Accessibility to Health Care in a GIS Environment: Synthesis and a Case Study in the Chicago Region. *Environ. Plan. B Plan. Des.* **2003**, *30*, 865–884. [CrossRef] [PubMed]
25. Luo, W.; Qi, Y. An enhanced two-step floating catchment area (E2SFCA) method for measuring spatial accessibility to primary care physicians. *Health Place* **2009**, *15*, 1100–1107. [CrossRef] [PubMed]
26. Dai, D. Black residential segregation, disparities in spatial access to health care facilities, and late-stage breast cancer diagnosis in metropolitan Detroit. *Health Place* **2010**, *16*, 1038–1052. [CrossRef]

27. Audretsch, B.D.; Belitski, M. The limits to open innovation and its impact on innovation performance. *Technovation* **2023**, *119*, 102519. [CrossRef]
28. Xu, C.; Haase, D.; Pribadi, D.O.; Pauleit, S. Spatial variation of green space equity and its relation with urban dynamics: A case study in the region of Munich. *Ecol. Indic.* **2018**, *93*, 512–523. [CrossRef]
29. De Haas, W.; Hassink, J.; Stuiver, M. The Role of Urban Green Space in Promoting Inclusion: Experiences From the Netherlands. *Front. Environ. Sci.* **2021**, *9*, 618198. [CrossRef]
30. Jennings, V.; Bamkole, O. The Relationship between Social Cohesion and Urban Green Space: An Avenue for Health Promotion. *Int. J. Environ. Res. Public Health* **2019**, *16*, 452. [CrossRef]
31. Kimpton, A. A spatial analytic approach for classifying greenspace and comparing greenspace social equity. *Appl. Geogr.* **2017**, *82*, 129–142. [CrossRef]
32. Comber, A.; Brunsdon, C.; Green, E. Using a GIS-based network analysis to determine urban green space accessibility for different ethnic and religious groups. *Landsc. Urban Plan.* **2008**, *86*, 103–144. [CrossRef]
33. Anderson, V.; Gough, W.A.; Agic, B. Nature-Based Equity: An Assessment of the Public Health Impacts of Green Infrastructure in Ontario Canada. *Int. J. Environ. Res. Public Health* **2021**, *18*, 5763. [CrossRef]
34. Diao, Y.; Hu, W.; He, B.J. Analysis of the Impact of Park Scale on Urban Park Equity Based on 21 Incremental Scenarios in the Urban Core Area of Chongqing, China. *Adv. Sustain. Syst.* **2021**, *5*, 2100171. [CrossRef]
35. Yu, Y.; Zhou, L.; Zhou, W.; Ren, H.; Kharrazi, A.; Ma, T.; Zhu, B. Decoupling environmental pressure from economic growth on city level: The Case Study of Chongqing in China. *Ecol. Indic.* **2017**, *75*, 27–35. [CrossRef]
36. Shi, Y.; Feng, C.C.; Yu, Q.; Guo, L. Integrating supply and demand factors for estimating ecosystem services scarcity value and its response to urbanization in typical mountainous and hilly regions of south China. *Sci. Total Environ.* **2021**, *796*, 149032. [CrossRef]
37. Locatelli, B.; Lavorel, S.; Sloan, S.; Tappeiner, U.; Geneletti, D. Characteristic trajectories of ecosystem services in mountains. *Front. Ecol. Environ.* **2017**, *15*, 150–159. [CrossRef]
38. Sil, Â.; Rodrigues, A.P.; Carvalho-Santos, C.; Nunes, J.P.; Honrado, J.; Alonso, J.; Marta-Pedroso, C.; Azevedo, J.C. Trade-offs and Synergies Between Provisioning and Regulating Ecosystem Services in a Mountain Area in Portugal Affected by Landscape Change. *Mt. Res. Dev.* **2016**, *36*, 452–464. [CrossRef]
39. Mohajeri, N.; Gudmundsson, A. Street networks in relation to landforms: Implications for fast-growing cities. *J. Geogr. Sci.* **2014**, *24*, 363–381. [CrossRef]
40. Anguelovski, I.; Connolly, J.J.T.; Masip, L.; Pearsall, H. Assessing green gentrification in historically disenfranchised neighborhoods: A longitudinal and spatial analysis of Barcelona. *Urban Geogr.* **2017**, *39*, 458–491. [CrossRef]
41. Miralles-Guasch, C.; Dopico, J.; Delclos-Alio, X.; Knobel, P.; Marquet, O.; Maneja-Zaragoza, R.; Schipperijn, J.; Vich, G. Natural Landscape, Infrastructure, and Health: The Physical Activity Implications of Urban Green Space Composition among the Elderly. *Int. J. Environ. Res. Public Health* **2019**, *16*, 3986. [CrossRef] [PubMed]
42. Davis, A.Y.; Belaire, J.A.; Farfan, M.A.; Milz, D.; Sweeney, E.R.; Loss, S.R.; Minor, E.S. Green infrastructure and bird diversity across an urban socioeconomic gradient. *Ecosphere* **2012**, *3*, 1–18. [CrossRef]
43. Dai, D. Racial/ethnic and socioeconomic disparities in urban green space accessibility: Where to intervene? *Landsc. Urban Plan.* **2011**, *102*, 234–244. [CrossRef]
44. Wright Wendel, H.E.; Zarger, R.K.; Mihelcic, J.R. Accessibility and usability: Green space preferences, perceptions, and barriers in a rapidly urbanizing city in Latin America. *Landsc. Urban Plan.* **2012**, *107*, 272–282. [CrossRef]
45. Chen, S.; Zhang, L.; Huang, Y.; Wilson, B.; Mosey, G.; Deal, B. Spatial impacts of multimodal accessibility to green spaces on housing price in Cook County, Illinois. *Urban For. Urban Green.* **2022**, *67*, 127370. [CrossRef]
46. Tran, D.X.; Pla, F.; Latorre-Carmona, P.; Myint, S.W.; Caetano, M.; Kieu, H.V. Characterizing the relationship between land use land cover change and land surface temperature. *ISPRS J. Photogramm. Remote Sens.* **2017**, *124*, 119–132. [CrossRef]
47. Jalal, H.; Buchanich, J.M.; Roberts, M.S.; Balmert, L.C.; Zhang, K.; Burke, D.S. Changing dynamics of the drug overdose epidemic in the United States from 1979 through 2016. *Science* **2018**, *361*, eaau1184. [CrossRef]
48. ESRI. How Hot Spot Analysis (Getis-Ord Gi*) Works? 2016. Available online: http://proarcgis.com/en/pro-app/tool-reference/spatial-statistics/h-how-hot-spot-analysis-getis-ord-gi-spatial-stati.htm (accessed on 16 February 2016).
49. Krajter Ostoić, S.; Konijnendijk van den Bosch, C.C.; Vuletić, D.; Stevanov, M.; Živojinović, I.; Mutabdžija-Bećirović, S.; Lazarević, J.; Stojanova, B.; Blagojević, D.; Stojanovska, M.; et al. Citizens' perception of and satisfaction with urban forests and green space: Results from selected Southeast European cities. *Urban For. Urban Green.* **2017**, *23*, 93–103. [CrossRef]
50. Sharifi, F.; Nygaard, A.; Stone, W.M. Heterogeneity in the subjective well-being impact of access to urban green space. *Sustain. Cities Soc.* **2021**, *74*, 103244. [CrossRef]
51. Guo, A.; Yang, J.; Xiao, X.; Xia, J.; Jin, C.; Li, X. Influences of urban spatial form on urban heat island effects at the community level in China. *Sustain. Cities Soc.* **2020**, *53*, 101972. [CrossRef]
52. Fan, P.; Xu, L.; Yue, W.; Chen, J. Accessibility of public urban green space in an urban periphery: The case of Shanghai. *Landsc. Urban Plan.* **2017**, *165*, 177–192. [CrossRef]
53. Gu, X.; Tao, S.; Dai, B. Spatial accessibility of country parks in Shanghai, China. *Urban For. Urban Green.* **2017**, *27*, 373–382. [CrossRef]
54. Shen, Y.; Sun, F.; Che, Y. Public green spaces and human wellbeing: Mapping the spatial inequity and mismatching status of public green space in the Central City of Shanghai. *Urban For. Urban Green.* **2017**, *27*, 59–68. [CrossRef]

55. Chen, W.Y.; Hu, F.Z.Y. Producing nature for public: Land-based urbanization and provision of public green spaces in China. *Appl. Geogr.* **2015**, *58*, 32–40. [CrossRef]
56. Federica Gaglione, C.G.; Zucarob, F.; Cottrilla, C. Urban accessibility in a 15-minute city: A measure in the city of Naples, Italy. *Transp. Res. Procedia* **2022**, *60*, 378–385. [CrossRef]
57. Tan, P.Y.; Wang, J.; Sia, A. Perspectives on five decades of the urban greening of Singapore. *Cities* **2013**, *32*, 24–32. [CrossRef]
58. Li, H.; Liu, Y. Neighborhood socioeconomic disadvantage and urban public green spaces availability: A localized modeling approach to inform land use policy. *Land Use Policy* **2016**, *57*, 470–478. [CrossRef]
59. La Rosa, D.; Takatori, C.; Shimizu, H.; Privitera, R. A planning framework to evaluate demands and preferences by different social groups for accessibility to urban greenspaces. *Sustain. Cities Soc.* **2018**, *36*, 346–362. [CrossRef]
60. Gupta, K.; Roy, A.; Luthra, K.; Maithani, S.; Mahavir, S. GIS based analysis for assessing the accessibility at hierarchical levels of urban green spaces. *Urban For. Urban Green.* **2016**, *18*, 198–211. [CrossRef]
61. Wu, J.; He, Q.; Chen, Y.; Lin, J.; Wang, S. Dismantling the fence for social justice? Evidence based on the inequity of urban green space accessibility in the central urban area of Beijing. *Environ. Plan. B Urban Anal. City Sci.* **2018**, *47*, 626–644. [CrossRef]

land

MDPI

Article

Spatiotemporal Differentiation of Land Ecological Security and Optimization Based on GeoSOS-FLUS Model: A Case Study of the Yellow River Delta in China Toward Sustainability

Lu Han, Yanbo Qu, Shufeng Liang, Luyan Shi, Min Zhang and Haiyan Jia *

The School of Public Administration and Management, Shandong University of Finance and Economics, Jinan 250014, China; hanlu@sdufe.edu.cn (L.H.); quyanbo@sdufe.edu.cn (Y.Q.); shufengliang@mail.sdufe.edu.cn (S.L.); shiluyan@mail.sdufe.edu.cn (L.S.); zhangminzm@mail.sdufe.edu.cn (M.Z.)

* Correspondence: jiahaiyan@sdufe.edu.cn; Tel.: +86-18953188826

Abstract: Land ecological security (LES) is a crucial issue pertaining to the sustainability of landscapes or land systems. Nevertheless, previous studies largely neglected the impacts of landscape structure on LES. This study, taking the Yellow River Delta (YRD) as the case area, analyzed the spatiotemporal differentiation patterns of regional LES in 19 counties (cities and districts) from 2000 to 2020 through the application of landscape metrics and employed GeoSOS-FLUS to forecast the landscape structure and corresponding land ecological security in 2030 under four different scenarios to advance sustainability. The results show that, first, landscape fragmentation and irregularity are decreasing, while aggregation and connectivity are increasing. The primary landscape change involves the transition from farmland to developed areas. Second, the LES index is rising. In 2020, about 73.68% of the research units reached level IV, while most were at level III. Third, it is projected that by 2030, the LES index will rise under all scenarios, exhibiting an eastward-declining gradient in its geographic distribution, with higher values prevailing in the eastern regions and gradually tapering off towards the west. It is expected that in the business-as-usual scenario, the proportion of LES reaching level V will be the highest. This study not only reveals the impact mechanism of the landscape structure on land ecological security but also provides an important basis for formulating scientific and rational land use planning and management strategies from the perspectives of sustainable land governance and sustainable land use.

Keywords: sustainable development; land ecological security; multi-scenario simulation; landscape structure; Yellow River Delta

1. Introduction

Land is a crucial medium for regional sustainable development, and land ecological security is indispensable for achieving enduring economic growth and societal progress, and additionally, it serves as a fundamental requirement for ensuring food safety and social stability [1]. With the continuous advancement of urbanization, human endeavors continue to interfere with the stability of agricultural ecosystems, and the contradiction between the economy, resources, and the environment has become increasingly prominent, destroying land ecosystems [2]. Significant portions of the global population face land ecological insecurity, environmental threats, and developmental shortcomings in low-income countries that rely on agriculture. These issues stem from failed agricultural development driven by factors such as a dearth of innovative practices, over-reliance on traditional methods, insufficient planned management, a lack of local community involvement, few economically viable alternatives, food shortages, unemployment, and low living standards, compounded by a lack of awareness [3]. At the same time, drought as well as the release of greenhouse gases, particularly carbon dioxide, into the atmosphere are also significant factors impacting land ecological security [4].

The ecosystem is built upon a specific landscape structure, and this landscape exerts a substantial effect on the comprehensive environmental safeguards for the land. Therefore, the land ecosystem and the landscape structure blend, thus showing characteristics of complexity [5]. A changing landscape structure is a critical aspect of landscape ecology research, as it affects landscape heterogeneity [6] and ecosystem stability [7]. As the landscape structure evolves, the patterns of material cycling and energy flow undergo corresponding changes, which may result in the degradation of regional ecosystem services [8]. This degradation not only threatens land ecological security but also poses severe challenges to sustainable land governance and land use through pathways such as environmental damage and decreased agricultural productivity. Understanding landscape structure change is vital for reaching landscape ecological balance and sustainable development [9].

The Yellow River Delta (YRD), the most recent and expansive wetland in China, is crucial for sustaining the ecological equilibrium of the area and moderating the local climate [10]. However, because of the inherent limitations and anthropogenic activities within the delta, the region's landscape structure and function have been compromised, resulting in inefficient land use, severe soil salinization, and wetland degradation, which demand urgent attention. The swift enlargement of construction land has further exacerbated these problems. In the context of landscape ecology, changes in the landscape pattern of the YRD not only impact the equilibrium and biodiversity of ecosystems but also have a direct bearing on the ecological safety of the region. With advancing urbanization along the Yellow River basin, the swift enlargement of construction land has exacerbated the already sensitive and fragile ecological environment issues. Consequently, investigations into the ecological safety of land are highly significant for optimizing the regional eco-security pattern and achieving sustainable progress in the Yellow River Delta.

Land ecological stability occupies a central position in landscape ecology and sustainable land utilization, as it not only concerns the environmental security of specific regions but also serves as the cornerstone for national security and socially sustainable development. In 1941, Aldo Leopold introduced the notion of "land health" [11]. Barthouse integrated the concept of safety into ecological and environmental conservation, conducting research on ecological systems [12]. The Food and Agriculture Organization of the United Nations (FAO) delineated the definition and essence of land assessment in 1976. It created the natural and social–economic evaluation indexes of land utilization in 1993 [13], which laid a robust theoretical basis for evaluating land eco-security. Regarding land eco-security, early research focused on land quality evaluation, land ecological security connotation, and qualitative regulation [14]. Given the growing tension and conflicts arising from the interactions between people's activities and land resources, ecosystems and the environmental circumstances are deteriorating. As a result, there is a growing body of research on terrestrial ecosystem protection. Research mainly focuses on the configurations of land ecological integrity [15,16], the evolution over space and time, factors affecting land ecosystem resilience [17,18], and the forecast of sustainable land ecosystem conditions [19,20]. The investigative techniques are varied yet predominantly focused on utilizing the Pressure–State–Response (PSR) framework, Driving–Pressure–State–Impact-Response (DPSIR) approach, the Economic–Environment–Social (EES) model, and the Natural–Economic–Social (NES) framework [21,22] to construct the indicator system and use principal component analysis, the ecological footprint method, landscape structure analysis, the TOPSIS method, and the matter component model [23–25] to evaluate land ecological security.

Landscape structure research includes special terrain areas, nature reserves, watersheds, provinces, cities, and other scales [8,26]. Initial investigations concentrated on the arrangement and pattern of landscape elements [27–29]. At the same time, recent research has expanded to landscape fragmentation, heterogeneity characteristics, land use, land evolution [30], landscape configuration, driving factor analysis, and landscape configuration simulation and optimization [31]. Research methods include a land use transfer matrix, a dynamic model [32], and a landscape pattern index [9]. A significant research domain that integrates terrestrial ecosystem patterns with sustainable land ecology centers on comprehending the impact of shifts in ecosystem patterns on the health of the area's ecological systems.

The interaction between landscape structure change and the ecological sustainability of land is subject to a trifecta of ecological protection, economic drive, and social development, which is the result of many factors [33], as shown in Figure 1. On the one hand, the landscape structure and function have experienced significant changes due to internal and external factors. These changes have led to other transformations in the arrangement of landscape elements and the range of landscape configuration, causing the landscape structure to deviate in its order and stability. At this state, the change in the landscape structure has a coercive effect on the safety of the "natural–social–economic system" and direct or indirect consequences on LES [34]. On the other hand, the landscape structure can resist external damage. Its changes in health can enhance the system's stability to a certain extent. This can then contribute positively to the safety of the "natural–social–economic system", thereby improving the degree of terrestrial ecosystem safety [35].

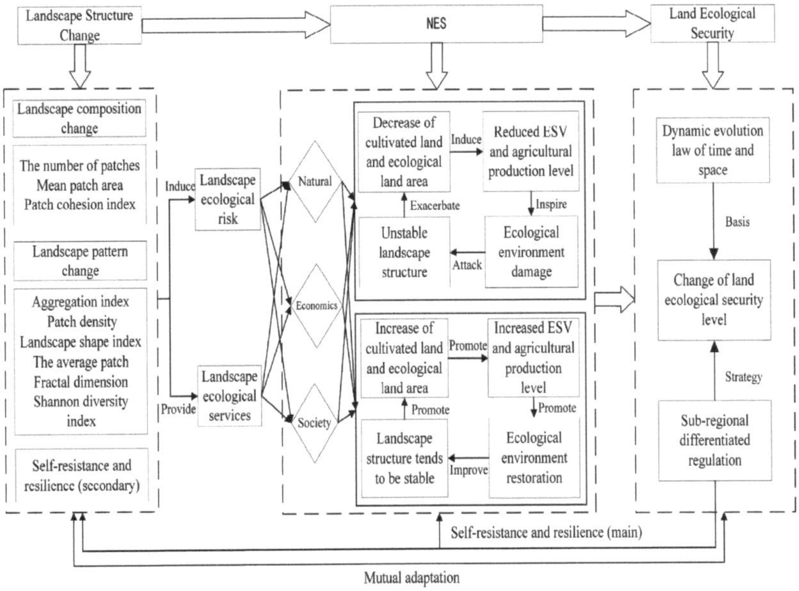

Figure 1. The mechanisms of interaction between the landscape structure change and LES.

This research direction explores the ecological impact of changes in landscape configuration, including the assessment of ecological security under different landscape models, the assessment of landscape ecological risks, and the assessment of the configuration vulnerability of the landscape. The ecological security assessment under differing landscape configurations encompasses a range of metrics, including, but not restricted to, the patch size, the number of fragments, a landscape diversity metric, an index of landscape dominance, a measure of landscape evenness, and a fragmentation assessment of the landscape. These indexes serve as quantitative tools for examining the evolving landscape alterations and elucidating the ecological security status within various landscape configurations [31]. To evaluate ecological hazard in the landscape, a model of ecological risk and security assessment is devised, incorporating indices of landscape disruption and vulnerability inherent in the landscape structure. Moreover, it explores how the landscape structure impacts ecological systems in wetlands, watersheds, and other research areas [36]. Constructing a landscape ecological vulnerability assessment model utilizing landscape sensitivity and the landscape fitness index allows for evaluating the vulnerability of landscape patterns. Similarly, spatial analysis and statistical methods examine geographical and temporal variations characterizing the region [37]. Scenario simulation and the prediction of the future ecological safety of land at the regional level will provide references for local managers and policymakers in territorial spatial planning. Currently, some more advanced models for land use research have gradually replaced traditional ones [38,39]. For example, the Verburg team used the CLUE-S model to transform changes in population, economy, and policies in Europe into changes in land use patterns through scenario simulations, thereby defining the trajectory of land use change in Europe [40]; the Yang Zitao team utilized the PLUS model to conduct three future scenario simulations for land use in the province by 2030 and evaluated the results of the development scenario [41]. The GeoSoS-FLUS model, formulated by the Li Xia team [42], is an effective model for geographic spatial modeling and spatial optimization, and aiding decision-making. This model can consider the transformation rules of land use types, spatial proximity effects, the impact of driving factors, and land use constraints [43], generating high-precision simulation results of land use changes [44].

These research findings provide valuable theoretical foundations and empirical cases of landscape ecological security patterns, sustainable land use resources, and sustainable land governance, yet vast research space remains to be explored. Firstly, despite the YRD boasting the largest land resources and most abundant wetlands in China, research has scarcely examined the spatiotemporal variations in land ecological security while adopting a landscape structural lens. Secondly, the landscape structure can indicate the condition of the ecosystem. Similarly, the value of ecosystem services directly reflects the condition of ecological security. By incorporating the two aspects within the measurement framework, a more impartial and comprehensive assessment of LES can be accomplished. However, most current evaluation studies overlook the influence of the two dimensions on the assessment outcomes. Thirdly, current studies aimed at forecasting the ecological stability of land primarily rely on time series analysis, and research on the trend of spatial evolution needs to be expanded. In contrast, research on multi-scenario prediction of land ecological security is rare.

This study explored the association between landscape structure and ecological land security in the YRD, China. A "natural–economic–social coupling system" (NES) model was developed to integrate the value of ecosystem functions into the framework for assessing land ecological security. Multiple scenarios were established, encompassing spontaneous growth, cultivated land conservation, economic growth, and ecological conservation. The GeoSOS-FLUS model was employed to forecast the landscape structure of the YRD in 2030, and the geospatial and temporal change trajectory of ecological land security was explored to evaluate the transformations of landscape structure in different situations and propose control measures.

2. Materials and Methods

The area included in this research is the Yellow River Delta region, characterized by fragile ecological foundations and landscapes exhibiting varying degrees of fragmentation. To foster national efforts in ecological preservation and sustainable development within the Yellow River basin, this research utilizes five phases of Normalized Difference Vegetation Index (NDVI) data (with a resolution of 1 km) and remote sensing land use monitoring data (also with a granularity of 1 km) spanning from 2000 to 2020, sourced from the Resource and Environment Science Data Center of the Chinese Academy of Sciences. Additionally, Digital Elevation Model (DEM) spatial distribution data (with a granularity of 90 m) from the National Geospatial Data Cloud, as well as socio-economic and lifestyle statistics from municipal and county-level statistical yearbooks (2000–2020) and the "National Economic and Social Development Statistics Bulletin" (2000–2020), are employed. By adopting the landscape pattern index method, utilizing the landscape change trajectory model, CRITIC–entropy weight combination weighting model, and GeoSOS-FLUS model, this research establishes a framework for assessing land ecological security to investigate the connection between changes in landscape structure and the ecological security of land in the Yellow River Delta area. Utilizing four scenario plans—natural development, farmland protection, economic development, and ecological conservation—this research simulates the spatiotemporal evolution trends of landscape structure and land ecological security under multiple scenarios to contribute to the environmentally sound development of this region.

The ecological integrity and safety of the YRD are discussed with regard to landscape composition, and research is conducted according to the subsequent steps. First, data are required to study the landscape framework change of the YRD utilizing landscape pattern metrics and the landscape change trajectory model. Secondly, the framework of land ecological security assessment is established, and the temporal and geographic distribution patterns of ecological stability within the YRD are examined by utilizing a combined CRITIC–entropy weighting model and a holistic appraisal technique. Subsequently, the GM (1,1) model and the GeoSOS-FLUS model are utilized to recreate multiple scenarios of LES in the YRD, describing the spatiotemporal attributes of LES across diverse scenarios. Finally, drawing on the aforementioned findings, the corresponding conclusions are drawn, and tangible recommendations are proposed to enhance ecological conservation and sustainable growth in the YRD (Figure 2).

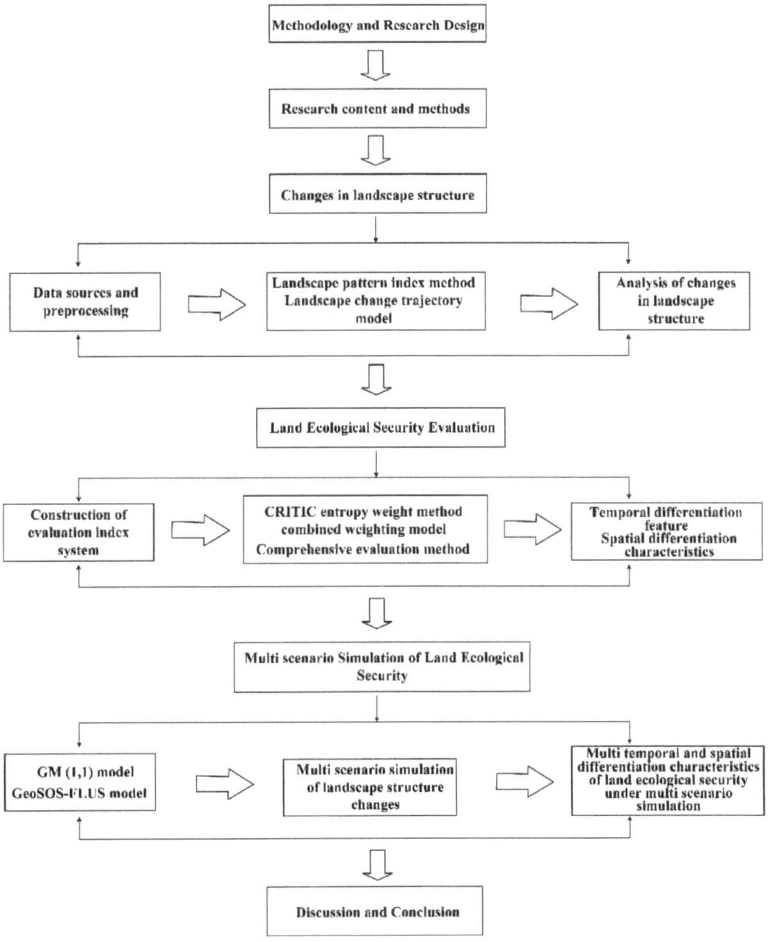

Figure 2. The research framework.

2.1. Study Area

The Yellow River Delta, designated as the YRD, is one of the largest river deltas with the greatest development potential in China (Figure 3). It is positioned at the juncture where the Yellow River meets the ocean, between 116°54′ E–120°18′ E and 36°25′ N–38°16′ N, and includes 19 counties (cities and districts) of Dongying, Binzhou, Weifang, Dezhou, Zibo and Yantai. The entire area under consideration spans approximately 2.65×10^4 km^2. Since introducing the Yellow River Delta Efficient Ecological Economic Zone Development Plan, the construction of the regional ecological civilization has achieved remarkable results. However, due to the fragile ecological foundation of the region and the increasing severity of climate change and anthropogenic impacts, the landscape exhibits varying degrees of fragmentation, which affects the stability and integrity of the landscape structure. Many problems with ecological and water resources such as soil salinization, untimely treatment of pollution sources and severe situations with water security need to be solved urgently, which limits the safety of the land ecosystem. Selecting this area for study holds significant practical value for advancing the national initiative on ecological preservation and sustainable, high-caliber growth in the Yellow River Basin.

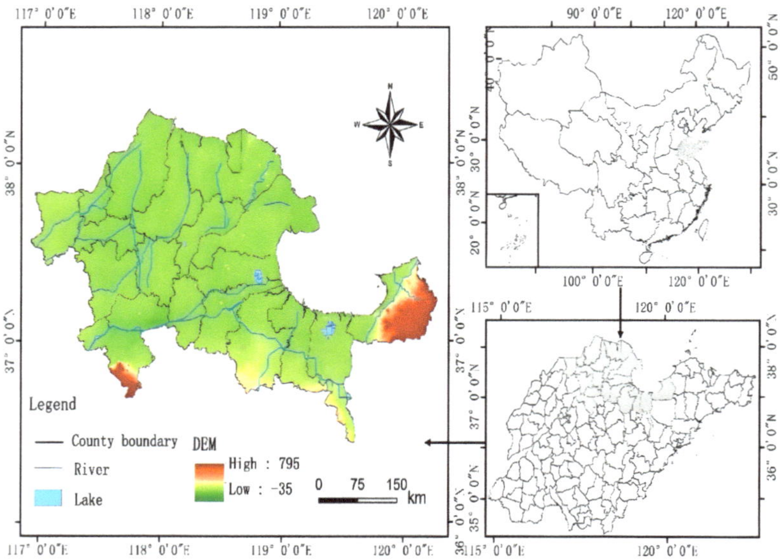

Figure 3. The location of the YRD.

2.2. Data Sources and Preprocessing

Within this research, the land use classifications within the designated region were delineated as categories of landscape. In accordance with the Land Use Status Classification (2010–2017) and the landscape features of the research region, landscape types have been categorized into six distinct classifications: cultivated land, forest land, grassland, wetland, construction land, and unused land. Normalized Difference Vegetation Index (NDVI) values and land utilization satellite imagery, both featuring a 1 km spatial accuracy, were obtained from the Resource and Environmental Science Data Center of the Chinese Academy of Sciences (http://www.resdc.cn, accessed on 6 November 2024) for 2000–2020. Digital Elevation Model (DEM) spatial distribution data (90 m resolution) originate from the National Geospatial Data Cloud, and elevation, gradient, and other data were derived based on DEM data. Socio-economic and living statistics are sourced from the statistical yearbooks of cities, counties, and districts (2000–2020) and from the Statistical Bulletin of National Economic and Social Development (2000–2020).

2.3. Methods

2.3.1. Landscape Pattern Index

In this research, fluctuations in landscape configuration in the YRD were studied, focusing on two key aspects: landscape composition and landscape pattern. This study utilized Fragstats4.2 software and a landscape metrics approach. The number of patches (NP), mean patch area (MPS), and patch cohesion index (COHESION) were chosen to characterize changes in landscape composition in the YRD [45]. The average fractal dimension of the patch (FRAC-MN), patch density (PD), maximum patch index (LPI), landscape shape index (LSI), Shannon diversity index (SHDI), contagion index (CONTAG), and aggregation index (AI) [20] were chosen to examine the transformation of the regional landscape pattern.

2.3.2. Landscape Change Trajectory Model

Combining the change trajectory model with the landscape configuration can more clearly show the characteristics of the landscape's composition change [46]. To reveal the features of evolving landscape structural dynamics, the present research employs a pathway analysis model to examine the evolution of six terrain categories within the YRD region and identifies the spatial transformation features among these varieties [47]. The mathematical expression is delineated below:

$$T = A \times 10^\wedge(n-1) + B \times 10^\wedge(n-2) + \cdots + E \times 10^\wedge(n-n), \quad (1)$$

where T is the newly generated landscape change trajectory code; $A, B, \ldots, E$ is the landscape grid trajectory coding of each year, which can be used to characterize the change in landscape types; and n is the number of time nodes.

2.3.3. CRITIC–Entropy Weight Method Combination Weighting Model

The CRITIC weighting approach determines the objective weight by estimating the contrast and conflict between the indices [48], which may reduce the impact of the correlation indices on weight distribution but cannot measure the degree of dispersion between indices. To compensate for the deficiency of this point and based on standardization, this study comprehensively uses the CRITIC–entropy weight method combination weighting model to ascertain the weight W_j for each indicator to determine the weight more objectively. The calculation formula is described in detail in [49].

2.3.4. Comprehensive Evaluation Method

Using spatial analysis and grid calculation, the standardized index data are superimposed with weights. The land ecological security (LES) index is obtained through a comprehensive assessment method as follows:

$$LES = \sum_{j=1}^{m} Y_{ij} W_j, \quad (2)$$

where Y_{ij} denotes the normalized value for every indicator, W_j represents the j-th index's weighting coefficient, and the LES range is [0, 1]. To more intuitively understand the present state of ecological land security within the research zone, this study refers to [9] and divides the LES index into five levels founded on the actual traits of the region (Table 1).

Table 1. The evaluation criteria used to determine the level of LES.

Safety Level	LES Level	Safety Zone
I	Severe	$LES \leq 0.42$
II	At-Risk	$0.42 < LES \leq 0.48$
III	Sensitive	$0.48 < LES \leq 0.54$
IV	Relatively secure	$0.54 < LES \leq 0.6$
V	Secure	$LES > 0.6$

2.3.5. GM (1,1) Model

The GM (1,1) model is suitable for predicting small sample data and can establish differential equations to predict future trends through the degree of dissimilarity between factors [50,51]. This study utilizes the GM (1,1) forecasting model to compute the forecasted figures for metrics within the ecosystem security assessment index system for the year 2030.

1. The modeling process

Let the primitive number be listed as $x^{(0)} = \left\{x^{(0)}(1), x^{(0)}(2), \cdots x^{(0)}(n)\right\}$ with the aim of accumulating a new series $x^{(1)} = \left\{x^{(1)}(1), x^{(1)}(2), \cdots x^{(1)}(n)\right\}$, where $x^{(1)}(k) = \sum_{i=1}^{k} x^{(0)}(i)$.

The gray differential equation is satisfied between the accumulated generated sequence and the original sequence as follows:

$$x^{(0)}(k) + \alpha z^{(1)}(k) = \mu \tag{3}$$

α denotes the growth gray number, μ represents the internal control gray number, and $z^{(1)}(k) = \left(0.5x^{(1)}(k) + x^{(1)}(k-1)\right)$.

Using the least square method for estimating the parameters, we can obtain the following:

$$[\alpha, \mu]^T = \left(B^T B\right)^{-1} B^T Y \tag{4}$$

where $B = \begin{bmatrix} -z^{(1)}(2)\ 1 \\ -z^{(1)}(3)\ 1 \\ \vdots \\ -z^{(1)}(n)\ 1 \end{bmatrix}, Y = \begin{bmatrix} x^{(0)}(2) \\ x^{(0)}(3) \\ \vdots \\ x^{(0)}(n) \end{bmatrix}$.

The generating sequence $x^{(1)}$ satisfies the derivative formula of the common GM (1,1) model and is presented below:

$$\frac{dx^{(1)}}{dt} + ax^{(1)} = \mu \tag{5}$$

The forecasted value from the GM (1,1) model's derivation proceeds as outlined below:

$$x\hat{}((1))(k) = (x\hat{}((0))(1) - \mu/\alpha)e\hat{}(-a(k-1)) + \mu/\alpha \tag{6}$$

2. Verification of simulation accuracy

In this study, the precision of the simulation was assessed using the posteriori error test, which tested the statistical characteristics of the residual distribution. The formula is as follows:

$$C = \frac{S_2}{S_1} \tag{7}$$

where C is the posterior difference ratio, S_1 represents the standard deviation of the initial data sequence, and S_2 is the residual standard deviation. A lower value of C corresponds to a higher simulation accuracy. When $C < 0.35$, the model accuracy is high; when $C < 0.5$, the model is moderately accurate; when $C < 0.65$, the simulation effect is fair; when $C > 0.65$, the simulation effect is poor. The test results show that $C = 0.329$, so the test simulation accuracy is high.

2.3.6. GeoSOS-FLUS Model

The GeoSOS-FLUS model is an upgraded model that draws its foundation from the principle of traditional cellular automation [52], which has high simulation accuracy and closely mirrors the actual patterns of land use. Utilizing the GeoSOS-FLUS model as a foundation, this study conducted a comprehensive multi-scenario simulation to analyze potential modifications to the landscape structure of the delta of the YRD by 2030, further delving into the implications of these alterations on land ecological security across various scenarios. It projected a geographical distribution of land ecological security across four different scenarios.

1. Selection of driving factors of landscape structure change

This study selected 12 driving factors as independent variables of landscape structure change, including annual average precipitation, annual mean precipitation, slope, elevation, public financial expenditures, average investments in fixed assets, population density, GDP per capita, aggregate power of farm machinery, distance from residential areas, distance from coastlines, and distance from roads [53]. Then, the standardized driving factor data were resampled to a 1 km resolution.

2. Accuracy verification analysis

To confirm the precision of the simulated results, utilizing the landscape structure data from the YRD for the year 2015, changes in landscape structure in 2020 were simulated, and the modeling outcomes were contrasted with the real values in 2020. The results showed that the kappa value reached 0.834, demonstrating the model's high precision.

3. Multi-scenario setting

Since 2009, the YRD has been acknowledged as part of the country's strategic growth initiatives, leading to a rise in regional economic development. With it being an important grain-producing area, its agricultural development has attracted much significance. Upon the implementation of sustainable protection and progressive development strategies within the Yellow River Basin, undeniable importance is placed on ecological protection and wetland protection in the region. To investigate how alterations in landscape structure affect the progression of land ecological security across various scenarios, this research incorporated the distinctive characteristics of the study area and refers to pertinent scholarly works to forecast the anticipated shifts in the landscape structure of the YRD by the year 2030 [9,54] (Table 2).

Table 2. Multi-scenario simulation mode and setting principle of the landscape structure change mode.

Scene Mode	Scenario Setting Principles
Business-as-usual	Do not ponder the restrictive effect of any planning policy on changing the landscape structure and do not impose any human interference and restrictions, so that the landscape structure naturally develops under the current real conditions.
Cultivated land protection	Enhance the safeguarding of cultivated land, comprehensively consider the proportion of cultivated land, construction land, ecological land, and other landscape structures, tightly regulate the rise in total construction land, and limit the conversion of cultivated land into other types of landscapes.
Economic development	Take economic development as the goal of optimization, follow the requirements of urban development, facilitate the development of new urban facilities, and expand the construction land area, while reducing the likelihood of the conversion of construction land into other landscapes, and constantly improve the level of urbanization.
Ecological protection	According to the "Outline of the Yellow River Basin Ecological Protection and High-quality Development Plan" and based on ensuring that the cultivated land area is not reduced, take the ecological protection priority as a goal of optimization; enhance the safeguarding of forest land, grassland, and wetland; and appropriately reduce the construction land to foster the development of ecological civilization and high-quality development in the YRD.

2.4. Land Ecological Security Evaluation Index System

Compared to traditional research models, the NES model more accurately captures the level of harmony among the land ecosystem and natural, economic, and social systems. The foundation of the land ecological security assessment index framework for the area is based on the features of the landscape composition and the actual state of socio-economic progress in the YRD (Table 3).

Table 3. The assessment criteria used to determine land ecological security in the YRD. "+" signifies a positive correlation, "−" signifies a negative correlation.

Target Layer	Criterion Layer	Indicator Layer	Trend	W_1	W_2	W_j
Land ecological security	Natural (N)	C1: NDVI (%)	+	0.066	0.036	0.051
		C2: Average annual precipitation (mm)	+	0.05	0.034	0.042
		C3: Proportion of saline–alkali land area (%)	−	0.035	0.025	0.030
		C4: Landscape matrix stability index (%)	+	0.076	0.034	0.055
		C5: Shannon diversity index (%)	+	0.064	0.054	0.059
		C6: Slope (°)	−	0.056	0.030	0.043
		C7: Area of cultivated land per capita (hm2/person)	+	0.05	0.048	0.049
		C8: Landscape fragmentation (%)	−	0.048	0.026	0.037
		C9: Structural stability index (%)	+	0.065	0.035	0.050
		C10: Wetland area (km^2)	+	0.042	0.134	0.088
		C11: Ecosystem service value (CNY)	+	0.042	0.090	0.066
	Economy (E)	C12: GDP per capita (CNY/person)	+	0.039	0.061	0.050
		C13: Share of tertiary industry in GDP (%)	+	0.046	0.036	0.041
		C14: Rural dwellers' average net earnings (CNY)	+	0.06	0.074	0.067
		C15: Fiscal expenditure on environmental protection (CNY)	+	0.043	0.083	0.063
	Society (S)	C16: Natural population growth rate (‰)	−	0.043	0.027	0.035
		C17: The level of urbanization (%)	+	0.049	0.039	0.044
		C18: Population density (Person/km^2)	−	0.02	0.062	0.041
		C19: Agricultural mechanization level (kw/hm^2)	+	0.046	0.042	0.044
		C20: Road network density (%)	−	0.060	0.030	0.045

The natural dimension reflects the impact of the landscape structure and environmental conditions on the LES. The landscape matrix stability index, structure stability index, and Shannon diversity index mirror the influence of landscape structure stability and diversity on land ecological security. Landscape fragmentation serves as an indicator of the adverse effects caused by the division and disruption of natural areas. Saline–alkali land and wetlands are typical for the YRD. The proportions of saline–alkali land and wetland areas are selected to better reflect the research area characteristics. The average cultivated land area per person shows the pressure of the regional population on cultivated land. Meanwhile, the value of an ecosystem service represents the benefits and well-being provided by the land ecosystem. The NDVI indicates that the greater the vegetation coverage, the more favorable the terrestrial ecosystem, and the average annual precipitation and slope reflect the impact of natural conditions on the safety of the land ecosystem [24,55].

The economic dimension reflects the influence that economic growth has on the LES. The greater the GDP per capita, the greater the beneficial influence on the standard of LES. The percentage of the tertiary sector in GDP indicates the level of regional economic modernization, and the greater the proportion, the more it can alleviate the strain that primary and secondary industries exert on the land ecosystem. The average GDP per rural resident can reflect the income of farmers' cultivated land production. The greater the index value, the more robust the region's capacity for cultivated land production. Fiscal expenditure on environmental protection reflects the level of regional environmental management and contributes positively to enhancing land ecological security [37,56].

The social dimension reflects the role of social development in land ecological security. The rate of the natural rise in the population and the population density shows the burden of the population on the land ecosystem. The concentration of the road network and the level of urbanization indicate the effect of urban development on the terrestrial ecosystem. The level of agricultural machinery demonstrates the influence of agricultural modernization on the land ecosystem [57].

3. Results

3.1. Analysis of the Landscape Structure Change in the YRD

3.1.1. Overall Change Characteristics

The overall attributes of the landscape structural alterations within the YRD between 2000 and 2020 are illustrated in Table 4 and Figure 4. The change in landscape structure in the YRD between 2000 and 2020 was primarily characterized by cultivated land, comprising over 60% of the entire surface. Construction land had an area ratio of 15% to 20%, followed by grassland and unused land. The expanse of cultivated land grew from 16,039 km^2 to 16,536 km^2 between 2000 and 2005, but decreased slightly after 2005 due to urban development and policies such as returning cultivated land to forests. Construction land increased by 24.5% between 2000 and 2015, but decreased to 4079 km^2 in 2020. Its change trend is primarily due to the continuous replenishment of water supplies in the YRD wetland in recent years. This effectively promotes the natural restoration of the ecosystem in the lower Yellow River area, the YRD wetland, and the nearby sea area. Forest land saw a slow decline, with grassland and unused land decreasing. The wetland developed rapidly in 2015, forming the main natural corridor of the YRD.

Table 4. The extent of various landscape categories in the YRD from 2000 to 2020.

Landscape Types	2000		2005		2010		2015		2020	
	Area/km^2	Percentage/%	Area /km^2	Percentage/%	Area /km^2	Percentage/%	Area /km^2	Percentage/%	Area /km^2	Percentage/%
Cultivated	16,039	62.20	16,536	64.48	16,369	63.62	16,204	62.59	16,039	61.96
Forest	245	0.95	262	1.02	268	1.04	280	1.08	190	0.73
Grassland	1917	7.43	1322	5.15	1289	5.01	1277	4.93	356	1.38
Wetland	1823	7.07	1620	6.32	1726	6.71	1923	7.43	4870	18.81
Construction	4001	15.52	4556	17.76	4856	18.87	4982	19.24	4079	15.76
Unused	1762	6.83	1351	5.27	1221	4.75	1224	4.73	352	1.36
Total	25,787	100.00	25,647	100.00	25,729	100.00	25,890	100.00	25,886	100.00

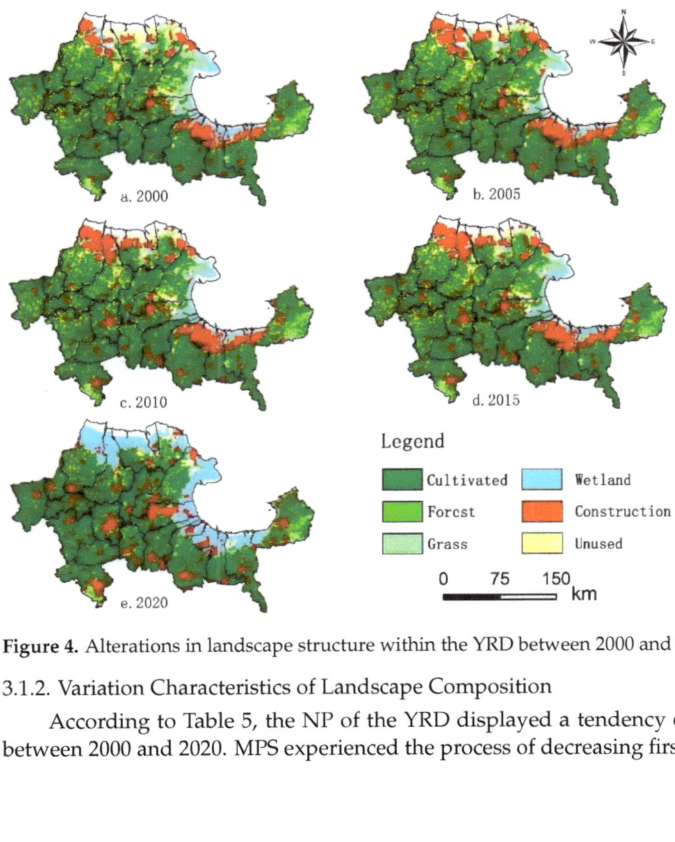

Figure 4. Alterations in landscape structure within the YRD between 2000 and 2020.

3.1.2. Variation Characteristics of Landscape Composition

According to Table 5, the NP of the YRD displayed a tendency of ups and downs between 2000 and 2020. MPS experienced the process of decreasing first and then increasing over the period under investigation. It is evident that from 2000 to 2015, the level of landscape fragmentation in the YRD escalated, indicating that human activities and urbanization progress had an important influence on the landscape structure of the YRD, leading to ongoing fragmentation of landscape patches and a steady rise in patch count. After 2015, land and resource conservation garnered increased attention. As a result, the changes in NP and MPS in the YRD exhibited a pattern of decrease and increase, respectively. And the level of landscape fragmentation was reduced. The COHESION value can indicate the extent of landscape aggregation and dispersion. During the research period, COHESION was greater than 98%.

Table 5. Characteristic index of landscape composition change.

Indicator Name	Unit	Year				
		2000	2005	2010	2015	2020
NP	-	2534	2517	2549	2579	1828
MPS	/km^2	10.17	10.18	10.09	10.03	14.16
COHESION	%	98.45	98.55	98.54	98.52	98.70

3.1.3. Characteristics of the Landscape Pattern Change

Table 6 shows that the PD in the YRD was stable at 0.1 per km^2 from 2000 to 2015, and then it decreased from 2015 to 2020, which meant that the degree of landscape fragmentation also diminished. The LPI initially rose and subsequently fell, reaching the highest value of 59.57% in 2005. It continued to decline from 2005 to 2020, indicating that human activity's impact on the landscape was intensifying, and the significance of dominant landscape types began to weaken after 2005. LSI exhibited a fluctuating decline from 28.74% to 24.12%, indicating that the plaque irregularity decreased gradually. The SHDI stabilized at around 1.15 between 2000 and 2015, and experienced a decline from 2015 to 2020, suggesting a reduction in the diversity of landscape types, and various types of patches are prone to uneven distribution in the landscape. CONTAG and AI can indicate the degree of interconnectivity and aggregation of landscapes, both of which have displayed a rise–fall–rise pattern, suggesting that the degree of agglomeration for various patch types has decreased from 2005 to 2015, while the distribution of patch types showed a trend towards increased concentration from 2015 to 2020. During the study period, FRAC-MN was stable at 1.02, and the trend of change was not obvious, indicating that the change in the general intricacy of the YRD landscape is not significant. Overall, throughout the research timeframe, the landscape fragmentation of the YRD decreased, connectivity and agglomeration gradually increased, and the landscape ecosystem was generally stable.

Table 6. Characteristic index of landscape pattern change.

Indicator Name	Unit	Year				
		2000	2005	2010	2015	2020
PD	Pcs/km^2	0.10	0.10	0.10	0.10	0.07
LPI	%	56.63	59.57	58.68	57.78	57.23
LSI	%	28.74	28.60	28.80	29.24	24.12
SHDI	/	1.19	1.12	1.13	1.15	1.06
FRAC-MN	/	1.02	1.02	1.02	1.02	1.02
CONTAG	%	40.41	43.08	42.73	41.87	50.28
AI	%	66.64	66.75	66.58	66.18	72.58

This study analyzed 199 change trajectory codes within the YRD, focusing on the primary types of changes (Table 7). Urbanization has shifted from cultivated land to construction land, affecting 2561.73 km^2. Cultivated land has been transformed into wetlands and grasslands, threatening food security. In 2015, the central policy document emphasized the conservation of cultivated land, leading to the conversion of 430 km^2 of wetland, 77 km^2 of forest land, 1740 km^2 of construction land, and 680 km^2 of grassland to cultivated land. In addition to cultivated land, the main direction of grassland landscape changes includes forest land, wetland, construction land, and unused land, which comprise 7.16% of the total expanse. The transformation of wetlands and construction land is the predominant feature of changing landscape patterns throughout the study duration, and the standard codes are 44445, 44554, 45554, and 55554. Most of the unused land was used for wetland protection and complemented the shortage of cultivated land, of which 696 km^2 was converted to wetland, 659 km^2 to cultivated land, and 143 km^2 to construction land.

Table 7. The landscape change trajectory in the YRD between 2000 and 2020.

Track	Area/km^2	Track	Area/km^2
Cultivated land to construction land	2561.73	Grassland to construction land	159
Cultivated land to wetland	764	Grassland to unused land	137
Cultivated land to grassland	118	Wetland to construction land	123
Wetland to construction land	430	Unused land to wetland	696
Forest land to cultivated land	77	Unused land to cultivated land	659
Construction land to cultivated land	1740	Unused land to construction land	143
Grassland to cultivated land	680	Wetland to wetland	182
Grassland to forest land	63	Construction land to wetland	1239
Grassland to wetland	462.44	Cultivated land to cultivated land	89

3.2. Spatiotemporal Fluctuations in Land Ecological Security Within the YRD

3.2.1. Temporal Characteristics of Land Ecological Security

The global LES index of the YRD improved significantly between 2000 and 2020, with a continued upward trend (Table 8). This improvement is primarily due to the stabilization of the landscape structure and the expansion of wetland areas, ensuring stable ecological services. Technological advancements and increased awareness of ecological land protection also improved LES levels. From 2000 to 2015, LES levels in administrative units were mostly level III, with some level II due to human activities and environmental destruction. In 2020, most counties and districts were level IV, accounting for 73.68% of administrative units. Due to the great improvement in the level of agricultural implements, the NDVI, and the utility of ecosystem services in Kenli and Changyi, the level of LES reached level V.

Table 8. The land ecological security index of the YRD between 2000 and 2020.

District \ Year	2000	2005	2010	2015	2020
Dongying	0.459	0.491	0.501	0.532	0.565
Hekou	0.442	0.488	0.513	0.520	0.575
Kenli	0.452	0.493	0.521	0.553	0.607
Lijin	0.459	0.497	0.512	0.546	0.539
Guangrao	0.499	0.537	0.538	0.555	0.564
Bincheng	0.488	0.514	0.521	0.541	0.551
Zhanhua	0.488	0.502	0.525	0.535	0.576
Zouping	0.503	0.496	0.515	0.539	0.517
Huimin	0.493	0.502	0.501	0.533	0.547
Yangxin	0.466	0.484	0.512	0.526	0.534
Wudi	0.478	0.496	0.516	0.536	0.592
Boxing	0.508	0.539	0.554	0.561	0.563
Laoling	0.479	0.492	0.520	0.540	0.559
Qingyun	0.486	0.510	0.528	0.544	0.543

Table 8. *Cont.*

District \ Year	2000	2005	2010	2015	2020
Gaoqing	0.499	0.527	0.520	0.541	0.554
Hanting	0.482	0.485	0.494	0.501	0.580
Shouguang	0.498	0.515	0.516	0.532	0.592
Changyi	0.512	0.528	0.553	0.563	0.603
Laizhou	0.466	0.482	0.518	0.514	0.562
Yellow River Delta	0.481	0.504	0.520	0.537	0.565

3.2.2. Spatial Attributes of Land Ecological Security

Software was employed to depict the geographical distribution of ecological land security across the YRD from 2000 to 2020 (Figure 5). The geographical distribution changed from "low in the north and high in the south" to "high in the east and low in the west" in 2000. Level I areas were distributed in Hekou due to the large saline–alkali area, low ecosystem services, and forest cover. Level II regions were dispersed in the northern section, while level III areas were clustered in the southern segment. Level IV areas were scattered throughout the counties. Regions like Zhanhua, Lijin, and Kenli transitioned from level II to level III in 2005, leading to an increase in level III area by 4364.79 km^2. The area of level IV also expanded and accounted for 13.52%. In 2010, the level of LES of all administrative units improved, and the areas at level III and level IV continued to expand, accounting for 71.86% and 21.49% of the entire research area, respectively. In 2015, the geographical pattern of LES in the YRD was differentiated and roughly characterized by a spatial distribution of "high in the east and low in the west". The level IV area changed greatly and was mainly distributed in Dongying, Guangrao, Boxing, and Changyi in the east. The proportion of level II areas increased and was predominantly located in Hekou due to the limited expanse of wetland and grassland and a comparatively low level of agricultural machinery. In 2020, the level of LES in the western area saw a significant improvement. Except for the southern part of Zouping, which is in level II, most of the area was converted to level IV, suggesting that against the backdrop of ecological civilization development, the land ecosystem of the YRD tends to be stable, with an ongoing enhancement in the condition of land ecological security. The level V region mainly occurs in the eastern Bohai Sea area due to the large number of wetlands and grasslands with more important ecological functions.

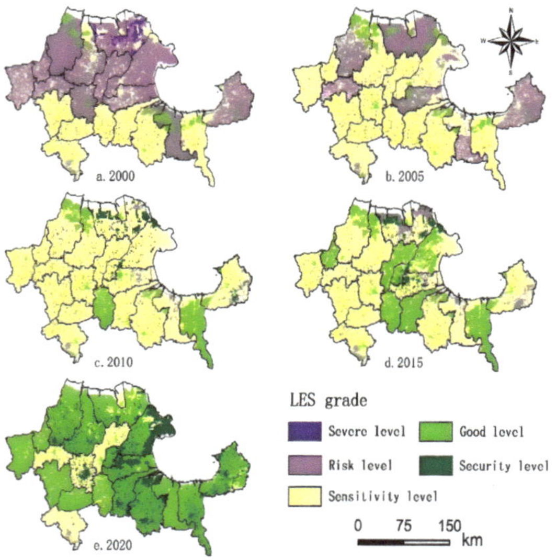

Figure 5. The geographical pattern of the land ecological security status in the YRD from 2000 to 2020.

3.3. Multi-Scenario Simulation of Land Ecological Security in the YRD

3.3.1. Multi-Scenario Simulation of Landscape Structure Change

The results of the simulation of different types of landscapes according to different scenarios are presented in Table 9. In 2030, the business-as-usual scenario shows the trend of expansion of cultivated land and wetland to construction land, grassland, and other landscapes, reflecting the necessary trajectory influenced by ecological conservation and the pursuit of high-quality development within the Yellow River Basin. The cultivated land protection scenario shows an increasing trend in cultivated land and unused land. In contrast, the economic growth scenario diminishes the likelihood of land use change, resulting in the continued occupation of cultivated land and grassland. The ecological conservation scenario enhances the landscape configuration, leading to a greater share of wetland, forest land, and grassland.

Table 9. The projected outcomes of diverse landscape types across various future scenarios by 2030.

Landscape Types	Business-as-Usual (Area/km^2)	Cultivated Land Protection (Area/km^2)	Economic Development (Area/km^2)	Ecological Protection (Area/km^2)
Cultivated	16,249	16,520	15,835	16,039
Forest	185	184	144	213
Grassland	274	356	281	365
Wetland	4926	4302	4812	5017
Construction	3939	4079	4501	3939
Unused	313	445	313	313

3.3.2. Spatial–Temporal Characteristics of Land Ecological Security Under Multi-Scenario Simulation

Drawing from the outcomes of the simulation of multi-scenario landscape structures in 2030, the LES of the YRD in multiple scenarios was evaluated (Figure 6). The LES indices under the four scenarios of spontaneous progression, preservation of cultivated land, growth of the economy, and conservation of the ecosystem are 0.5841, 0.6224, 0.6123, and 0.6341, respectively, which are between level IV and level V. The results showed significant differences in the LES index based on different scenarios, with the LES index showing increasing trends due to ongoing economic and societal advancement and the reinforcement of environmental conservation measures. The geographical distribution of LES in four scenarios was "eastern elevation and western depression". The ecological protection scenario saw a significant increase in level V areas, primarily enhancing the safeguarding of forest land, grassland, and wetland and promoting ecosystem services, landscape fragmentation, and landscape stability. In scenarios prioritizing the preservation of cultivated land and economic expansion, the level V area experiences further enlargement in contrast to the standard operating scenario, accounting for 74.64% and 70.68%, respectively. The level V area was focused on the northeast region of the YRD. The area of level IV and level III was reduced. The level of the LES of these five counties has not yet changed to level V. In line with the ecological conservation scenario, the area of level III is only 1 km^2, and the level IV area accounted for 11.45%. The level V area increased significantly, accounting for 88.54%, an increase of 19,109.04 km^2 compared to 2020, showing a spatially concentrated distribution. This is chiefly the result of the enhanced protection of forest land, grassland, and wetland under this scenario, which increases ecosystem service values, reduces landscape fragmentation, improves the stability of the landscape structure, and continuously expands the high-value area of LES within the research area.

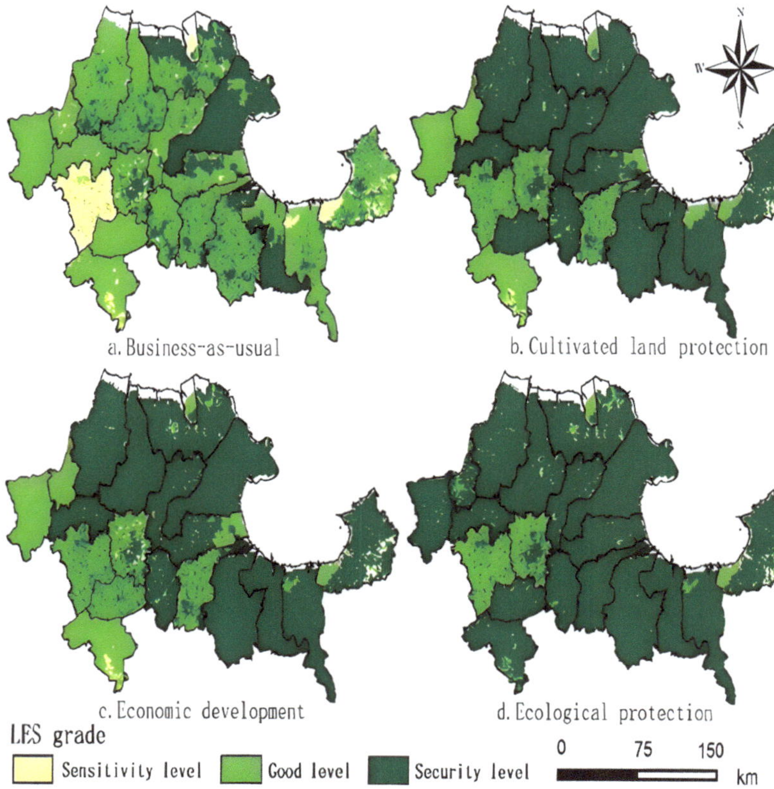

Figure 6. The multi-scenario simulation of land ecological security in the YRD in 2030.

4. Discussion

4.1. Contribution

This study examined the ecological land security of 19 counties in the YRD from 2000 to 2020, focusing on alterations in landscape structure. This study expanded the research horizons concerning the Yellow River Basin, offering a new viewpoint on ecological conservation and sustainable development in the region. The investigation integrated measures of landscape configuration and the value of ecosystem services into the assessment framework, yielding a more holistic view of how the land ownership system influences the security of ecological land. In contrast to existing studies on land ecological security [58,59], this study not only relied on the preliminary assessment predicated on stress, condition, and reaction but also evaluated the alterations in terrestrial ecological integrity over the study's timeframe, while accounting for shifts in landscape configuration. This method provides in-depth insight into how the land system intrinsically affects land ecological security and allows for a multi-dimensional assessment of LES. The GeoSOS-FLUS model was employed to simulate the landscape structure, and the GM (1,1) model was applied to simulate several ecological land security scenarios. The CA-Markov model frequently serves as a tool in current studies for multi-scenario simulation of LES [60,61]. These types of research only perform a numerical prediction of the time series within a single scenario. This research, however, employed the GeoSOS-FLUS model to simulate the landscape structure. Using these simulation outcomes, the GM (1,1) model was utilized to simulate the multiple scenarios of land ecological security, which better reflected the influence of the landscape structure on land ecological security. This research also produced a geographical and chronological evolution of ecological land security trends in the YRD, thereby expanding the research scope of multi-scenario simulation. The study proposes optimization and control countermeasures derived from the current state of the YRD and simulation results, offering guidance for the sustainable development and stimulating inquiry into excellence in additional domains.

This study displayed that the configuration of the landscape exerts varying levels of influence on the landscape of each county, with some counties experiencing greater fragmentation and lower stability due to human activities and urbanization. The research results are roughly the same as those of previous studies [62]. However, as ecological protection efforts intensify, the landscape structure stabilizes, characterized by an annual growth in the wetland area. This trend ensured a relatively consistent provision of various ecological service functions and contributed to the overall increase in the LES index of the YRD.

This research considered the impact of the landscape configuration on ecological land security in multiple scenarios, offering a foundation for land utilization strategy and the safeguarding of ecological integrity within the research region. The no-intervention business-as-usual scenario is expected to encroach on cultivated land, grassland, and forest land, resulting in the lowest LES index. The preservation scenario for cultivated land favors the enlargement of cultivated land and grassland, but the wetland area is expected to decrease, leading to a growth in unused land and construction land. The economic development scenario favors urbanization around existing infrastructure, increasing construction land and negatively affecting cultivated land, forest land, wetland, and the overall LES index. The scenario that focuses on ecological protection considers an increase in forest land, grassland, and wetland. Compared to other scenarios, there is no clear evidence of the increased cultivated and construction land. In this scenario, the YRD obtains the largest safety level and the highest LES index. This study recommends prioritizing cultivated land protection and ecological safeguards to achieve a balanced progression in economic growth, environmental conservation, and food security.

4.2. Limitations

This research had limitations: First, constructing a land ecological safety evaluation index framework and selecting indicators are key to assessment research. Due to insufficient data and the availability of some data, some important indicators were not selected, such as the severity of soil erosion, which might exert a certain influence on the evaluation outcomes. Second, the GeoSOS-FLUS model that was utilized to forecast the trend in the evolution of the landscape structure and land ecological security expected to occur in 2030 lacked a long-term sequence simulation analysis. In addition, in the current process of scenario simulation research, the assumption that policy regulation methods remain unchanged may lead to biases in the comprehensiveness and timeliness of the research results. In subsequent studies, basic data, control standards, policy regulations, and other elements will be further incorporated into the model operations.

4.3. Policy Implications

Research on ecological land security regulation and condition assessment is crucial for consolidating ecological barriers in the YRD, ensuring the protection of ecology and culture alongside high-quality development. This study offers a range of recommendations to foster the integrated advancement of landscape structure and ecological land security, thereby fulfilling the nation's ecological conservation and sustainable goals:

(1) Adhere to the boundary of permanently protected essential cultivated land, optimize the landscape structure, strengthen the conservation and restoration of cultivated land, strengthen the rise in agricultural land, promote the continuous development of cultivated land, and build a "trinity" pattern of cultivated land protection.
(2) Stop the increase in construction land and strengthen the management of forest land, grassland, and wetlands. The wetland area has experienced different degrees of degradation. In the future, it will be crucial to strictly abide by the ecological red

line, improve wetland restoration, monitor, and assess systems, implement wetland classification management, and increase the value of its ecosystem services.

(3) Actively develop unused land and establish a three-dimensional and long-term monitoring platform for to stock unused land resources. For areas with more unused land resources, we should clarify their area and layout, undertake reasonable planning, determine the use of unused land, and determine unreasonable land use changes promptly during the development process.

(4) Adhere to the ecological priority, strengthen the public awareness of environmental protection, increase investment in land ecological environment infrastructure, continue to explore new technologies for alkaline land management in the YRD, and strictly prohibit the blind and uncontrolled development of natural resources, so as to improve the ecological environment protection system and develop efficient ecological agriculture.

(5) Optimize the industrial structure of the region and continue to advance the new industrialization strategy, centered around a green circular economy, to reduce the environmental impact caused by industrial activities. This measure aims to foster the balanced growth of local economic prosperity and terrestrial ecological safety, to promote the formation of the YRD. The endogenous driving force with innovation as the core realizes the new development pattern of internal and external linkage.

(6) Establish a regional integration model by leveraging the influence of the central city, which has abundant resources and a comparatively high level of ecological land security. Neighboring regions should focus on improving their environmental development, leveraging their geographic advantages, and fostering cooperation and interconnectivity between development industries. The government should prioritize development issues, provide specific financial support, improve transportation networks, foster cooperation, and establish integrated development models in the YRD.

5. Conclusions

Currently, numerous studies on land ecological security focus on spatial patterns, driving factors, regulatory responses, or numerical predictions across time series under a single scenario, with limited integration of landscape patterns into multi-scenario simulations of land ecological security. In response, incorporating the landscape structure into land ecological security evaluation indicators and conducting simulations and predictions for different scenarios can provide more intuitive and comprehensive advice for local land policy formulation. This study examined the change in landscape structure and ecological land security in 19 counties (cities and districts) in the YRD. The landscape configuration metric approach, the NES model, and the GeoSOS-FLUS model were utilized to investigate spatiotemporal changes. Furthermore, four scenarios—business-as-usual, cultivated land protection, economic expansion, and ecological protection—were established to project future trends. This study found that cultivated land was the dominant landscape type between 2000 and 2020. The regions of cultivated land, forest, and construction land demonstrated an upward surge at first, with a subsequent fall. The landscape structure shifted from fragmentation to patchy irregularities, with trajectories of landscape change mainly moving from cultivated to construction land. The LES index exhibited an upward trend within the study duration and the spatial variance features of land ecological security evolved from "northernly depressed and southernly elevated" to "eastern elevation and western depression". However, its differences were narrowed. In 2030, significant differences were shown in the LES in different scenarios, and an upward trend was observed compared to that in 2020. This prediction underscores the importance of formulating scientific and reasonable land management strategies to ensure the sustainable exploitation of land resources and the protection of the ecological environment. In summary, this study not only reveals the intrinsic link between landscape structural changes and ecological land security but also integrates the concepts of sustainable land use and governance, providing an important basis for formulating scientific and reasonable strategies for sustainable land use and governance.

Author Contributions: L.H.: Conceptualization, methodology, software, validation, formal analysis, resources, data curation, writing—original draft, writing—review and editing, funding acquisition; Y.Q.: conceptualization, methodology, software, validation, formal analysis, supervision, advice; S.L.: conceptualization, formal analysis, data curation, writing—review and editing; L.S.: methodology, conceptualization, formal analysis, data curation, writing—review and editing, supervision; M.Z.: methodology, formal analysis, data curation, writing—review and editing; H.J.: methodology, conceptualization, resources, validation. All authors have read and agreed to the published version of the manuscript.

Funding: This research was funded by the general project of the National Natural Science Foundation of China: Grant No. 42077434 and National Natural Science Foundation of China: Grant No. 42471295.

Data Availability Statement: All relevant data sets in this study are described in the manuscript.

Conflicts of Interest: The authors declare no competing interests.

References

1. Zhang, Y.; Hu, Z.; Zhou, H.; Cui, G.; Li, J. Spatial and Temporal Distribution Characteristics on Land Ecological Security Along Main Stream of Yellow River in Shaanxi Province. *Bull. Soil Water Conserv.* **2022**, *42*, 258–267. [CrossRef]
2. Liu, Y.; Jing, Y.; Han, S. Ecological function zoning of Nansi Lake Basin in China based on ecosystem service bundles. *Environ. Sci. Pollut. Res.* **2023**, *30*, 77343–77357. [CrossRef]
3. Xiong, J. Review, Difficulties and Thinking of Land Ecological Security Evaluation. *Geog-Raphy Geo-Inf. Sci.* **2018**, *34*, 71–76. [CrossRef]
4. Hu, X.; Dong, C.; Zhang, Y. Impacts of Cropland Utilization Patterns on the Sustainable Use Efficiency of Cropland Based on the Human–Land Perspective. *Land* **2024**, *13*, 863. [CrossRef]
5. Lv, Y.; Chen, L.; Fu, B. Analysis of the Integrating Approach on Landscape Pattern and Ecological Processes. *Prog. Geogr.* **2007**, *26*, 1–10. [CrossRef]
6. Su, H.; Wu, C. Land use eco-risk assessment and its spatio-temporal variation of agroforestry ecotone based on landscape structure. *Econ. Geogr.* **2017**, *37*, 158–165. [CrossRef]
7. Wang, B.; Pan, K. Analysis of the landscape pattern and ecological security of Yinchuan City from 1980 to 2018. *J. Lanzhou Univ.* **2022**, *58*, 27–38. [CrossRef]
8. Xu, Q.; Wang, P.; Wang, S.; Ding, M.; Zhang, H. Influence of landscape structures on river water quality at multiple spatial scales: A case study of the Yuan river watershed. *China Ecol. Indic.* **2021**, *121*, 107226. [CrossRef]
9. Li, Y.; Geng, H. Evolution of Land Use Landscape Patterns in Karst Watersheds of Guizhou Plateau and Its Ecological Security Evaluation. *Land* **2022**, *11*, 2225. [CrossRef]
10. Zhang, Z.; Han, J.; Zhang, Y.; Sun, Y.; Sun, Z. Connotation, status, and governance of land ecological security in China's new urbanization: Recent advances and future prospects. *Environ. Sci. Pollut. Res. Int.* **2023**, *30*, 119654–119670. [CrossRef]
11. Yao, L.; Halike, A.; Yao, K.; Wei, Q.; Tang, H.; Tuheti, B.; Luo, J.; Duan, Y. Dynamic simulation of landscape ecological security and analysis of coupling coordination degree: A case study of Bole. *PLoS ONE* **2024**, *55*, 1932–6203. [CrossRef] [PubMed]
12. Wu, Y.; Qiu, L. Evaluation of Land Ecological Security in the Guanzhong Plain Urban Agglomeration. *Soil Water Conserv. China* **2024**, *2*, 52–56. [CrossRef]
13. Han, L.; Song, Z.; Zhang, M.; Ding, L. Visual Analysis of China's Land Ecological Security Based on CiteSpace. *J. Yunnan Agric. Univ.* **2022**, *16*, 141–149. [CrossRef]
14. Fu, B.J.; Liu, S.H.; Lu, Y.H. Comparing the soil quality changes of different land uses determined by two quantitative methods. *Res. Environ. Sci.* **2003**, *15*, 167–172. [CrossRef]
15. He, N.; Zhou, Y.; Wang, L.; Li, Q.; Liu, J.Y.; Li, M.Y. Spatiotemporal evaluation and analysis of cultivated land ecological security based on the DPSIR model in Enshi autonomous prefecture, China. *Ecol. Indic.* **2022**, *145*, 109619. [CrossRef]
16. Xu, S.; Yang, G.; Zhang, C.; Sun, D.; Sun, X. Study on evaluation of urban land ecological environment security based on PSR model: A case study of Qingdao city. *Ecol. Eng.* **2019**, *37*, 199–204. [CrossRef]
17. Wang, H.; Qin, F.; Zhang, X. A spatial exploring model for urban land ecological security based on a modified artificial bee colony algorithm. *Ecol. Inform.* **2019**, *50*, 51–61. [CrossRef]
18. Yu, W.; Cai, H.; Zhang, Y.; Lyu, B. The Early-warning and Regulation of Land Ecological Security: A Case Study of Hubei Province. *Environ. Sci. Technol.* **2018**, *41*, 189–196. [CrossRef]
19. Zhang, N.; Shi, S.; Li, B.; Li, H.; Li, S. Evaluation and Prediction of Land Ecological Security Based on PSR Model—Take Shenyang as an Example. *Chin. J. Soil Sci.* **2022**, *53*, 28–35. [CrossRef]
20. Huang, L.J.; Yang, P. Identification and Optimization of Land Ecological Security Pattern in Wuhan Metropolitan Area Based on Machine Learning. *J. Ecol. Rural. Environ.* **2020**, *36*, 862–869. [CrossRef]

21. Ye, H.; Bai, D.; Tan, S.; Shao, D.; Wang, J. Evaluating the ecological security of land resources based on multi-source data in the Altay Region of China. *J. Resour. Ecol.* **2021**, *12*, 757–765. [CrossRef]
22. Ye, H.; Ma, Y.; Dong, L. Land ecological security assessment for Bai autonomous prefecture of Dali based using PSR model–with data in 2009 as case. *Energy Procedia* **2011**, *5*, 2172–2177. [CrossRef]
23. Cheng, H.; Zhu, L.; Meng, J. Fuzzy evaluation of the ecological security of land resources in mainland China based on the Pressure-State-Response framework. *Sci. Total Environ.* **2022**, *804*, 150053. [CrossRef] [PubMed]
24. Liu, S.; Xu, L.; Zhang, J. Spatiotemporal change of land ecological security in Xinjiang. *Acta Ecol. Sin.* **2019**, *39*, 3871–3884.
25. Feng, Y.; Liu, Y.; Liu, Y. Spatially explicit assessment of land ecological security with spatial variables and logistic regression modeling in Shanghai, China. *Stoch. Environ. Res. Risk Assess.* **2017**, *31*, 2235–2249. [CrossRef]
26. Tran, D.; Pearson, D.; Palmer, A.; Lowry, J.; Gray, D.; Dominati, E. Quantifying spatial non-stationarity in the relationship between landscape structure and the provision of ecosystem services: An example in the New Zealand hill country. *Sci. Total Environ.* **2022**, *808*, 152126. [CrossRef]
27. Xiao, D.; Zhao, Y.; Sun, Z.; Zhang, G. Study on the variation of landscape pattern in the west suburbs of Shenyang. *Chin. Appl. Ecol.* **1990**, *1*, 75–84.
28. Hagen-Zanker, A. A computational framework for generalized moving windows and its application to landscape pattern analysis. *Int. J. Appl. Earth Obs. Geoinf.* **2016**, *44*, 205–216. [CrossRef]
29. Xu, J.; Yue, W.; Tan, W. A Statistical Study on Spatial Scaling Effects of Urban Landscape Pattern: A Case Study of the Central Area of the External Circle Highway in Shanghai. *Acta Ecol. Sin.* **2004**, *6*, 1058–1067. [CrossRef]
30. Fu, F.; Liu, W.; Wu, D.; Deng, S.; Bai, Z. Research on the spatiotemporal evolution of land use landscape pattern in a county area based on CA-Markov model. *Sustain. Cities Soc.* **2022**, *80*, 103760. [CrossRef]
31. Gao, Y.; Wu, Z.; Lou, Q.; Huang, H.; Cheng, J.; Chen, Z. Landscape ecological security assessment based on projection pursuit in Pearl River Delta. *Environ. Monit. Assess* **2012**, *184*, 13. [CrossRef] [PubMed]
32. Jiao, G.Y.; Yang, X.Z.; Huang, Z.Q.; Zhang, X.; Lu, L. Evolution characteristics and possible impact factors for the changing pattern and function of "Production-Living-Ecological" space in Wuyuan county. *J. Nat. Resour.* **2021**, *36*, 1252–1267. [CrossRef]
33. Fujihara, M.; Kikuchi, T. Changes in the landscape structure of the Nagara River Basin, central Japan. *Landsc. Urban Plan.* **2005**, *70*, 271–281. [CrossRef]
34. Zhang, W.; Yi, S.; Fan, X.; Zhang, X.; Liu, X.; Fang, Y.; Ma, H.; Liu, F.; Liang, Y. Relationship between ecosystem multifunctionality and plant diversity in different dimensions in the Yellow River Delta. *Wetl. Sci.* **2023**, *21*, 1–8. [CrossRef]
35. Ouyang, Z.; Zhu, C.; Yang, G.; Xu, W.; Zheng, H.; Zhang, Y.; Xiao, Y. Gross ecosystem product: Concept, accounting framework and case study. *Acta Ecol. Sin.* **2013**, *33*, 6747–6761. [CrossRef]
36. Xun, L.; Wang, Z. Landscape ecological risk assessment and its driving factors of multi-mountainous city. *Ecol. Indic.* **2023**, *146*, 146. [CrossRef]
37. Zhang, R.; Lu, J. Simulation of Land Use Pattern Evolution from a Multi-Scenario Perspective: A Case Study of Suzhou City in Anhui Province, China. *Int. J. Environ. Res. Public Health* **2021**, *18*, 921. [CrossRef]
38. Talukdar, S.; Singha, P.; Mahato, S.; Pal, S.; Liou, Y.A.; Rahman, A. Land-use land-cover classification by machine learning classifiers for satellite observations—A review. *Remote Sens.* **2020**, *12*, 1135. [CrossRef]
39. Wang, X.; Bao, Y. Research methods on the dynamic change of land use. *Prog. Geogr. Sci.* **1999**, *1*, 83–89.
40. Verburg, P.H.; Van Berkel, D.B.; Van Doorn, A.; van Eupen, M.; van den Heiligenberg, H.A. Trajectories of land use change in Europe: A model-based exploration of rural futures. *Landsc. Ecol.* **2010**, *25*, 217–232. [CrossRef]
41. Hu, F.; Zhang, Y.; Guo, Y.; Zhang, P.; Lyu, S.; Zhang, C. Spatial and temporal changes in land use and habitat quality in the Weihe River Basin based on the PLUS and InVEST models and predictions. *Arid. Land Geogr.* **2022**, *45*, 1125–1136. [CrossRef]
42. Liu, X.; Liang, X.; Li, X.; Xu, X.; Ou, J.; Chen, Y.; Pei, F. A future land use simulation model (FLUS) for simulating multiple land use scenarios by coupling human and natural effects. *Landsc. Urban Planning* **2017**, *168*, 94–116. [CrossRef]
43. Ji, Z.; Wei, H.; Xue, D.; Liu, M.; Cai, E.; Chen, W.; Feng, X.; Li, J.; Lu, J.; Guo, Y. Trade-off and projecting effects of land use change on ecosystem services under different policies scenarios: A case study in central China. *Int. J. Environ. Res. Public Health* **2021**, *18*, 3552. [CrossRef]
44. Yang, Z.; Yu, W.; He, P. Ecological Security Assessment of Jingzhou City Based on GeoSOS-FLUS Model. *Tianjin Agric. Sci.* **2024**, *30*, 47–54. [CrossRef]
45. Yang, Q.; Pu, L.; Huang, S. Review of Urbanization-Associated Farmland Research in China: A Sustainability Perspective. *Land* **2024**, *13*, 534. [CrossRef]
46. Liu, C.; Wu, X.; Wang, L. Analysis on land ecological security change and affect factors using RS and GWR in the Danjiangkou Reservoir area, China. *Appl. Geogr.* **2019**, *105*, 1–14. [CrossRef]
47. Liu, J.; Wang, D.; Sun, R.; Wang, F.; Hu, B.; Cheng, J.; Sun, Z. Study on spatial relevance of ecological-land loss based on change trajectory analysis method. *Geogr. Res.* **2020**, *39*, 103–114. [CrossRef]
48. Simic, V.; Gokasar, I.; Deveci, M.; Karakurt, A. An integrated CRITIC and MABAC based type-2 neutrosophic model for public transportation pricing system selection. *Socio-Econ. Plan Sci.* **2022**, *80*, 101157. [CrossRef]
49. Fu, W.Z.; Chu, L.P. Research on the evaluation of high-quality development of manufacturing industry from the perspective of integration of the Yangtze River Delta—TOPSIS evaluation model based on improved CRITICAL-Entropy weight method. *J. Ind. Tech. Ecol.* **2020**, *39*, 145–152. [CrossRef]

50. Alipbeki, O.; Alipbekova, C.; Sterenharz, A.; Toleubekova, Z.; Makenova, S.; Aliyev, M.; Mineyev, N. Analysis of Land-Use Change in Shortandy District in Terms of Sustainable Development. *Land* **2020**, *9*, 147. [CrossRef]
51. Kong, F.Z.; Cui, W.L.; Xi, H.H. Spatial–temporal variation, decoupling effects and prediction of marine fishery based on modified ecological footprint model: Case study of 11 coastal provinces in China. *Ecol. Indic.* **2021**, *132*, 108271. [CrossRef]
52. Pan, Z.; He, J.; Liu, D.; Wang, J. Predicting the joint effects of future climate and land use change on ecosystem health in the Middle Reaches of the Yangtze River Economic Belt, China. *Appl. Geogr.* **2020**, *124*, 102293. [CrossRef]
53. Wang, H.; Zhang, C.; Yao, X.; Yun, W.; Ma, J.; Gao, L.; Li, P. Scenario simulation of the tradeoff between ecological land and farmland in black soil region of Northeast China. *Land Use Policy* **2022**, *114*, 105991. [CrossRef]
54. Zhang, M.; Bao, Y.; Xu, J.; Han, A.; Liu, X.; Zhang, J.; Tong, Z. Ecological security evaluation and ecological regulation approach of East-Liao River basin based on ecological function area. *Ecol. Indic.* **2021**, *132*, 108255. [CrossRef]
55. Gao, X.; Wang, G.; Innes, J.L.; Zhao, Y.; Zhang, X.C.; Zhang, D.H.; Mi, F. Forest ecological security in China: A quantitive analysis of twenty-five years. *Glob. Ecol. Conserv.* **2021**, *32*, e01821. [CrossRef]
56. Su, Y.; Liu, G.; Zhao, J.; Guo, L.; Lin, F.; Niu, J. Land Ecological Security Early Warning Measure and Spatial-Temporal Pattern Evolution in Fenhe River Basin During 2006–2018. *J. Soil Water Conserv.* **2021**, *41*, 144–151. [CrossRef]
57. Wu, Y.; Jiang, M.; Yang, S. Security Measurement and Early Warning of Regional Innovation Ecosystem in Shaanxi Province from Ecological Niche Perspective. *Bull. Soil Water Conserv.* **2024**, *44*, 86–96. [CrossRef]
58. Wang, N.; Zhu, P.; Zhou, G.; Xing, X.; Zhang, Y. Multi-Scenario Simulation of Land Use and Landscape Ecological Risk Response Based on Planning Control. *Int. J. Environ. Res. Public Health* **2022**, *19*, 14289. [CrossRef]
59. Yu, Z.; Li, M.; Qian, Y.; Peng, R.; Yang, G. Land Use Change and Its Ecological Effects in Three Gorges Reservoir Area Basedon CA_Markov Model with Multiple Scenarios Simulation. *Res. Soil Water Conserv.* **2024**, *31*, 363–372. [CrossRef]
60. Zhao, L.; Liu, G.; Xian, C.; Nie, J.; Xiao, Y. Simulation of Land Use Pattern Based on Land Ecological Security: A Case Study of Guangzhou, China. *Int. J. Environ. Res. Public Health* **2022**, *19*, 9281. [CrossRef]
61. Zhang, H.; Yan, Q.; Xie, F.; Ma, S. Evaluation and Prediction of Landscape Ecological Security Based on a CA-Markov Model in Overlapped Area of Crop and Coal Production. *Land* **2023**, *12*, 207. [CrossRef]
62. Han, L.; Song, Z.Y. Dynamic Evaluation and Obstacle Factor Diagnosis of Cultivated Land Ecological Security-A Case Study at Counties Along Lower Yellow River. *Bull. Soil Water Conserv.* **2021**, *41*, 148–156+164. [CrossRef]

land

MDPI

Article

How Part-Time Farming Affects Cultivated Land Use Sustainability: Survey-Based Assessment in China

Xinwei Pei [1], Xinger Zheng [1,2] and Cong Wu [3,*]

[1] School of International Affairs and Public Administration, Ocean University of China, Qingdao 266100, China; 2023104@ouc.edu.cn (X.P.); z18059386033@163.com (X.Z.)
[2] School of Geography and Ocean Science, Nanjing University, Nanjing 210023, China
[3] School of Sociology and Law, Shanxi Normal University, Taiyuan 030031, China
* Correspondence: wucong2022@sxnu.edu.cn

Abstract: Part-time farming is a widespread phenomenon associated with the long-term global trend of urbanization, especially in China since its reform and opening-up in 1978. The shift of agricultural labor to non-agricultural sectors has significantly impacted cultivated land use activities, yet the connection between part-time farming and cultivated land use sustainability (CLS) remains understudied. Here, we construct an index system for assessing CLS that integrates ecological, economic, and social sustainability. Using survey data from seven Chinese villages across three provinces, we analyze the impact pattern and mechanism of part-time farming on CLS. We find the following: (1) The impact of part-time farming on CLS presents an inverted U-shape, peaking negatively at a 45% inflection point; (2) Spatial heterogeneity exists in the effect of part-time farming on CLS; (3) A household's non-agricultural workforce size and the gender of the household head significantly moderate the link between part-time farming and CLS; (4) CLS strongly hinges on various factors including the household head's health, other family members' education levels, commercial insurance, and agricultural skills training. Our findings provide empirical insights into governing part-time farming for sustainable cultivated land use and, eventually, rural human–land system sustainability.

Keywords: sustainable development; part-time farming; cultivated land use sustainability; food security; rural household

1. Introduction

Land is the essential link that binds the interactions of ecological elements and provides the space for human production and life [1]. And cultivated land is the most fundamental resource required for human survival and development [2]. Faced with daunting global challenges such as the escalation of climate warming and food insecurity, the imperative to attain cultivated land use sustainability (CLS) has emerged as a critical and pressing concern of the contemporary era. Since the concept of sustainable land use was officially established in 1990, a plethora of studies on CLS has been conducted by experts and scholars worldwide [3]. Previous studies on CLS have primarily focused on the following three aspects. First, several studies have measured the level of sustainable use of cultivated land in countries or regions through the construction of specific evaluation indicators [4,5]. For instance, in 1997, the International Conference on Sustainable Land Use Management and Information Systems divided the evaluation indicators of sustainable use of cultivated land into environmental and technical indicators, economic indicators, and social indicators. In accordance with the United Nations 2023 Sustainable Development Goals, Lu et al. selected 43 indicators from three dimensions to construct an evaluation index system for sustainable land use [4]. Second, some studies have focused on specific regions and the analysis of countermeasures. Scholars have examined issues regarding the sustainable use of cultivated land in plain areas [6], plateau regions [7], tropical zones [8], and ecologically fragile areas [9] and have offered optimization recommendations. Third, studies have analyzed the influencing factors of the sustainable use of cultivated land. These studies have explored the impacts of urbanization [10], agricultural labor transfer [11], land tenure [12], cultivated land fragmentation [13], and food production efficiency [14] on sustainable use. Within the realm of factors influencing CLS, existing research has identified the latent effect of part-time farming on CLS.

Part-time farming is a widespread phenomenon associated with the long-term global trend of urbanization, especially in China since its reform and opening-up in 1978. As a result of the shift in agricultural labor to non-farming employment resulting from urbanization [15], coupled with the influences of seasonal agricultural production and the instability of non-farming employment opportunities [16], part-time farming is increasingly becoming the predominant production and management strategy adopted by the majority of farmers [17]. And it has long been a rational choice for Chinese farming households to mitigate risks and maximize income. Among the impact of part-time farming on CLS, it has been found that part-time farming has an impact on fertilizer and pesticide application [18], agricultural labor productivity [19], agricultural irrigation [20], cultivated land protection behavior [21], willingness to return straw to the land [22], farmland transfer [23], and the abandonment of cultivated land [24]. Research indicates that part-time farming can promote the adoption of machinery by farmers, and mechanical tillage can significantly reduce the intensity of chemical fertilizer application [18]. The reduction implies a more rational control of agricultural pollution [25], effectively enhancing the level of CLS. Moreover, part-time farming may exacerbate land abandonment through changes in household income [26], thereby impacting the economic dimension of CLS.

Protecting and enhancing the sustainability of cultivated land is of significant importance for ensuring food security and promoting the sustainable development of society. Currently, the global population has reached 8 billion people; however, nearly 800 million people still face hunger issues [27], leading to sustained pressure on cultivated land production. Despite having vast areas of arable land in China, the per capita arable land area is less than 40% of the global average, meaning that China needs to feed nearly 20% of the world's population with only 9% of the global arable land. With rapid economic development and urbanization, food production faces the dual risks of tightening land resources and declining cultivated land quality. In light of the increasingly severe international food security situation, exploring scientific pathways to enhance China's cultivated land sustainability (CLS) is essential. It is also important to clarify the relationship between part-time farming and CLS, as well as to elucidate the mechanism that links part-time farming and CLS.

The previous research efforts in cultivated land sustainability have yielded substantial findings. However, there still exist areas that require further investigation: (1) Currently, there is still a lack of studies concerning the holistic correlation between part-time farming and CLS. Most of the land sustainability studies have focused on the impact of part-time farming on a single dimension of CLS, including food production efficiency, agricultural non-point source pollution, and agricultural irrigation efficiency, but lack a broader understanding of the impact of part-time farming on CLS. (2) Existing studies have only partially revealed the mechanisms through which part-time farming affects the sustainability of cultivated land, leaving much to be explored in understanding the intricate dynamics of this relationship. (3) Previous studies have paid insufficient attention to the impact of spatial variables and have not effectively explored the spatial differentiation in the impact effect.

In this study, based on the data collected from a field survey in seven villages across three provinces in China, we constructed an index system for evaluating CLS and analyzed the impact and mechanisms of part-time farming on CLS. The main objective of this research is to investigate the effect of part-time farming on CLS in different regions of China. Our research provides empirical evidence that validates the relationship between part-time farming and CLS. Furthermore, the study's findings will assist the government in developing better plans and strategies to achieve socio-environmental sustainability at local, regional, and global levels by promoting land use sustainability.

2. Framework and Hypotheses

2.1. Analytical Framework of How Farmers' Part-Time Farming Affects CLS

Sustainability is a people-oriented concept that consists of the following three elements: ecological sustainability, economic sustainability, and social sustainability. Sustainability signifies the responsible management of resources to satisfy the current ecological, economic, and social needs without compromising the ability to meet future requirements [28]. Cultivated land is the material foundation of agricultural production activities. The sustainable utilization of cultivated land calls for the effective use of the existing land resources, the mitigation or eradication of detrimental environmental impacts, and the enhancement of yields while augmenting the flow of natural capital [29]. This involves the multi-dimensional attributes of environmental, economic, and social dimensions. The farmer's perspective is an essential vantage point for comprehending CLS, with the influence pathways encompassing both crowding-out and income effects.

Regarding the crowding-out effects, as the key agents of cultivated land conservation, the willingness of farmers to invest in protective measures significantly influences the quality of the cultivated land [30]. The structural changes in family labor are also closely linked to land management practices [31]. The nature of part-time farming, as a process of the reallocation of family labor resources between the agricultural and non-agricultural production sectors, is characterized by a significant displacement of labor and a rapid shift of the young and middle-aged labor force to the non-agricultural sector. Consequently, on the one hand, part-time farming leads to a large-scale transfer of labor to non-agricultural sectors, affecting the transfer of cultivated land, the scale of land management [19], and the forms of agricultural organization [19]. On the other hand, it also exacerbates the phenomenon of agricultural aging. The elderly labor force's capacity to accept new technologies and equipment, as well as their willingness to invest, will further influence the protective investment in cultivated land, the input of agricultural production materials [32], and the efficiency of agricultural production [33]. In the context of the income effect, agricultural income is a significant factor that influences the farmers' behavior in protecting their cultivated land [34]. The income of farmers directly determines the level of investment in agricultural production and management. Consequently, the pluralistic employment behavior of farmers, which leads to changes in family income and its structure, including the increase in non-agricultural income and the expansion of income channels, will affect the livelihood strategies of farmers [17], the agricultural production methods [35], and the application of technological equipment [36], thereby impacting CLS.

At present, China's labor supply is experiencing the "Lewis turning point", where the labor market has shifted from a long-term surplus to a state of general scarcity, and labor wages have transitioned from long-term stability to a phase of general increase [37]. The phase differences in agricultural labor and income will lead to stage differences in CLS. Furthermore, industrialization and urbanization are the primary factors driving the phenomenon of farmers' pluralistic employment [15]. Regional socio-economic disparities also influence the process by which farmers' pluralistic employment affects CLS. Based on this, the analytical framework is proposed in Figure 1.

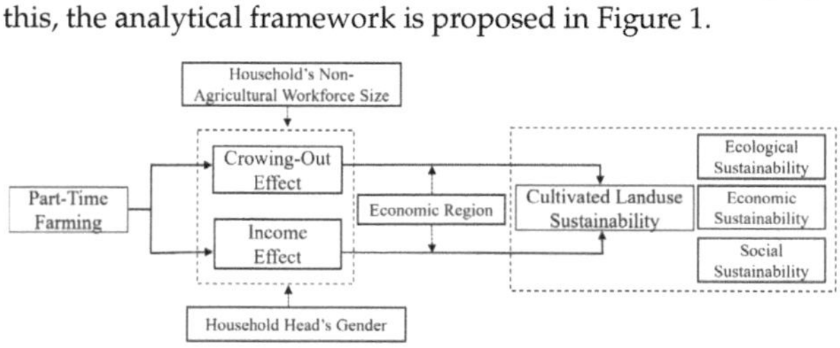

Figure 1. Analytical framework of this study.

2.2. Hypothesis

The crowding-out and income effects on CLS manifest in the multiple dimensions of ecology, economy, and society, with varying impacts. Regarding the positive influence of part-time farming, on the one hand, the crowding-out of agricultural labor by pluralistic employment can optimize the allocation of labor in terms of division of labor and specialization, intensifying the degree of labor specialization in agricultural production [38]. This is conducive to accelerating land transfer, improving the concentration of plots, promoting the scale operation of land, and simultaneously fostering the development of agricultural mutual aid organizations [19], thereby enhancing the social sustainability of CLS. On the other hand, the increase in non-agricultural income brought about by a higher degree of multiple employment opportunities will provide farmers with additional income channels, significantly alleviating their financial constraints. This enables farmers to compensate for the shortage of labor by improving the level of agricultural mechanization and purchasing specialized services [39], thereby enhancing agricultural production efficiency [33], and serving the needs of economic sustainability.

Additionally, the increase in non-agricultural income may diminish the enthusiasm of farmers for engaging in agricultural activities, thereby reducing the use of pesticides and fertilizers [18], which will continue to contribute to the green production of cultivated land. In terms of the adverse effects of part-time farming, such behaviors are likely to result in a lack of sufficient labor within farming households for agricultural production. To mitigate these negative impacts on agricultural production, the households engaged in multiple employment may increase the input of pesticides and fertilizers [32] and adopt more extensive irrigation methods [40], which could pose hazards to the cultivated land ecosystem. In addition, the shortage of labor may inhibit the adoption of new technologies and equipment, thereby restricting the mechanization of agriculture [41] and the scale of land operations [42], reducing agricultural production efficiency [43]. The shift in young and middle-aged labor to non-agricultural industries may also exacerbate the aging phenomenon of the agricultural labor force [44]. Typically, the elderly labor force has a lower capacity for using water-saving technologies and a lower willingness to invest in agricultural water conservancy facilities, leading to lower irrigation efficiency [40] and a low rate of straw return to the fields [45]. The increase in non-agricultural income may also lead to a lack of emphasis on the role of cultivated land by farmers, potentially resulting in land abandonment. Accordingly, hypotheses H1 and H2 are proposed in this study.

H1: *Part-time farming affects CLS through crowding-out effects.*

H2: *Part-time farming affects CLS through income effects.*

Part-time farming is bound to alter agricultural production and management activities, thereby causing changes in CLS. Before the agricultural labor force reaches the "Lewis turning point", part-time farming by farmers will not have a negative impact on agricultural production [46]. Specifically, the impact of part-time farming on CLS is mainly divided into the following two stages: First, when the degree of part-time farming is low, farmers have a strong dependence on and an emotional attachment to the land. Although part-time farming may lead to the transfer of some agricultural labor, this period is mainly characterized by seasonal farming or working outside during the agricultural off-season, which does not significantly negatively affect grain production. At the same time, part-time farming increases the level of family income, and farmers may offset the resultant labor shortage by improving the level of mechanization. The farmers who are more exposed to modern concepts are also more likely to adopt more cultivated land protection behaviors, enhancing the level of CLS. Second, when the degree of part-time farming is high, the main source of household income for farmers comes from non-agricultural industries and the economic and emotional dependence on cultivated land decreases. The labor force engaged in agricultural production gradually becomes both scarce and aging, compounded by the lower environmental awareness and weaker acceptance of new things among the elderly, which is very likely to lead to low irrigation efficiency, low straw return rate, less arable land protection behaviors, and a significant area of abandoned land. At the same time, to alleviate the negative impact of labor shortages during this period, farmers will increase the use of agricultural chemicals such as pesticides and fertilizers, thereby reducing the ecological level of CLS. Furthermore, despite a notable rise in farmers' household income, the diminished status of agriculture may result in a suboptimal investment of production factors by farmers. Consequently, the substitution effect of agricultural mechanization and intermediate organizations has not fully compensated for the loss of the agricultural labor force, thereby dampening the overall level of CLS. In summary, hypothesis H3 is proposed in this study.

H3: *The impact pattern of the degree of part-time farming on CLS exhibits an inverted U-shape.*

Additionally, due to the influence of factors such as geographical location, natural resources, national policies, and social culture, there are significant regional disparities in the levels of industrialization and urbanization in China, with the eastern regions showing higher levels compared to the central regions. In turn, the central regions surpass the western regions in the development of industrialization and urbanization. The regional disparities in these respective levels are bound to lead to regional differences in the degree of part-time farming, as well as regional disparities in the crowding-out and income effects, which in turn will cause regional differences in CLS. In summary, hypothesis H4 is proposed in this study.

H4: *There exists a significant regional heterogeneity in the impact of part-time farming on CLS.*

3. Methodology and Data

3.1. Statistical Modeling

After examining the scatter plot of the sample data and evaluating several models for data fitting, the researchers determined that the quadratic regression model exhibited the best goodness of fit.

In this study, the correlation between part-time farming and CLS was analyzed using an OLS model, defined in Equation (1):

$$Y_i = \alpha_0 + \alpha_1 x_i + \alpha_2 x_i^2 + \alpha_s Controls_i + \varepsilon_i \quad (1)$$

where, Y_i in (1) is CLS score of farmer i; x_i is the independent variable; the degree of part-time farming; x_i^2 stands for the square number of the independent variable; $Controls_i$ represents the set of control variables; α_0 and ε_i represent the intercept and error terms, respectively.

In addition, this study conducted regional heterogeneity and moderation effect analysis on the model results. We employed truncation handling, robust regression, and ridge regression analysis to perform robustness checks on the OLS model results.

The moderation effect model is defined by the following equation:

$$Y_i = \beta_0 + \beta_1 x_i + \beta_2 Mod_i + \beta_3 x_i * Mod_i + \beta_s Controls_i + \varepsilon_i \quad (2)$$

Mod_i in Equation (2) is the moderating variable, with the meanings of other variables being the same as in Equation (1).

3.2. Data Collection

The data we collected were obtained from a field survey conducted from January to March 2024. According to the Economic Divisions of China, a total of seven villages were sampled for investigation from three representative provinces of China, including Shandong Province in the eastern region, Henan Province in the central region, and Shaanxi

Province in the western region. The characteristics of these seven villages are quite similar to the economic and social development status of their respective regions, allowing them to accurately mirror the diverse economic and social conditions found across China. The survey mainly covered the characteristics of farmers' individual and family structures, inputs of production factors for farmland cultivation, household income and expenses, agricultural production subsidies, as well as land transfer and family farmland area. To ensure the randomness of the sampling process, team members first developed a sampling frame based on the household registration list, then used a simple random method to select samples, and replaced households that were not available or refused to participate with random samples. In the end, a total of 246 households in seven villages were visited in this survey, and after validity testing, 237 valid questionnaires were collected. The KMO value in the validity test was 0.743, and the result of Bartlett's sphericity test was significant ($p < 0.001$), indicating good data quality for this survey. Additionally, Cronbach's alpha value of the data was 0.697, indicating good reliability of the questionnaire.

3.3. Index System for Assessing CLS

Based on the connotations of sustainability and considering the evaluation indicators for sustainable land use, this study subdivides CLS into ecological sustainability, economic sustainability, and social sustainability. Additionally, referring to the global indicator framework for the Sustainable Development Goals issued by the United Nations [47], and drawing from the research of scholars such as Lu [4], Ge [48], Ustaoglu [49], Lv [50], etc., this study establishes three-level indicators in the index system for CLS (Table 1). The specific reasons for establishing these indicators are as follows:

(1) Agricultural subsidy amount. Agricultural subsidy can be interpreted as providing certain preferential treatment or compensation in order to promote farmers' behavior in improving the quality of cultivated land [51]. The impact of the agricultural subsidy amount on farmers' behavior in protecting cultivated land triggers a chain reaction, ultimately affecting CLS.

(2) The number of mu (1 mu = 0.1647369 acres) of cultivated land for which agricultural insurance has been purchased. Agriculture is counted as one of the most vulnerable areas to disastrous weather; therefore, purchasing agricultural insurance is an essential approach to diversifying the risks of agricultural production and operation [52]. Purchasing agricultural insurance, a crucial strategy for enhancing farmers' resilience against risks, serves as a significant indicator in assessing CLS.

(3) Whether the farmer has joined a farmer's cooperative or other mutual agricultural production organization. As farmers' engagement in mutual agricultural production organizations drives improvements in the scale management of cultivated land and the degree of agricultural mechanization, it subsequently boosts the progress of agricultural labor efficiency [53]. The fully developed mutual agricultural production organization can prevent farmers from abandoning cultivated land and enhance CLS.

(4) The average annual grain yield per mu and net income per mu of grain. The average grain yield per mu is a direct factor in measuring the productivity of cultivated land, and the net income of grain per mu is closely related to the average grain yield per mu. Both of these are important indicators of CLS.

(5) Cultivation patterns of cultivated land. Within cultivation patterns, there are three distinct categories: fully mechanized, partially mechanized, and entirely reliant on manpower. The mechanized farming pattern can enhance agricultural labor productivity [54], thereby facilitating the development of CLS.

(6) Whether to hand over the cultivated land to others for management. Handing over cultivated land to others for management is the main manifestation of cultivated land circulation. This behavior enables the transfer of cultivated land to more productive farmer through the market [55], which helps to promote the scale management of cultivated land and prevent its abandonment [56].

(7) Whether the cultivation process incorporates at least one of the following practices for the protection of cultivated land. Adopting acts of cultivated land protection during cultivation can directly improve cultivated land productivity [57] and enhance CLS.
(8) The number of mu of cultivated land treated with the return of straw to the land. Straw returning can enhance the soil's carbon sequestration capacity, reduce carbon loss in the soil, and mitigate environmental pollution [58]. As the amount of cultivated land treated with straw returning increases, so does the potential for improving CLS.
(9) The main irrigation method and average irrigation time per mu. The use of scientific irrigation methods can improve the irrigation efficiency of cultivated land [59]. Provided that the irrigation effect is not disturbed, an increase in irrigation efficiency, manifesting as a reduction in the irrigation time per mu, positively correlates with an elevation in the degree of CLS.
(10) The amount of money spent on pesticides and fertilizer per mu of cultivated land in the last year. The excessive use of pesticides and fertilizer can lead to soil compaction, water pollution, and a decline in cultivated land fertility [60]. As the amount of pesticides and fertilizers purchased per mu increases, meaning the usage of pesticides and fertilizers is enhanced, it can result in damage to the ecological level of CLS.

After standardizing the positive and negative indicators, we constructed an indicator system for CLS using the entropy method (Table 1). In the existing research, it should be noted that farmland abandonment is an important indicator for measuring CLS. However, no instances of farmland abandonment were observed in China during the survey period. Hence, this study does not include farmland abandonment in the indicator system for CLS.

Table 1. A comprehensive index system for assessing CLS.

Level 1 Indicators	Level 2 Indicators	Level 3 Indicators	Weight Coefficient	Direction of Impact
Cultivated Land use Sustainability	Social Sustainability	Annual agricultural subsidy amount	2.31%	+
		The number of mu of land for which agricultural insurance has been purchased	9.03%	+
		Whether the farmer has joined a farmer's cooperative or other mutual agricultural production organization	16.05%	+
	Economic Sustainability	Net income per mu of grain	1.41%	+
		Cultivation patterns of cultivated land	18.69%	-
		Whether to hand over the farmland to others for management	18.25%	+
		Whether the cultivation process incorporates at least one of the following practices for the protection of cultivated land	11.78%	+
		The average annual grain yield per mu	0.09%	+
	Ecological Sustainability	The number of mu of cultivated land treated with straw returning	2.46%	+
		The main irrigation method	18.47%	-
		Average irrigation time per mu	0.73%	-
		The amount of money spent on pesticides per mu of cultivated land last year	0.24%	-
		The amount of money spent on fertilizer per mu of cultivated land last year	0.49%	-

Note: 1 mu = 0.1647369 acres.

3.4. Variables

3.4.1. Dependent Variables

The main focus of this study was to examine the impact of the degree of part-time farming behavior on CLS. Therefore, CLS served as the dependent variable in this research. Drawing on existing research findings, this study divided CLS into the following three positive sub-indicators: social sustainability, economic sustainability, and ecological sustainability. The researchers employed the entropy method to calculate the weights of each indicator and the scores of each level of indicators, ultimately determining CLS. A higher score for CLS indicated a higher level of sustainability.

3.4.2. Independent Variables

The degree of part-time farming was the independent variable in this study. According to the criteria of China's Third Agricultural Census, full-time farming households referred to farming households whose household income was derived entirely from agricultural sources, and farming households engaged in part-time farming referred to farming households whose household income was partially derived from non-agricultural sources [48]. Following the definition of part-time farmers and with reference to the existing research results [18], this study used the proportion of non-farming income to the annual net income of households to express the degree of part-time farming. In other words, the higher the proportion of non-farming income in the annual net income of households, the higher the degree of part-time farming.

3.4.3. Control Variables

To mitigate the potential biases in the estimation results of regression models caused by omitted variables, this study, apart from the variable of farmers' part-time farming, also took into account other variables that may affect the sustainability of cultivated land use as much as possible. Drawing on previous studies, this study included the following five control variables: (1) The health status of the household head; (2) Whether the household head has received agricultural skills training; (3) Whether the household head has other commercial insurance; (4) The total years of education for other family members besides the head of the household; (5) The distance between the rural area where the farmers reside and the nearest county seat.

3.4.4. Moderating Variables

This study conducted moderation analysis on the relationship between the core variables. The moderating variables were the number of non-agricultural labor participants in the family and the gender of the household head. Typically, the number of non-agricultural labor participants in the family directly affected the non-agricultural income, thereby influencing the degree of part-time farming. However, in contemporary China, with the upgrading of the industrial structure in major urban areas, the opportunities available for Chinese migrant workers are decreasing [61]. Whether the relationship between agricultural labor migration and non-agricultural income (degree of part-time farming) continues to be significant requires examination with the latest evidence. Additionally, the gender of the household head has a significant influence on the family; therefore, this study sets this as a moderating variable. The specific variable settings and descriptive statistics are presented in Table 2.

Table 2. Descriptive statistics of the dependent, independent, and control variables.

Category	Name	Definition	Notation	Min	Max	Mean	Standard Deviation
Dependent Variables	CLS	Score	SF	0	0.87	0.17	0.19
	Social sustainability	Score	SF1	0	0	0.04	0.066
	Economic sustainability	Score	SF2	0	1	0.09	0.120
	Ecological sustainability	Score	SF3	0	0	0.04	0.063
Independent Variable	Degree of farmer's part-time farming	%	PE	0	1	0.63	0.29
Control Variables	The health status of the household head	Very unhealthy = 1, very healthy = 5, 5 dimensions in total	C1	1	5	3.63	1.18
	Agricultural skills training	True = 1, false = 0	C2	0	1	-	-
	Other commercial insurance	True = 1, false = 0	C3	0	1	-	-
	Total years of education for other family members	Years	C4	0	60	19.17	10.638
	The distance between the rural area and the nearest county seat	Kilometer	C5	9	25	13.16	4.89

Table 2. *Cont.*

Category	Name	Definition	Notation	Min	Max	Mean	Standard Deviation
Categorical Variable	Province	Eastern region = 1; central region = 2, western region = 3	Prov	1	3	-	-
Moderating Variables	Non-farm labor force participants	Number	Re1	0	4	0.99	0.776
	The gender of the head of household	Male = 1, female = 0	Re2	0	1	-	-

4. Results

4.1. Impact Pattern Analysis

The analysis of the sample data revealed that the impact of the degree of part-time farming on the level of CLS presented an inverted U-shaped pattern. Therefore, in this study, we performed a curve fitting on the core variables and established a quadratic regression model.

From Model 1, it can be inferred that the coefficient of determination for the quadratic regression model was 0.369, indicating that the variance in CLS data that can be explained by the degree of part-time farming was approximately 36.9%. Model 1 also passed the F-test (F = 18.393, $p < 0.05$). Specifically, the regression coefficient for the degree of part-time farming (PE) was 0.904 ($p < 0.001$), and the regression coefficient for the square of the degree of part-time farming (PE^2) was -0.928 ($p < 0.001$). This implies that the degree of part-time farming has a significant "inverted U-shaped" effect on CLS. That is, in the initial stages of the increase, part-time farming continuously improves the level of CLS. However, after reaching a turning point, higher levels of the degree of part-time farming lead to lower levels of CLS (Table 3). From the model fitting of various sub-indicators of CLS, it is evident that the degree of part-time farming among farmers significantly impacts social sustainability, economic sustainability, as well as ecological sustainability, with the distribution of data mirroring that of the overall indicators, i.e., an inverted U-shape trend. (Models 2, 3, and 4).

Table 3. Quadratic regression of CLS against part-time farming with various model settings. Model 1: the regression results of part-time farming on CLS. Model 2: the regression results of part-time farming on the social level of CLS. Model 3: the regression results of part-time farming on the economic level of CLS. Model 4: the regression results of part-time farming on the ecological level of CLS.

	Model 1 (SF)	Model 2 (SF1)	Model 3 (SF2)	Model 4 (SF3)
PE	0.904 ***	0.146 *	0.586 ***	0.150 *
	(0.179)	(0.066)	(0.115)	(0.063)
PE^2	−0.928 ***	−0.155 **	−0.583 ***	−0.167 **
	(0.164)	(0.061)	(0.105)	(0.057)
_cons	0.047	0.025	0.002	0.023
	(0.042)	(0.015)	(0.027)	(0.015)
R^2	0.369	0.183	0.35	0.225
F	18.393 ***	4.067 *	16.316 ***	6.236 **

Note: * $p < 0.05$, ** $p < 0.01$, *** $p < 0.001$, standard errors are in parentheses.

The researchers utilized the U-shaped test method proposed by Lind et al. [62] to analyze the data, thereby obtaining the inflection point value of the degree of part-time farming on the "inverted U-shaped" impact on CLS. From the analysis results, the inflection point value was determined to be 0.45, indicating that when the degree of part-time farming exceeded 45%, the level of CLS began to decline. This validates hypothesis H3 (Table 4).

Table 4. Impact pattern of part-time farming on CLS (Inverted U-shaped).

	SF	SF1	SF2	SF3
Inflection point	0.45	0.47	0.46	0.45
Whether the inflection point is within the sample range	Yes	Yes	Yes	Yes
U-shaped test results	Inverted U-shaped	Inverted U-shaped	Inverted U-shaped	Inverted U-shaped

4.2. OLS Regression

This study separately established OLS regression models for CLS and three secondary indicators (SF1, SF2, SF3). Both independent and control variables were included in the models using the "enter" method, which helped to highlight the relationships between core variables more prominently in the data results. The results of Model 5, Model 6, Model 7, and Model 8 showed that after adding control variables, the variables in the degree of part-time farming (PE) and the square of the numerical value of the degree of part-time farming (PE^2) remained significant in each model. The explanatory efficacy of the models also demonstrated robust performance (Table 5).

Table 5. OLS regression of CLS against part-time farming with various model settings. Model 5: the regression results of part-time farming on CLS. Model 6: the regression results of part-time farming on the social level of CLS. Model 7: the regression results of part-time farming on the economic level of CLS. Model 8: the regression results of part-time farming on the ecological level of CLS.

	Model 5	Model 6	Model 7	Model 8
PE	1.021 *** (0.168)	0.173 ** (0.061)	0.649 *** (0.112)	0.172 ** (0.062)
PE^2	−1.077 *** (0.155)	−0.203 *** (0.056)	−0.659 *** (0.103)	−0.187 *** (0.057)
C1	0.024 ** (0.009)	0.006 (0.003)	0.014 * (0.006)	0.003 (0.004)
C2	0.073 ** (0.026)	0.016 (0.009)	0.035 * (0.017)	0.021 * (0.010)
C3	0.049 (0.032)	0.027 * (0.011)	0.033 (0.021)	−0.012 (0.012)
C4	0.001 (0.001)	0.002 *** (0.000)	0.001 (0.001)	−0.001 (0.00)
C5	0.009 *** (0.002)	0.003 *** (0.001)	0.004 ** (0.001)	0.002 (0.001)
_cons	−0.212 *** (0.059)	−0.075 *** (0.021)	−0.127 *** (0.039)	−0.003 (0.022)
R^2	0.269	0.238	0.204	0.097
F	12.031 ***	10.224 ***	8.370 ***	3.528 ***

Note: * $p < 0.05$, ** $p < 0.01$, *** $p < 0.001$, standard errors are in parentheses.

Model 5 revealed the effects of the control variables. Notably, it is clear that the better the health status of the householder head (C1), the higher the level of CLS. Additionally, households benefit from heads with agricultural skills training (C2), showing that expertise in farming positively impacts sustainability. Moreover, having other commercial insurance (C3) is shown to increase CLS, indicating the effectiveness of risk management strategies in supporting agriculture. Furthermore, the total years of education of non-head family members (C4) is a significant factor, with higher education correlating with better CLS. Lastly, the distance from the county seat (C5) also plays a role, with greater distances associated with higher levels of CLS. These results corroborate previous studies [63,64].

The reasons behind the results can be outlined as follows: a healthier householder typically demonstrates better management capabilities over cultivated land use. The heads of households who undergo agricultural skills training exhibit heightened levels of agricultural knowledge and demonstrate a greater inclination toward agricultural work. The households with commercial insurance generally enjoy higher incomes compared to those without, and higher education levels among other family members besides the head of the household often correlate with higher household incomes. Consequently, such households possess adequate financial resources to protect their cultivated land; for instance, through the adoption of practices like the application of organic fertilizers, they can sustain soil fertility [65]. The farther away the household is from the county seat, the higher the overall cost of working in the city and the greater the proportion of agricultural income in their household income. In this case, farmers' willingness to maintain the fertility of cultivated land and increase their income from growing grain will also be stronger. Therefore, the assumption is that H2 (income effect) is true. In turn, Model 6, Model 7, and Model 8 analyzed social sustainability, economic sustainability, and ecological sustainability, and the data results were essentially consistent with those of Model 5. However, the influence of the three control variables on the secondary indicators (SF1, SF2, SF3) was weakened by the physical health status of the head of household (C1), whether the head of household had other commercial insurance (C3), and the education level of other family members (C4) (Table 5).

4.3. Heterogeneity Analysis

Considering the long-standing regional disparities in China, particularly the existence of the developed manufacturing industry in the eastern region [66], while the central and western regions account for a larger proportion of China's total grain output and bear greater ecological protection responsibilities [67], it is necessary to conduct a test for regional heterogeneity in the relationships between the core variables. Based on common regional division methods, this study divided China into eastern, central, and western regions and sought to separately provide model estimates for each region (see Table 6 for details). The results vividly showed significant differences in the levels of CLS among the eastern region (Model 9), central region (Model 10), and western region (Model 11). From the eastern to central region and then to the western region, the level of CLS continuously and significantly decreased, with values of 1.465, 1.432, and 0.36. In all three regions, the square of the degree of part-time farming (PE^2) significantly influenced the level of CLS, with negative regression coefficients. This implies that beyond the inflection point value, the level of CLS in all three regions decreases as the degree of part-time farming continues to rise. However, before reaching the inflection point value, the degree of part-time farming had a significant positive impact on the level of CLS in the eastern and central regions (1.465 ***, 1.432 ***). Conversely, the western region did not exhibit the same relationship between variables. Based on this, hypothesis H4 is confirmed.

Table 6. Regression results of part-time farming on CLS in China's different regions. Model 9: Eastern China. Model 10: Central China. Model 11: Western China.

	Model 9 (Prov1)	Model 10 (Prov2)	Model 11 (Prov3)
PE	1.465 ***	1.432 ***	0.360
	(0.323)	(0.285)	(0.243)
PE^2	−1.505 ***	−1.470 ***	−0.434 +
	(0.295)	(0.257)	(0.232)
C1	0.017	0.051 ***	−0.007
	(0.026)	(0.013)	(0.015)
C2	0.106 *	0.073	0.036
	(0.044)	(0.047)	(0.041)
C3	0.154	0.044	0.040
	(0.079)	(0.057)	(0.040)
C4	0.000	0.003	0.006 **
	(0.002)	(0.002)	(0.002)
C5	0.006 *	0.019	0.001
	(0.003)	(0.011)	(0.005)
_cons	−0.145	−0.515 ***	−0.021
	(0.118)	(0.146)	(0.106)
R^2	0.405	0.439	0.184
F	7.015 ***	8.149 ***	2.195 *

Note: + $p < 0.1$, * $p < 0.05$, ** $p < 0.01$, *** $p < 0.001$, standard errors are in parentheses.

4.4. Moderation Effect Analysis

In this study, the moderating variable of household non-farming labor participation (Rel) was added to the model for moderating effect analysis (Table 7). Model 12 showed that the interaction term (PE* Rel) between the degree of part-time farming and the number of household non-farming labor participants was significantly at the $p < 0.05$ level. The results showed that the household labor structure of farming households had a significant moderating effect on the relationship between the degree of part-time farming and the level of CLS. The data analysis results indicate a significant negative effect of part-time farming on the overall level of CLS. Moreover, the involvement of more family members in non-agricultural labor intensifies this negative effect. This implies that as the number of family members engaged in non-agricultural labor increases, this negative effect becomes more pronounced (Figure 2a). Based on this, we assume that hypothesis H1 is valid.

Table 7. Moderation effect analysis results. Model 12: the moderated influence of part-time farming on CLS with the inclusion of the number of family members engaged in non-agricultural labor. Model 13: the moderated influence of part-time farming on CLS with the household head's gender as a moderating factor.

	Model 12	Model 13
C1	0.014 (0.010)	0.017 (0.010)
C2	0.070 * (0.028)	0.076 ** (0.028)
C3	0.042 (0.035)	0.040 (0.034)
C4	0.001 (0.001)	0.002 (0.001)
C5	0.007 ** (0.002)	0.008 ** (0.002)
PE	−0.193 ** (0.060)	−0.136 ** (0.043)
Rel	0.02 (0.023)	
PE* Rel	−0.145 * (0.062)	
Re2		−0.026 (0.024)
PE* Re2		0.223 ** (0.083)
_cons	0.003 (0.065)	−0.049 (0.052)
R^2	0.136	0.145
F	4.472 ***	4.850 ***

Note: * $p < 0.05$, ** $p < 0.01$, *** $p < 0.001$, standard errors are in parentheses.

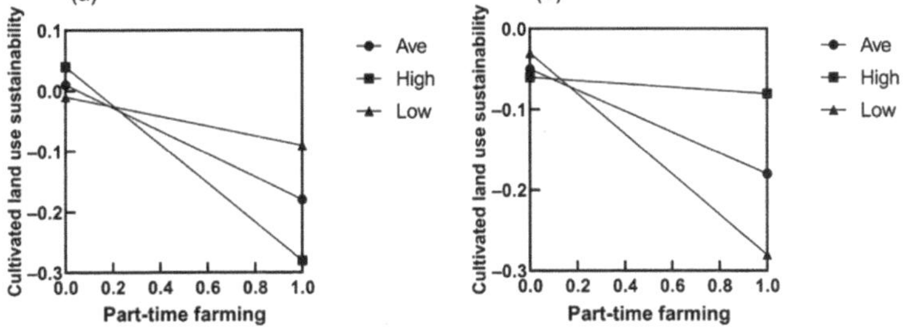

Figure 2. Moderating effect of household non-farming labor force participation and the gender of the household head. (**a**) The moderating effect of household non-farming labor force participation; (**b**) the moderating effect of the household head's gender.

As a large number of male laborers have migrated to urban areas, women have emerged as the primary workforce in Chinese agriculture. Compared to men, women exhibit significant differences in terms of ideology, physiological conditions, and off-farm employment opportunities. Therefore, we believe it is necessary to consider the influence of the household head's gender on the correlations between core variables. Through the introduction of the moderating variable of the household head's gender into the analysis, Model 13 demonstrates that the interaction term between the degree of part-time farming and the household head's gender (PE* Re2) was significant at the $p < 0.01$ level. This result indicates that the gender of the household head significantly moderates the relationship between part-time farming and the level of CLS. Overall, the degree of part-time farming among farmers has a significant negative effect on the level of CLS, and this effect is more pronounced in households where the head of the household is female. In other words, among two households with the same degree of non-agricultural employment, the level of CLS is lower in the household headed by a female compared to the one headed by a male, as illustrated in Figure 2b. This can be explained by the fact that female household heads engage less in land protection activities (Re2: Female = 0). This study suggests that households headed by females often indicate the absence or weakening of the husband's role, and such households usually face more difficult livelihoods. Therefore, these farmers lack both the willingness and the capability to undertake land protection activities. Additionally, the physiological characteristic of women being physically weaker than men may also contribute to the aforementioned result, as households headed by females generally have fewer adult male laborers.

4.5. Robustness Tests

In order to validate the robustness of the research findings, three methods of robustness tests were employed in this study (Table 8). The first method involved truncating the data (Model 14). Even after truncation (at both sides 1%), the results remained significant. Namely, the higher the degree of farmers' part-time farming activities, the lower the level of CLS. Regression coefficients, coefficient directions (positive/negative), significance levels, and collinearity indicators did not exhibit significant changes compared to the baseline model. The second method employed robust regression (Model 15), yielding consistent results. Considering the collinearity effects of the quadratic terms, the third method utilized ridge regression on core variables (Model 16). The model yielded a determination coefficient of 0.24, supported by an F-test ($F = 10.358$, $p < 0.001$). These three model results remained consistent with the OLS model, thus verifying the robustness of the results.

Table 8. Robustness test results. Model 14: results of the censoring process for the data. Model 15: results of the robust regression tests. Model 16: the findings from the ridge regression analysis of the key variables.

	Model 14 Truncating	Model 15 Robust Regression	Model 16 Ridge Regression
PE	1.021 ** (0.168)	−0.922 ** (0.146)	−0.618 ** (0.088)
PE^2	−1.077 ** (0.155)	0.893 ** (0.159)	0.524 ** (0.095)
C1	0.024 * (0.009)	0.011 (0.009)	0.021 * (0.010)
C2	0.073 ** (0.026)	0.072 ** (0.024)	0.069 ** (0.026)
C3	0.049 (0.032)	0.042 (0.030)	0.044 (0.032)
C4	0.001 (0.001)	0.002 (0.001)	0.002 (0.001)
C5	0.009 ** (0.002)	0.008 ** (0.002)	0.008 ** (0.002)
_cons	−0.212 ** (0.059)	−0.171 ** (0.056)	−0.097 (0.050)
R^2	0.269	0.250	0.24
F	12.031 ***	10.922 ***	10.358 ***

Note: * $p < 0.05$, ** $p < 0.01$, *** $p < 0.001$, standard errors are in parentheses.

5. Discussion

5.1. Directing Farmers to Maintain a Rational Proportion of Part-Time Income

Sustainable development is an important goal for the advancement of society and is the central theme of our era. In the process of achieving sustainable development, sustainable thinking is essential for scholars, government officials, social workers, and the general public. In this study, grounded in sustainability thinking, we aimed to understand and elucidate the impact of part-time farming on CLS. Our goal was to promote the sustainable use of cultivated land through the obtained insights. To achieve this, we constructed an index system for CLS and analyzed the effects of part-time farming on this sustainability, thereby bridging a gap in the existing research. The findings revealed that the impact of part-time farming on CLS follows an inverted U-shaped pattern. This pattern is also observed in the economic, social, and ecological dimensions of CLS (Table 3). Some studies have suggested that part-time farming leads to a shortage of agricultural labor, resulting in decreased agricultural production efficiency and the increased use of pesticides and fertilizers [68], which in turn undermines CLS. While this reasoning is partially valid, it overlooks the fact that many rural migrants move back and forth between urban and rural areas when the degree of part-time farming among farming households is low. Specifically, these migrants work in urban areas during the off-season and return to their hometowns during peak farming periods, thus mitigating the negative impact of the "crowding out effect" on CLS. During this period, the predominant influence is the "income effect", showing a positive correlation between the degree of part-time farming among farmers and CLS. As the degree of part-time farming surpasses 45%, the "crowding out effect" gradually intensifies, diminishing the impact of the "income effect" and consequently leading to a decline in CLS. Hence, enhancing subsidies for farming households engaged in grain cultivation and judiciously adjusting grain purchase prices should be the focal points of China's agricultural policy in order to effectively manage the proportion of part-time income within the total household income.

5.2. Facilitating Structured Reallocation of Agricultural Labor Force and Bolstering Engagement of Female Household Heads in Cultivated Land Conservation Efforts

Since the 1980s, China's industrialization, marketization, and urbanization have led to a significant migration of agricultural labor to cities, resulting in a demographic shift marked by an increase in the proportion of women and older workers in the agricultural sector [69]. This change indirectly affects CLS. Generally, households with a higher proportion of members engaged in non-agricultural work tend to demonstrate lower levels of attention to CLS among those with the same degree of part-time farming (Table 7). This is because more non-agricultural workers in a family increase the "crowding out effect" of part-time work on the labor force, hindering the sustainable use of cultivated land. Additionally, our findings suggest that households with female heads of household tend to demonstrate lower levels of CLS activity compared to those with male heads of household, within the same level of part-time employment (Table 7). The previous research further reinforces the arguments advanced in this paper and has revealed that women generally have lower psychological expectations regarding the value of cultivated land use compared to men, often due to disparities in education, policy advocacy, and economic and social development [70]. Additionally, women shoulder significant childcare responsibilities [71], leading them to prefer managing smaller, less resource-intensive cultivated land [72], and they exhibit less proactive behavior in conserving cultivated land. Consequently, households led by women tend to exhibit lower levels of CLS activity. Therefore, the Chinese government must strategically guide the transition of agricultural labor, encourage men to reengage in agricultural activities in their hometowns, optimize the demographic composition of rural communities, and mitigate the adverse impact of agricultural labor shortages on the sustainable use of cultivated land. In addition, the government can enhance women's awareness of cultivated land protection and their psychological expectations of cultivated land value through ideological education, policy publicity, economic incentives, etc., so as to effectively mobilize women's enthusiasm for cultivated land protection. For example, the Law of the People's Republic of China on the Protection of Women's Rights and Interests, revised in 2022, clearly emphasizes the protection of rural women's land and their related rights and interests, which can help raise women's psychological expectations of the value of cultivated land to encourage their effective protection of it [73].

5.3. Other Strategies to Enhance CLS

Influenced by geographical positioning, resource allocation, and government policies, China displays significant disparities in economic development among its regions, with the eastern region leading, followed by the central region, and then the western region [74]. Aligned with the level of economic development, employment opportunities and wage levels in the eastern region significantly outstrip those in the central and western regions. Consequently, in the early 21st century, a substantial rural population, particularly from the central and western regions, migrated to the urban areas in the eastern region. This phenomenon resulted in a pronounced shortage in the western region's young and middle-aged rural labor force, alongside a substantial conversion of high-quality cultivated land in the eastern region for non-agricultural purposes. In comparison with the eastern and central regions, the rural population's migration distances are greater in the western region. Consequently, commuting costs between urban and rural areas (i.e., between workplace and hometown) are higher, exacerbating the pronounced "crowding out effect" of part-time farming on the household labor force, with limited alleviation prospects. Therefore, at a lower degree of part-time farming, it has no significant positive impact on CLS in the western region. Furthermore, as the degree of part-time farming increases, the decline in CLS in the western region occurs at a notably slower rate (Table 6). Previous studies have also corroborated this finding, suggesting that women living in developed areas hold higher expectations regarding cultivated land [70]. Conversely, in the relatively underdeveloped western regions, female-headed households have lower expectations for cultivated land, resulting in fewer protective farming behaviors among farmer households in these areas. Hence, government policies should be tailored to the unique characteristics of different regions, including location, resource endowment, transportation conditions, and policy intensity, to implement region-specific strategies for enhancing CLS.

This study found a direct correlation between higher levels of education, improved physical health, and comprehensive agricultural skills training among farmers and the heightened CLS within their households (Table 5). Previous research has also demonstrated that higher levels of education are correlated with an increased likelihood of farmers adopting new technologies or methods to enhance CLS [75]. Moreover, initiatives such as the establishment of field schools [76] and providing agricultural skills training for farmers have been shown to encourage more proactive cultivated land conservation behaviors. In recent years, with improvements in the education and healthcare systems, the educational attainment of the Chinese population has been steadily increasing. As a result, a growing number of agricultural workers have received agricultural skill training, contributing to enhanced cultivated land-management practices. Additionally, there has been an overall improvement in the physical health of the population. Therefore, highly skilled professional farmers will become the primary participants in China's future agricultural sector. The advances in science and technology, coupled with enhancements in human capital, have mitigated the adverse effects of an aging agricultural labor force and rural depopulation on cultivated land conservation [77]. So far, an increasing number of regions in China have embraced and implemented smart agricultural practices, not only enhancing the mechanization level of agricultural production but also integrating advanced technologies such as IoT detection, artificial intelligence analysis, and drones. As a result, the pessimistic perspectives regarding the state of cultivated land protection in China are becoming increasingly untenable.

Drawing from empirical surveys, this study thoroughly explores the relationship between the degree of part-time farming and CLS in China, revealing statistically significant findings. These results contribute to advancing sustainable development practices at the land level. However, it should be noted that this study primarily relies on the proportion of non-farming income in household net income to assess farmers' part-time farming activities, lacking a deeper analysis regarding factors like location, timing, and the number of farmers engaged in part-time farming. Additionally, this study only examines the sustainability of cultivated land used for food crops, excluding the consideration of cultivated land for cash crops. Furthermore, this study does not analyze how farmers' part-time farming affects the sustainability of different types of cultivated lands. Despite concerns from some scholars about widespread cultivated land abandonment in China [78], none of the farmers surveyed reported such instances. This absence may be attributed to the long-standing protection of cultivated land by the Chinese government [79], leading to cultivated land abandonment being excluded from the study's sustainability index system. These are crucial areas for future research, as a thorough understanding of these issues can better promote sustainable development at the land level.

6. Concluding Remarks

Land relates to the sustainability of our coupled human–Earth system through various entry points. Our study takes a social–environmental system perspective on cultivated land resources and thus operationalizes CLS in a broader sense than the environmentally friendly use of cultivated land. Our research contributes by proposing a comprehensive indicator system for assessing CLS and shedding new light on the understudied connection between CLS and part-time farming, which is a long-term nationwide trend in rural China. We quantitatively assessed the impact pattern and moderating effects of part-time farming on CLS, using first-hand survey data obtained from village representatives in Eastern, Central, and Western China. Based on the rich empirical findings, we further proposed actionable suggestions for policymakers to advance CLS.

To recap, this study provides three key empirical insights. First, the impact pattern of the degree of part-time farming on CLS exhibits an inverted U-shape, with a clear turning point at 45%. When the degree of part-time farming is below 45%, CLS increases with the degree of part-time farming. However, once the degree of part-time farming exceeds 45%, CLS begins to decline with any further increases. Second, there exists significant regional heterogeneity in the impact of part-time farming on CLS. Prior to reaching the inflection point, the degree of part-time farming has a notably positive impact on CLS in the eastern and central regions, whereas no significant positive correlation is observed in the western region. Following the inflection point, CLS in all three regions experiences a significant decline with increasing degrees of farmers' part-time farming at varying magnitudes. Third, the involvement of households in non-farming activities and the gender of the household head significantly moderate the correlation between farmers' part-time farming and CLS. Additionally, factors such as the physical health of the household head, the cumulative years of education among family members excluding the head, access to commercial insurance, and participation in agricultural skills training play crucial roles in influencing CLS. Moreover, households with healthier heads and higher levels of education among family members (excluding the head) tend to exhibit greater awareness and protection of CLS. Similarly, farmers with access to commercial insurance and those who have undergone agricultural skills training demonstrate higher levels of understanding of and respect for CLS. We anticipate that these findings will support governments facing the long-term trend of part-time farming in formulating informed policies to advance rural and regional sustainability with CLS as an entry point.

Author Contributions: Conceptualization, X.P.; methodology, X.P. and C.W.; software, C.W.; validation, C.W.; formal analysis, C.W.; investigation, X.P. and C.W.; resources, X.P. and X.Z.; data curation, C.W.; writing—original draft preparation, X.P. and C.W.; writing—review and editing, X.P., C.W. and X.Z.; visualization, X.P. and C.W.; supervision, X.Z.; project administration, X.P. All authors have read and agreed to the published version of the manuscript.

Funding: This research was funded by the "Rural Revitalization" Research Special Fund of Ocean University of China (Project No: ZX2024003).

Data Availability Statement: The data presented in this study are available upon reasonable request from the corresponding author.

Acknowledgments: We appreciate the constructive suggestions and comments from the editor and anonymous reviewers.

Conflicts of Interest: The authors declare no conflicts of interest.

References

1. Foley, J.A.; DeFries, R.; Asner, G.P.; Barford, C.; Bonan, G.; Carpenter, S.R.; Chapin, F.S.; Coe, M.T.; Daily, G.C.; Gibbs, H.K.; et al. Global consequences of land use. *Science* **2005**, *309*, 570–574. [CrossRef]
2. Trewavas, A. Malthus foiled again and again. *Nature* **2002**, *418*, 668–670. [CrossRef]
3. Yang, Z.; Liu, Y.; Tao, W.; Xu, J.; Zhao, Q. Method for evaluating the degrees of land use sustainability of mountainous county and its application in Yunnan province, China. *J. Mt. Sci.* **2008**, *5*, 98–112. [CrossRef]
4. Lu, X.; Zhang, Y.; Lin, C.; Wu, F. Analysis and comprehensive evaluation of sustainable land use in China: Based on sustainable development goals framework. *J. Clean. Prod.* **2021**, *310*, 127205. [CrossRef]
5. Gadanakis, Y.; Bennett, R.; Park, J.; Areal, F.J. Evaluating the sustainable intensification of arable farms. *J. Environ. Manag.* **2015**, *150*, 288–298. [CrossRef]
6. Zhou, D.; Zhang, R.; Liu, L.; Gao, L. Evaluation of the sustainable land use status of the north China plain. *Int. J. Sustain. Dev. World Ecol.* **2009**, *16*, 253–259. [CrossRef]
7. Wang, X.; Zhou, D.; Jiang, G.; Peng, C. How can the sustainable goal of cultivated land use in the Qinghai-Tibet plateau be realized?—Based on a research framework of cultivated land use patterns. *Front. Environ. Sci.* **2023**, *11*, 1134136. [CrossRef]
8. DeFries, R.; Rosenzweig, C. Toward a whole-landscape approach for sustainable land use in the tropics. *Proc. Natl. Acad. Sci. USA* **2010**, *107*, 19627–19632. [CrossRef] [PubMed]
9. Niu, W.; Shi, J.; Xu, Z.; Wang, T.; Zhang, H.; Su, X. Evaluating the sustainable land use in ecologically fragile regions: A case study of The Yellow River Basin in China. *Int. J. Environ. Res. Public Health* **2022**, *19*, 3222. [CrossRef]
10. Yu, D.; Wang, D.; Li, W.; Liu, S.; Zhu, Y.; Wu, W.; Zhou, Y. Decreased landscape ecological security of peri-urban cultivated land following rapid urbanization: An impediment to sustainable agriculture. *Sustainability* **2018**, *10*, 394. [CrossRef]
11. Zhong, T.; Zhang, X.; Huang, X. Impact of labor transfer on agricultural land use conversion at rural household level based on logit model. *Chin. Geogr. Sci.* **2008**, *18*, 300–307. [CrossRef]
12. Xu, Y.; Huang, X.; Bao, H.X.H.; Ju, X.; Zhong, T.; Chen, Z.; Zhou, Y. Rural land rights reform and agro-environmental sustainability: Empirical evidence from China. *Land Use Pol.* **2018**, *74*, 73–87. [CrossRef]
13. Niu, S.; Lyu, X.; Gu, G.; Zhou, X.; Peng, W. Sustainable intensification of cultivated land use and its influencing factors at the farming household scale: A case study of Shandong province, China. *Chin. Geogr. Sci.* **2021**, *31*, 109–125. [CrossRef]
14. Zeng, Z. The empirical study of sustainable utilization of cultivated land based on grain production technical efficiency perspective. *Jiangsu Agric. Sci.* **2018**, *46*, 454–458. (In Chinese)
15. Wang, C.; Rada, N.; Qin, L.; Pan, S. Impacts of migration on household production choices: Evidence from China. *J. Dev. Stud.* **2014**, *50*, 413–425. [CrossRef]
16. Bouchakour, R.; Saad, M. Farm and farmer characteristics and off-farm work: Evidence from Algeria. *Aust. J. Agric. Resour. Econ.* **2020**, *64*, 455–476. [CrossRef]
17. Xie, H.; Jin, S. Evolutionary game analysis of fallow farmland behaviors of different types of farmers and local governments. *Land Use Pol.* **2019**, *88*, 104122. [CrossRef]
18. Lang, Y.; Wang, G.; Attipoe, S.G.; Han, D. Does off-farm employment contribute to chemical fertilizer reduction? New evidence from the main rice-producing area in Jilin province, China. *PLoS ONE* **2022**, *17*, e0279194. [CrossRef] [PubMed]
19. Jiang, X.; Zhong, S.; Huang, C.; Guo, X.; Zhao, J. Blessing or curse? The impacts of non-agricultural part-time work of the large farmer households on agricultural labor productivity. *Technol. Econ. Dev. Econ.* **2022**, *28*, 26–48. [CrossRef]
20. Yin, N.; Huang, Q.; Yang, Z.; Wang, Y. Impacts of off-farm employment on irrigation water efficiency in north China. *Water* **2016**, *8*, 452. [CrossRef]
21. Kousar, R.; Abdulai, A. Off-farm work, land tenancy contracts and investment in soil conservation measures in rural Pakistan. *Aust. J. Agr. Resour. Econ.* **2016**, *60*, 307–325. [CrossRef]
22. Zeng, J.; Lyu, J. Simultaneous decisions to undertake off-farm work and straw return: The role of cognitive ability. *Land* **2023**, *12*, 1599. [CrossRef]
23. Gao, J.; Song, G.; Sun, X. Does labor migration affect rural land transfer? Evidence from China. *Land Use Pol.* **2020**, *99*, 105096. [CrossRef]
24. He, Y.; Xie, H.; Peng, C. Analyzing the behavioural mechanism of farmland abandonment in the hilly mountainous areas in China from the perspective of farming household diversity. *Land Use Pol.* **2020**, *99*, 104826. [CrossRef]

25. Ren, K.; Xu, M.; Li, R.; Zheng, L.; Liu, S.; Reis, S.; Wang, H.; Lu, C.; Zhang, W.; Gao, H.; et al. Optimizing nitrogen fertilizer use for more grain and less pollution. *J. Clean. Prod.* **2022**, *360*, 132180. [CrossRef]
26. Wang, C.; Wang, Y.; Tian, Y.; Chen, S. Spatial patterns and determinants of village abandonment in the mountainous areas of china. *Soc. Indic. Res.* **2021**, *157*, 1111–1130. [CrossRef]
27. FAO; IFAO; UNICEF; WFP; WHO. *The State of Food Security and Nutrition in the World 2023*; FAO: Rome, Italy, 2023; p. 6.
28. Purvis, B.; Mao, Y.; Robinson, D. Three pillars of sustainability: In search of conceptual origins. *Sustain. Sci.* **2019**, *14*, 681–695. [CrossRef]
29. Lv, X.; Niu, S.; Gu, G.; Peng, W. Conceptual cognition and research framework on sustainable intensification of cultivated land use in China from the perspective of the "New Agriculture, Countryside and Peasants". *J. Nat. Resour.* **2020**, *35*, 2029–2043. (In Chinese)
30. Liu, H.; Luo, X. Understanding farmers' perceptions and behaviors towards farmland quality change in northeast China: A structural equation modeling approach. *Sustainability* **2018**, *10*, 3345. [CrossRef]
31. Deng, C.; Xie, B.; Pan, Y.; Li, Y. Influencing factor of cultivated land disposal of migration labor force in typical agricultural country: The case of Xinhua County in Hunan Province. *Econ. Geogr.* **2015**, *35*, 134–140. (In Chinese)
32. Ma, W.; Abdulai, A.; Ma, C. The effects of off-farm work on fertilizer and pesticide expenditures in China. *Rev. Dev. Econ.* **2018**, *22*, 573–591. [CrossRef]
33. Chang, H.-H.; Wen, F.-I. Off-farm work, technical efficiency, and rice production risk in Taiwan. *Agric. Econ.* **2011**, *42*, 269–278. [CrossRef]
34. Niu, H.; Xiao, D.; Zhao, S. Drivers of farmers' behavior toward compensation scheme for cultivated land protection in Chengdu pilot area, China. *Ecosyst. Health Sustain.* **2021**, *7*, 1978330. [CrossRef]
35. Qing, C.; Zhou, W.; Song, J.; Deng, X.; Xu, D. Impact of outsourced machinery services on farmers' green production behavior: Evidence from chinese rice farmers. *J. Environ. Manag.* **2023**, *327*, 116843. [CrossRef]
36. Zhou, S.; Herzfeld, T.; Glauben, T.; Zhang, Y.; Hu, B. Factors affecting chinese farmers' decisions to adopt a water-saving technology. *Can. J. Agric. Econ.-Rev. Can. Agroecon.* **2008**, *56*, 51–61. [CrossRef]
37. Zhang, Y.; Shao, T.; Dong, Q. Reassessing the Lewis turning point in China: Evidence from 70,000 rural households. *China World Econ.* **2018**, *26*, 4–17. [CrossRef]
38. Duc, L.N.; Grote, U.; Trung, T.N. Migration, crop production and non-farm labor diversification in rural Vietnam. *Econ. Anal. Policy* **2019**, *63*, 175–187.
39. Gartaula, H.; Niehof, A.; Visser, L. Shifting perceptions of food security and land in the context of labour out-migration in rural Nepal. *Food Secur.* **2012**, *4*, 181–194. [CrossRef]
40. Yang, J.; Wang, H.; Jin, S.; Chen, K.; Riedinger, J.; Peng, C. Migration, local off-farm employment, and agricultural production efficiency: Evidence from China. *J. Prod. Anal.* **2016**, *45*, 247–259. [CrossRef]
41. Guo, L.; Song, Y.; Zhao, S.; Tang, M.; Guo, Y.; Su, M.; Li, H. Dynamic linkage between aging, mechanizations and carbon emissions from agricultural production. *Int. J. Environ. Res. Public Health* **2022**, *19*, 6191. [CrossRef]
42. Viira, A.-H.; Poeder, A.; Vaernik, R. The determinants of farm growth, decline and exit in Estonia. *Ger. J. Agric. Econ.* **2013**, *62*, 52–64. [CrossRef]
43. Kawagoe, T.; Hayami, Y.; Ruttan, W.R. The intercountry agricultural production function and productivity differences among countries: Reply. *J. Dev. Econ.* **1988**, *28*, 125–126. [CrossRef]
44. Larcher, M.; Engelhart, R.; Vogel, S. Agricultural professionalization of Austrian family farm households: The effects of vocational attitude, social capital and perception of farm situation. *Ger. J. Agric. Econ.* **2019**, *68*, 28–44. [CrossRef]
45. Zhou, W.; Yang, Y.; He, J.; Xu, D. Does labor aging inhibit farmers' straw-returning behavior? Evidence from rural rice farmers in southwest China. *Land* **2023**, *12*, 1816. [CrossRef]
46. Yang, S.; Fan, G. An effect analysis of the impact of part-time farming on rural eco-environment. *J. S. China Univ.* **2018**, *17*, 52–63. (In Chinese)
47. Inter-Agency and Expert Group on SDG Indicators. SDG Indicators. 2017. Available online: https://unstats.un.org/sdgs/indicators/indicators-list/ (accessed on 27 May 2024).
48. Ge, D.; Kang, X.; Liang, X.; Xie, F. The impact of rural households' part-time farming on grain output: Promotion or inhibition? *Agriculture* **2023**, *13*, 671. [CrossRef]
49. Ustaoglu, E.; Collier, M.J. Farmland abandonment in Europe: An overview of drivers, consequences, and assessment of the sustainability implications. *Environ. Rev.* **2018**, *26*, 396–416. [CrossRef]
50. Lv, Y.; Zhang, C.; Ma, J.; Yun, W.; Gao, L.; Li, P. Sustainability assessment of smallholder farmland systems: Healthy farmland system assessment framework. *Sustainability* **2019**, *11*, 4525. [CrossRef]
51. Xie, H.; Wang, W.; Zhang, X. Evolutionary game and simulation of management strategies of fallow cultivated land: A case study in Hunan Province, China. *Land Use Pol.* **2018**, *71*, 86–97. [CrossRef]
52. Zhang, Y.; Ju, G.; Zhan, J. Farmers using insurance and cooperatives to manage agricultural risks: A case study of the swine industry in China. *J. Integr. Agric.* **2019**, *18*, 2910–2918. [CrossRef]
53. Zhong, Z.; Jiang, W.; Li, Y. Bridging the gap between smallholders and modern agriculture: Full insight into China's agricultural cooperatives. *J. Rural Stud.* **2023**, *101*, 103037. [CrossRef]

54. Yi, Q.; Chen, M.; Sheng, Y.; Huang, J. Mechanization services, farm productivity and institutional innovation in China. *China Agric. Econ. Rev.* **2019**, *11*, 536–554.
55. Qiu, T.; Luo, B.; Li, S.; He, Q. Does the basic farmland preservation hinder land transfers in rural China? *China Agric. Econ. Rev.* **2020**, *12*, 39–56. [CrossRef]
56. Qiu, T.; Zhang, D.; Zhu, M. Do land transfers reduce the effects of labour costs on agricultural land abandonment? *Appl. Econ.* **2024**, 1–4. [CrossRef]
57. Lu, H.; Chen, Y.; Huan, H.; Duan, N. Analyzing cultivated land protection behavior from the perspective of land fragmentation and farmland transfer: Evidence from farmers in rural China. *Front. Environ. Sci.* **2022**, *10*, 901097. [CrossRef]
58. Yang, C.; Wang, X.; Li, J.; Zhang, G.; Shu, H.; Hu, W.; Han, H.; Liu, R.; Guo, Z. Straw return increases crop production by improving soil organic carbon sequestration and soil aggregation in a long-term wheat- cotton cropping system. *J. Integr. Agric.* **2024**, *23*, 669–679. [CrossRef]
59. Song, C.; Oxley, L.; Ma, H. What determines irrigation efficiency when farmers face extreme weather events? A field survey of the major wheat producing regions in China. *J. Integr. Agric.* **2018**, *17*, 1888–1899. [CrossRef]
60. Sun, B.; Zhang, L.; Yang, L.; Zhang, F.; Norse, D.; Zhu, Z. Agricultural non-point source pollution in China: Causes and mitigation measures. *AMBIO* **2012**, *41*, 370–379. [CrossRef]
61. Wang, J.; Hu, Y.; Zhang, Z. Skill-biased technological change and labor market polarization in China. *Econ. Model.* **2021**, *100*, 105507. [CrossRef]
62. Lind, J.T.; Mehlum, H. With or without u? The appropriate test for a u-shaped relationship. *Oxf. Bull. Econ. Stat.* **2010**, *72*, 109–118. [CrossRef]
63. Aregay, F.A.; Zhao, M.; Xu, T. Knowledge, attitude and behavior of farmers in farmland conservation in China: An application of the structural equation model. *J. Environ. Plan. Manag.* **2018**, *61*, 249–271. [CrossRef]
64. Loureiro, M.L. Farmers' health and agricultural productivity. *Agric. Econ.* **2009**, *40*, 381–388. [CrossRef]
65. Li, X.; Wu, X. The impact of social norms on rice farmers' behavior of organic fertilizers application: Mediating effect of value perception and moderating effect of education level. *Int. J. Low-Carbon Technol.* **2021**, *16*, 1492–1503. [CrossRef]
66. Han, Z.; Liu, Y.; Guo, X.; Xu, J. Regional differences of high-quality development level for manufacturing industry in China. *Math. Biosci. Eng.* **2022**, *19*, 4368–4395. [CrossRef] [PubMed]
67. Xie, K.; Ding, M.; Zhang, J.; Chen, L. Trends towards coordination between grain production and economic development in China. *Agriculture* **2021**, *11*, 975. [CrossRef]
68. Ke, N.; Zhang, X.; Lu, X.; Kuang, B.; Jiang, B. Regional disparities and influencing factors of eco-efficiency of arable land utilization in China. *Land* **2022**, *11*, 257. [CrossRef]
69. Asadullah, M.N.; Kambhampati, U. Feminization of farming, food security and female empowerment. *Glob. Food Secur.-Agric. Policy* **2021**, *29*, 100532. [CrossRef]
70. Yan, Z.; Wei, F.; Deng, X.; Li, C.; He, Q.; Qi, Y. Feminization of agriculture: Do female farmers have higher expectations for the value of their farmland?-Empirical evidence from China. *Agriculture* **2022**, *12*, 60. [CrossRef]
71. Pilgeram, R.; Amos, B. Beyond "inherit it or marry it": Exploring how women engaged in sustainable agriculture access farmland. *Rural Sociol.* **2015**, *80*, 16–38. [CrossRef]
72. Ball, J.A. Women farmers in developed countries: A literature review. *Agric. Hum. Values* **2020**, *37*, 147–160. [CrossRef]
73. China NPC Network. The Law of the People's Republic of China on the Protection of Woman's Rights and Interests. 2022. Available online: http://www.npc.gov.cn/npc/c2/c30834/202210/t20221030_320091.html (accessed on 27 May 2024).
74. Li, B.; Li, T.; Yu, M.; Chen, B. Can equalization of public services narrow the regional disparities in China? A spatial econometrics approach. *China Econ. Rev.* **2017**, *44*, 67–78. [CrossRef]
75. Begho, T.; Glenk, K.; Anik, A.R.; Eory, V. A systematic review of factors that influence farmers' adoption of sustainable crop farming practices: Lessons for sustainable nitrogen management in south asia. *J. Sustain. Agric. Environ.* **2022**, *1*, 149–160. [CrossRef]
76. Waddington, H.; Snilstveit, B.; Hombrados, J.; Vojtkova, M.; Phillips, D.; Davies, P.; White, H. Farmer field schools for improving farming practices and farmer outcomes: A systematic review. *Campbell Collab.* **2012**, *10*, 336. [CrossRef]
77. Liu, Y.; Zou, L.; Wang, Y. Spatial-temporal characteristics and influencing factors of agricultural eco-efficiency in China in recent 40 years. *Land Use Pol.* **2020**, *97*, 104794. [CrossRef]
78. Song, W. Mapping cropland abandonment in mountainous areas using an annual land-use trajectory approach. *Sustainability* **2019**, *11*, 5951. [CrossRef]
79. Yan, J.; Yang, Z.; Li, Z.; Li, X.; Xing, L.; Sun, L. Drivers of cropland abandonment in mountainous areas: A household decision model on farming scale in southwest China. *Land Use Pol.* **2016**, *57*, 459–469. [CrossRef]

land

MDPI

Article

Understanding the Rise and Fall of Rural Specialty Agriculture from Social–Ecological Land System Perspective: A Longitudinal Case Study in China

Xue Wang, Mingyu Wang, Lianbing Gong and Chuangchuang Yu *

School of International Affairs and Public Administration, Ocean University of China, Qingdao 266100, China; wangxue1056@ouc.edu.cn (X.W.); wangmingyu3525@stu.ouc.edu.cn (M.W.); gLb@ouc.edu.cn (L.G.)
* Correspondence: yuchuangchuang@stu.ouc.edu.cn

Abstract: Specialty agriculture, promoted in China to enhance rural revitalization and farmers' wellbeing, depends heavily on local social–ecological conditions and land transformations. However, existing studies often emphasize socio-institutional successes while oversimplifying ecological complexity and neglecting failures. This study examines the rise and fall of rural specialty agriculture through a social–ecological land system perspective, using the cherry tomato industry in C Town, China, as a case study. Drawing on qualitative data from field investigations and interviews, we adapt an extended Social–Ecological Systems Framework (SESF) to analyze interactions among governance subsystems, land resources, and external shocks, and their role in social–ecological transitions. Our findings show that the shift from farmer-led self-organization to government-driven and enterprise-dominated governance significantly undermined system resilience and adaptability. Although external interventions initially spurred industrialization and scaling, they marginalized farmers, leading to the system's collapse under stressors like market fluctuations and environmental challenges. Our findings highlight the importance of inclusive governance for sustaining land-based agricultural systems, and point to the need of prioritizing farmer autonomy, sustainable land use planning, and adaptive governance to ensure the resilience and long-term sustainability of specialty agriculture systems.

Keywords: specialty agriculture; rural sustainability; governance dynamics; self-organization; social–ecological systems; land system science

1. Introduction

In recent years, specialty agriculture has been actively advocated by the Chinese government across China's vast rural areas in order to enhance farmers' incomes, alleviate rural poverty, and implement the Rural Revitalization Strategy [1–3]. Specialty agriculture, as in the Chinese context, refers to place-based and economically competitive agricultural practices that take advantage of a region's unique social–ecological conditions such as soil, climate, and other natural endowments [1]. In contrast to the often disorganized subsistence agriculture, specialty agriculture is developed to meet market and societal needs that typically requires economies of scale and organization of various agents [2,4,5]. On one hand, the development of specialty agriculture can yield a range of positive impacts on rural sustainability, e.g., increased income, restored ecosystems, and enhanced social capital [6–9]. On the other hand, however, specialty agriculture brings higher production costs and bears greater exposure to biophysical, socio-economic, and even political risks [6,10]. For example, studies have found specialty agriculture to be more vulnerable to changes in climate [11], water resources [12], and labor supply [13]. Moreover, the intensification of specialty agriculture often leads to the excessive monocropping of crops, which can contribute to land and environmental degradation, a decline in biodiversity, and adverse effects on environmental health, food security, and the long-term sustainability of agricultural systems [14–16].

Major advances have been made in the existing literature to explain the successful development of specialty agriculture, which highlights two development models in China: the top-down, government-led model versus the bottom-up [17], endogenous development model [18]. The former model is more commonly studied in the literature, with recognized governmental influences behind various successful cases [19,20]. Such governmental contribution is attributed to local government's motivation rooted in top-down political pressure to fulfill administrative duties and indigenous economic pressure to increase fiscal revenues [21]. Specifically, local governments, following the general principles of national policies, have implemented various measures to stimulate the growth of specialty agriculture. These measures include special funding, financial policy support, strategic planning, establishment of demonstration agricultural zones, infrastructure development, facilitation of land access, technical support, information services, assistance in brand promotion, and regional marketing initiatives [1,22,23]. Notably, the government-led model also involves collaboration with the private sector in varying forms, e.g., by offering generous subsidies, favorable loans, tax exemptions, and land integration and agricultural support [24]. In short, China's local governments have the motivation and capability to mobilize agribusinesses to enter the specialty agriculture sector. As a result, the traditional disorganized, scattered, subsistence-based agriculture in selected villages can transform into a more specialized, large-scale, market-oriented specialty agriculture, in a relatively short period of time [25,26].

In contrast, the latter model, which is less common in existing studies, emphasizes the leading role of farmers, growers, and the cooperatives they establish in initiating and promoting specialty agriculture [27–29]. Such successful cases often involve, in one way or another, a key agent other than the government—rural elites. Driven by economic, social, and/or political incentives [18] and thanks to their outstanding entrepreneurial skills and strong social capital, some of the rural elites, by chance or intentionally, take the lead in developing rural specialty agriculture by aligning local resource endowments with market demands and policy supports [30]. That said, farmers often form cooperatives with those rural elites as a core to take collective actions in mitigating the limitations of small-scale production and taking advantages of the scale economy associated with agricultural industrialization and commercialization [28,29]. For instance, in central Kenya, smallholder banana farmers have successfully facilitated the flow of information, the adoption of innovative technologies, and the upgrading of production systems through forming and participating in cooperatives. This has also helped strengthen the connection between farmers and emerging high-value market chains [31]. In Costa Rica's Central Valley and Coto Brus, participation in cooperatives has had a positive impact on farmers' decisions to grow specialty coffee and secure higher prices [32]. In Sedomyun, South Korea, local agricultural cooperatives play a pivotal role in organizing the export system for cherry tomatoes. The cooperatives collect the tomatoes from farmers and sell them to exporters [33]. These cooperatives can serve as a platform for smallholder farmers to cooperate, share information and risks, and address common challenges [34]. They often offer technical assistance and training, supply fertilizers and pesticides, provide seeds and seedlings, engage in processing, packaging, and marketing of products, and other information and services related to production technologies and marketing [17,35]. Such collective actions not only help farmers increase production efficiency and improve the quality and market value of specialty agricultural products [29,34], but also build rural community and sometimes contributing to improved eco-environment as well [27,34].

Notwithstanding the insightful research progresses, existing studies seem still insufficient in revealing the whole story about the development of China's specialty agriculture. To begin with, specialty agriculture, by definition, is strongly dependent on the relative comparative advantages rooted in a village's favorable social–ecological conditions. Such conditions can be in various forms of natural endowments, usually relating to land in one way or another. However, the ecological dimension of this story—particularly the roles of land in the development of specialty agriculture—remains surprisingly oversimplified and remarkably understudied. In fact, as the spatial nexus of social–ecological interactions and a productive base for agriculture in general [36,37], land is deeply involved in the development of China's specialty agriculture. Specifically, first, China's unique land tenure systems impacts the extent and type of involvement of small-scale farmers and enterprises in specialty agriculture. As a result, various agribusiness modes coexist, including contracted management, land leasing, and independent cooperatives [17,25,38]. Second, land transfer is a critical foundation for achieving economies of scale in specialty agriculture [38,39]. Promoting the concentration of land use can provide the necessary space for the development of specialty agriculture [40], thus improving the production efficiency and ensuring their long-term sustainability [39,41]. Third, land consolidation helps eliminate land fragmentation, expand the area of arable land, improve planting conditions, and enhance land quality and productivity, thus facilitating the expansion of agricultural production and the development of multifunctional agriculture [42–44]. Finally, the modes of specialty agricultural production also transform the use of farmland, affecting the farmland functions, production efficiency, and property rights realization methods. This shift leads to a transition from decentralized to more concentrated modes of production [45–47].

Another insufficiency of the existing more social-focused studies, which often take a socio-institutional or governance perspective, is their obsession with explaining the successful cases (e.g., [17–19,23]). Equally important, however, if not more, is the need of understanding why some cases failed. Additionally, the existing studies seem satisfied with their oversimplification of a dichotomy of stereotypes, namely, the top-down, government-led model versus the bottom-up, farmer-led model. The fact is that the governance subsystem itself often evolved over time along with the development of specialty agriculture. In other words, further needed is a dynamic instead of static perspective on the roles of the governance subsystem (and land resource subsystem) in the rise and fall and China's specialty agriculture.

To help fill the above-noted two research gaps, here we present a longitudinal case study in Shandong Peninsula of China to reveal how the governance subsystem changed over time at different stages of the specialty agriculture development—from its endogenous emergence to rapid development, then to regional proliferation, and lastly to abrupt decline—while highlighting the involved land transformations and land-related sustainability impacts. In this sense, we explicitly take a dynamic social–ecological land system perspective, with our analytical framework adapted from the Social–Ecological Systems Framework (SESF) originally proposed by Ostrom and her colleagues [48–50]. The remainder of the paper is organized as follows: Section 2 details our research design; Section 3 describes the four development stages of our case and analyzes the social–ecological land transitions between the stages; Section 4 discusses the overarching social–ecological mechanisms underlying the rise and fall of the specialty agriculture, elaborates the roles played by the land resource subsystem, and reflects on policy implications for sustainability; and Section 5 concludes in relation to the broader theme of this Special Issue, "Advancing

Sustainability through Land-Related Approaches: Theoretical Foundations and Empirical Investigations".

2. Methodology and Materials

2.1. Framework for Analyzing Transition of Social–Ecological Land Systems

As modern society has evolved, there has been a growing recognition of the universality and complexity of interactions between ecological dynamics and social processes, as well as the formidable challenges they pose [51,52]. Over the past two decades, the idea of Social–Ecological Systems (SESs) has developed into a leading interdisciplinary research field that examines interdependent linkages between social and environmental change and how these linkages influence the achievement of sustainability goals [53–58]. Among various frameworks, the SESF [48,59] developed by Ostrom and others stands out as arguably the most comprehensive and widely recognized [57], offering a general tool for understanding and analyzing SES sustainability [48,52]. While SESF is applicable to different resource systems at varying scales [53], it has been particularly influential in analyzing small-scale public resource systems [60], with case-based qualitative studies being a common research method [57].

A SES is a complex adaptive system shaped by the intricate interactions and interconnections between social and ecological systems. In this context, an ecological system refers to a network of interdependent organisms or biological units, while a social system is a network in which organisms are similarly interdependent [61]. Ecological systems provide essential resources and services that sustain the functioning of social systems, while social systems, through management and utilization activities, influence the structure and function of ecological systems. The defining feature of SESs lies in the dynamic interaction effects and feedback loops between social and ecological systems [55,62–64]. Agriculture is a complex social–ecological system [52,64–66], and its evolution is the result of dynamic interactions and feedback loops between the social subsystem and ecological subsystem [56,67–69]. This has led to an increasing number of scholars employing SESF to study agricultural development issues. For example, Kopainsky and colleagues used SESF to analyze Switzerland's agricultural system [70], while Dwyer and others employed an adjusted SESF to evaluate sustainable agriculture policies and initiatives across Europe [71]. Zhu and Wang, on the other hand, applied the SESF indicator system to explore the relationship between cooperatives and collective action among farmers [72].

Our study adapts the SESF with additional focus on its dynamic flows to explore how interactions between governance subsystems and land resource subsystem change over the course of specialty agriculture development and associated implications of such transitions. According to SESF (see Supplementary Figure S1a), the first tier of the specialty agriculture SESs comprises four subsystems: Resource Systems (RSs) (the resource system for specialty agriculture); Resource Units (RUs) (natural resource units, including land, soil, and water; biological resource units, such as crops, insects, and microorganisms; and technological resource units, such as greenhouses, irrigation systems, and drainage facilities, etc.); Actors (As) (government, enterprises, cooperatives, and farmers); and Governance Systems (GSs) (the governance network or institutions formed by these actors). These subsystems interact with each other, collectively influencing the Interactions (Is) and Outcomes (Os) in specific action situations, while being indirectly influenced by external factors. Additionally, they are embedded within and interact with larger social, economic, political (S), and ecological (ECO) environments [48]. According to Ostrom's SES diagnostic framework [50] (see Supplementary Table S1), we developed an key variable inventory for each stage of the specialty agriculture development to analyze connections between specific variables. Importantly, our extension of SESF is focusing on the dynamic nature of the specialty agriculture SESs (see Supplementary Figure S1b), investigating the potential transition of social–ecological interactions over time [49].

2.2. Case Study Area: C Town of P City in Shandong Province of China

Case study as a widely used method requires the focal case to serve research purpose and needs researchers to collect comprehensive information on the focal case. Thus, we used the specialty agriculture of C Town in eastern China as our empirical case, which present typical rise and fall of development and allows us best available first-hand data. C Town is administratively subordinate to P City, a county-level city (cf. a county-level city in China corresponds to a county in the USA, in terms of its administrative level) under the jurisdiction of Qingdao, Shandong Province, situated in the southwestern part of P City (Figure 1). As of 2024, C Town spans an administrative area of 214 square kilometers, with 209,000 mu (1 mu = 0.1647369 acres) of arable land and a population of approximately 78,000. The terrain of C Town is predominantly flat, with the eastern area higher than the west, and the northern part higher than the south. Some low-lying regions feature weakly alkaline groundwater and soil. C Town experiences a warm temperate monsoon climate, characterized by abundant sunlight and frequent summer thunderstorms and floods. Historically, C Town has long been practicing traditional agriculture, predominantly cultivating crops such as wheat, corn, and cotton. Since the top-down movement of the Targeted Poverty Alleviation program in 2014, C Town was classified as an economically disadvantaged town by the Qingdao government. In 2016, C Town began to promote cherry tomato cultivation, and by 2019, it was recognized as a strong agricultural town by Shandong Province. In 2022, the town was honored as a National "One Village, One Product" Demonstration Town by the Ministry of Agriculture and Rural Affairs of China.

Q Village, located 6 km northeast of C Town, is named for its low-lying terrain. In the past, Q Village primarily cultivated wheat and cotton. However, due to the alkaline soil and water, crop yields in Q Village were consistently lower than surrounding villages, with the average annual income per capita hovering around CNY 8000 to 9000, leaving many villagers in poverty. In 2008, under the leadership of the village head, a few proactive villagers began cultivating cherry tomatoes. This was the origin of the cherry tomatoes industry in C Town. By 2016, C Town had established a tomato poverty alleviation industrial park in Q Village. By 2019, over ten major investment enterprises, along with tens of hundreds of small agricultural companies, got involved in the local cherry tomato industry. However, by 2023, most of these enterprises had withdrawn due to years of financial losses, leaving increasing numbers of abandoned greenhouses behind.

The cherry tomatoes cultivated in C Town are primarily of the "Busan 88" variety, a herbaceous crop with a sugar–acid ratio of 12:3, offering a richer taste and higher vitamin content. Due to the region's alkaline soil and mineral-rich water sources, the resulting tomatoes are notably sweeter and have a superior flavor. In 2020, the unique taste and high quality of the cherry tomatoes, influenced by the local environment, were recognized by the Chinese Ministry of Agriculture as a national geographical indication protected agricultural product.

In terms of cultivation practices, cherry tomatoes are grown in greenhouses. These greenhouses were initially basic structures but were later upgraded to winter-heated plastic greenhouses with investments from both the government and enterprises. The greenhouses are equipped with drip irrigation systems, and there are government-built irrigation wells, roads, and simple roadside drainage channels nearby. Local practices do not include crop rotation, intercropping, or organic farming. Fertilizers and pesticides are applied frequently during the growing season, particularly in the summer when pests and diseases are most prevalent. Given China's inadequate phytosanitary system—especially for non-export crops—phytosanitary products have never been used. Before farmers in C Town began growing cherry tomatoes independently, they consulted agricultural experts to test and analyze the local soil and climate and conducted a basic market survey. Based on this information, they chose to cultivate cherry tomatoes, experimenting with various varieties, ultimately selecting the "Busan 88" variety as the most suitable. When enterprises became involved, agricultural planting companies assumed responsibility for market demand analysis and planting decisions. However, neither the government nor the enterprises provided systematic training for the farmers, who instead relied on the exchange of cultivation experiences among themselves.

Unless seedlings die completely due to climatic factors or diseases, cherry tomatoes are typically grown in a single cycle per year. Planting begins at the end of August, with fruit production starting in November and continuing until June of the following year. Once harvested, the cherry tomatoes are initially sold by the cooperative to local buyers. After the enterprises became involved, the tomatoes are directly sold to large supermarkets in nearby cities. Most of the tomatoes are sold fresh, while lower-quality tomatoes are processed into jams.

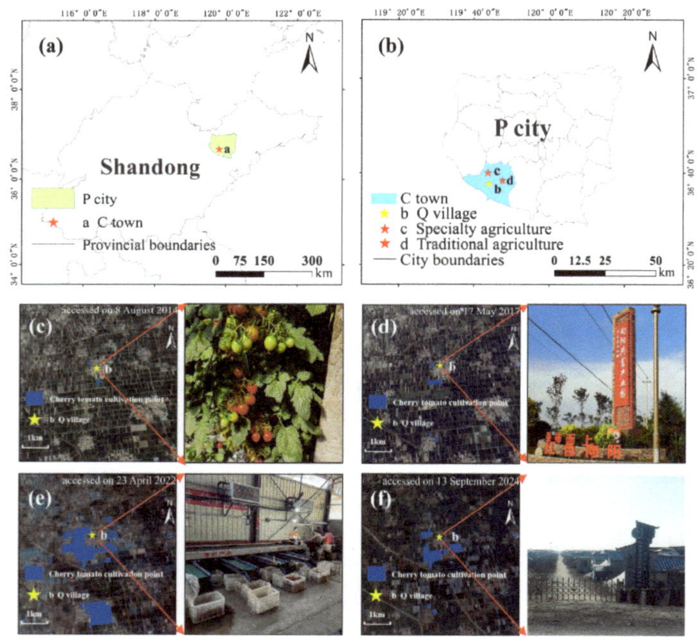

Figure 1. Study area and evolution of its specialty agriculture. (**a**) C Town in P City, a county-level city in Shandong Province of China. (**b**) C Town is situated in southwestern P City. While agricultural land in the southeast is dedicated to traditional farming, that in the northwest was developed into a specialty tomato cultivation zone. The development of specialty tomato agriculture in C Town has progressed through four phases (with Q Village as a case, which is located 6 km northeast of the C Town government): (**c**) from 2008 to 2015, Q Village farmers initiated by themselves to establish greenhouses for cherry tomato production; (**d**) from 2016 to 2018, governmental poverty alleviation funds were invested to develop an industrial park, expanding cultivation; (**e**) from 2019 to 2022, significant commercial investments facilitated industrialization and commercialization; and (**f**) from 2023 to 2024, persistent financial losses led to the withdrawal of enterprises, causing widespread greenhouse abandonment and extensive specialty agricultural land fallow.

2.3. Field Investigation and Interview

To track the long-term development of cherry tomato cultivation in C Town, we conducted four field investigations in P City, C Town, and Q Village: in August 2021, September 2022, July 2023, and May 2024. In addition to on-site observations and visits to local industrial parks and greenhouses, we conducted in-depth interviews with 20 individuals, including officials from the P City's Agricultural and Rural Bureau, government staff from C Town, village head of Q Village, representatives from Huaken Agricultural Company, Ping Agricultural Tourism Group, Shi Agricultural Company, and local farmers and villagers (see Supplementary Table S2 for the interview list). Most of the interviewees were revisited during the four investigations, particularly government officials and key actors. The interviews were semi-structured, lasting over half an hour, with some key interviews extending beyond two hours. Additionally, we reviewed all available official documents, statistical yearbooks, and government reports, along with collecting printed and online promotional materials and news articles.

During the first two investigations, we focused primarily on the development trajectory of the cherry tomato industry and were particularly interested in understanding the roles of and interactions among the government, enterprises, cooperatives, and farmers as well as related land transformation in successfully developing this specialty agriculture. At the third field investigation, however, we recognized serious challenges facing this local industry. In fact, the situation deteriorated far more rapidly than anticipated—during the fourth investigation, conducted less than a year later, we observed large numbers of abandoned and idle greenhouses, and several of the companies we had interviewed the previous year had already left. We made efforts to reconnect with the original enterprises and farmers to understand why the local specialty agriculture experienced such a dramatic decline in such a short period.

3. Results

Based on consensus among key actors and consistency with second-hand data, the specialty agricultural development of the cherry tomato industry in C Town can be divided into four stages: endogenous emergence (2008–2015), rapid development (2016–2018), regional proliferation (2019–2022), and abrupt decline (2023–2024). The division into these four stages is primarily based on the involvement of different actors and the resulting significant changes in planting scale and industrial upgrading. For instance, after the government's involvement, the planting area was expanded, and a specialized industrial park was established. The participation of enterprises contributed to a rapid expansion of the industry scale. We then employed the SESF, along with secondary variables, to conduct a detailed analysis of each stage, exploring how interactions between actors, governance systems, resource units, and resource systems, within specific political, economic, social, and ecological contexts, led to the evolution of the specialty agriculture and land transformations.

3.1. Endogenous Emergence (2008–2015), Characterized by Farmers' Self-Governance

In this phase, under the political, economic, social, and ecological conditions of the time, a small group of farmers, led by the village head, initiated land consolidation and self-organizing activities that laid the foundation for the initial development of cherry tomato agriculture. Figure 2 illustrates the dynamics of the SES in this phase, while Supplementary Table S3 provides empirical evidence for the secondary variables.

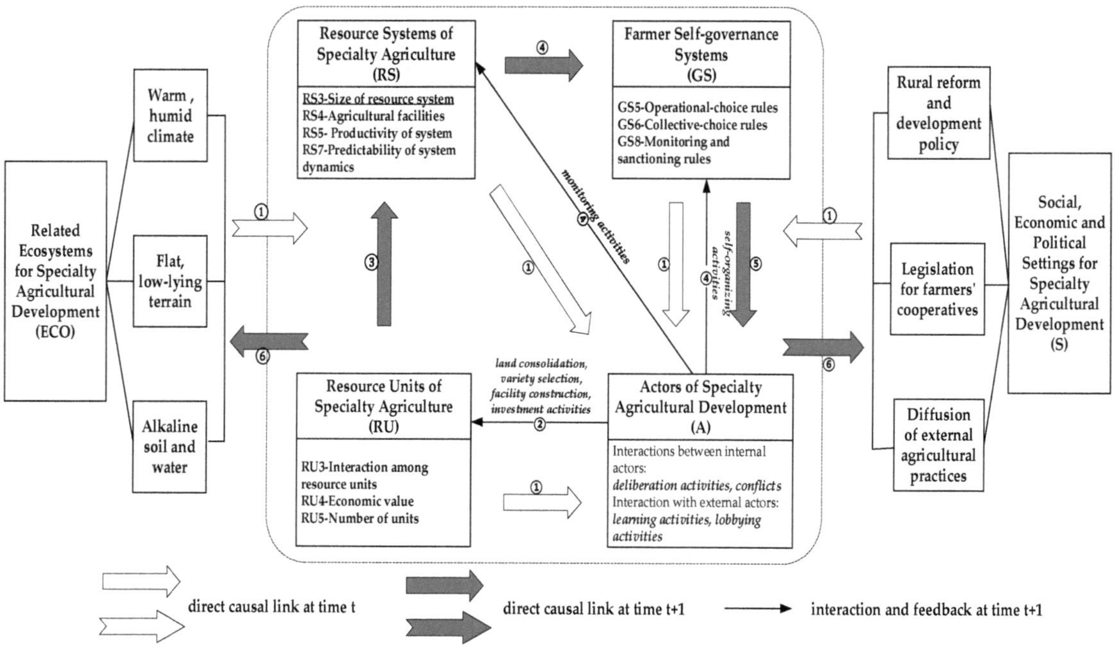

Figure 2. Endogenous emergence (2008–2015). Farmers in the region engaged in land consolidation for greenhouse construction through plot exchanges with other villagers. Given the area's slightly alkaline soil conditions, they carefully selected a variety of cherry tomatoes that was most suited to these environmental factors. These measures not only promoted the early-stage emergence of specialty agriculture but also enhanced the efficiency of land use.

① Ostrom identified ten key secondary variables influencing self-governance within SESs [48] (marked with * in Supplementary Table S3). The interplay of these variables contributed to the collective actions and self-organizing efforts of Q Village's farmers to cultivate cherry tomatoes. Before 2008, Q Village had relatively abundant farmland, with 3423 mu allocated to individual households after the household responsibility system was implemented (RS3). However, the village's low-lying terrain, alkaline soil, and frequent summer storms (ECO) made crop yields highly dependent on weather conditions (RS7). As a result, agricultural production was low (RS5), and many villagers lived in poverty. In 2004, Ju, an entrepreneurial and leadership-minded individual, was elected as the village head (A5). Upon taking office, he was deeply concerned about the poverty and underdevelopment in the village. In 2008, the Central Committee of the Communist Party of China issued the "Decisions on Several Major Issues Concerning Rural Reform and Development", which called for the active promotion of modern agriculture (S4a). This reform direction ignited Ju's entrepreneurial zeal. After learning about the successful greenhouse vegetable model in Shouguang, Shandong Province (S5), he quickly mobilized 18 villagers to start cultivating vegetables in greenhouses (A1).

② Once the proactive farmers were organized, they engaged in a series of collective actions involving interactions among themselves, external actors, and land and other resource units. Initially, the farmers undertook learning, consultation, and lobbying with external actors to obtain the necessary knowledge, experience, and resources. In 2008, the proactive farmers visited Shouguang to learn cultivation techniques. Ju also invited agricultural experts from Qingdao to assess the local soil and water quality. The tests revealed that the local soil and water are slightly alkaline, making the area particularly suitable for growing cherry tomatoes (I2). Ju then actively sought local government policies and loan support (I6). Furthermore, in order to consolidate land, the farmers negotiated land exchanges with other villagers on a 1:1.3 ratio (I3). Secondly, the farmers held collective discussions and made decisions about the development of specialty agriculture. After deliberation, they decided to grow cherry tomatoes (I3). Thirdly, they engaged with land and other resources. The proactive farmers consolidated 140 mu of contiguous land by exchanging land with other villagers (I10), built 18 greenhouses (I5, I10), and trial-planted tomato varieties suited to the soil conditions (I10).

③ The interaction between farmers and resources led to changes. By consolidating land, building greenhouses, and cultivating tomatoes, the economic value of land resource unit and their crops increased (RU4), their quantity grew (RU5), and their spatial distribution became more concentrated (RU7). In terms of interactions between resource units, compared to traditional farming, cultivating cherry tomato varieties that are better suited to the local soil and water conditions optimizes the use of land resources. Greenhouse cultivation also safeguards specialty crops from extreme weather, thereby stabilizing agricultural production. However, cherry tomatoes are water-intensive and susceptible to pests and diseases, requiring significant amounts of water, fertilizers, and pesticides (RU3) [73,74]. The above interactions, in turn, facilitated the expansion of the resource system. The farmers planted 18 greenhouses on the 140 mu of consolidated land (RS3, RS4), resulting in higher crop yields and greater economic returns (RS5). However, the predictability of the resource system decreased (RS7). During the first trial, the chosen cherry tomato variety failed due to extreme weather conditions and severe market homogenization, leading to lower yields and prices and significant losses for the farmers. Nevertheless, most farmers persevered.

④ The dynamic changes in the resource system led to new interactions. According to Ostrom's research, a slight expansion in the resource system strengthens the motivation for self-organization, while decreased predictability can lead to larger-scale or higher-degree self-organization [50]. In this case, the changes in the resource system indeed reinforced the farmers' self-organizing efforts and increased the level of organization within the self-governance system. After the first trial failure, farmers conducted market research (I9) and consulted experts (I2) to identify a more suitable cherry tomato variety, "Busan 88", and after successful cultivation, they achieved higher economic returns. This success attracted more villagers to join, but Ju and others soon noticed that some farmers engaged in unhealthy competition and substandard practices (I9), resulting in conflicts (I4). As a result, they continued learning from the cooperative model in Shouguang. Based on the Farmers' Professional Cooperative Law enacted in 2006 (S4b), the villagers, after months of deliberation (I3), established the "Q Village Fruit and Vegetable Professional Cooperative" in 2014 (I7), with Ju as the chairman and hundreds of local farming households as main members.

⑤ These self-organizing efforts led to the expansion, organization, and standardization of the self-governance system. The cooperative established a more comprehensive management and service system, guiding and regulating the actions of an increasing number of participants. The cooperative primarily provided services such as vegetable cultivation and sales, greenhouse management and maintenance, and the promotion of agricultural technologies to its members. It also addressed competitive chaos by directly liaising with buyers, overseeing operations, and disciplining members for improper behaviors. Moreover, the cooperative stipulated that farmers could join by contributing their land's contractual rights at CNY 800 per mu per year. The cooperative would then collectively lease the land for greenhouse construction (GS5, GS6, GS8).

⑥ The dynamic interactions between the four subsystems produced noticeable social and ecological outcomes. Farmers in Q Village, leveraging local land resources and market demand, began developing greenhouse agriculture to cultivate cherry tomatoes, a high-value crop. This not only increased the villagers' economic income but also significantly enhanced their self-governance effectiveness. The local specialty industry began to take shape, and the villagers' economic income and self-governance effectiveness significantly improved. At the same time, the development of facility agriculture transformed land use, improving both land and production efficiency (O1). Compared to traditional farming, cultivating a single variety of cherry tomatoes in greenhouses is more likely to result in soil degradation, nutrient depletion, and loss of biodiversity. The increased use of plastic films, fertilizers, and pesticides also contributes to soil, water, and air pollution (O2). These factors create hidden risks to the sustainable development of specialty agriculture from the emergence.

3.2. *Rapid Development (2016–2018), Characterized by Governments Interventions*

In accordance with the national strategy to overcome poverty, local governments took proactive steps to promote the development of specialty agriculture rooted in local characteristics, leading to the swift expansion of cherry tomato cultivation in C Town. However, this also introduced new challenges. Figure 3 illustrates the dynamics of the SES in this phase, while Supplementary Table S4 provides empirical evidence for secondary variables.

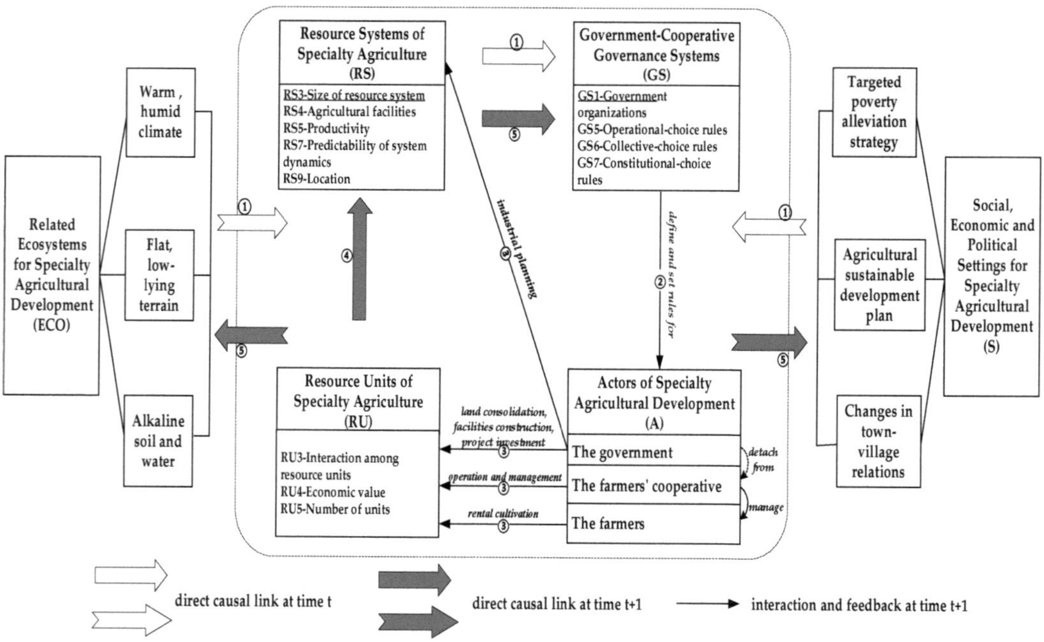

Figure 3. Rapid development (2016–2018). The government harnessed poverty alleviation projects to implement land transfers and consolidation, improving infrastructure such as roads, drainage channels, and irrigation wells, while also establishing a poverty alleviation industrial park. These efforts not only expanded the scale of specialty agriculture but also optimized the allocation of land resources.

① At the end of 2015, the Central Committee of the Communist Party of China and the State Council made landmark decisions concerning the national poverty alleviation strategy, mandating that "local party committees and governments at all levels must treat poverty alleviation as a major political task". They also proposed specific strategies for target poverty alleviation and for developing specialty industry-based poverty alleviation (S4a). These politically driven poverty alleviation tasks and their rigid targets were passed down the vertical power structure, exerting pressure at all levels of government. Qingdao City, P City, and C Town governments rushed to identify and nurture potential industries for poverty alleviation within their locales. The rapidly expanding tomato industry in Q Village quickly caught the attention of the local government.

② In their effort to succeed in the national "poverty alleviation competition [75]", P City and C Town governments strategically embedded themselves within the local specialty agriculture governance system, fostering its rapid development. This resulted in the creation of a new governance framework dominated by government leadership, with the cooperative at the core (G1). In this framework, local governments, utilizing their political clout and poverty alleviation resources, became the central decision-makers for the development of specialty agriculture, directly formulating plans and strategies for its macro-development (GS7). The cooperatives, acting under government guidance, assumed the role of operational managers, handling government investment projects, executing development strategies, and overseeing day-to-day operations (GS6, GS5).

③ Within the governance system of "Government-Cooperative" and its guiding principles, the government, cooperative, and farmers undertook specific actions to further develop specialty agriculture, engaging with resource units and systems. In 2016, C Town pooled CNY 5.4 million in poverty alleviation funds allocated by higher levels of government for six impoverished villages (I5) and consolidated over 260 mu of project land in Q Village to establish 36 high-standard winter greenhouses (I10), aiming to create a poverty alleviation industrial park (I11). The completed industrial park was then handed over by the government to the cooperatives for management. The cooperative was tasked with leasing out greenhouses, hiring workers, and providing leasing farmers with production plans, seeds, technical guidance, and market access service. Under the cooperative's management and services, the farming model adopted by the farmers remained largely consistent with that of the initial stage (I11). Land transfers and consolidation, driven by government initiatives and supported by substantial poverty alleviation funding, provided the necessary land resources and improved agricultural infrastructure, both of which were crucial to the expansion of specialty agriculture. A significant portion of agricultural land, especially fields previously abandoned by farmers due to land salinization, was repurposed for protected agriculture, thereby increasing its economic value. As a result, the cultivation area of cherry tomatoes expanded, and the construction and utilization of new greenhouses significantly enhanced both the quality and yield per unit area, as well as the economic returns (RU3). By 2017, the cherry tomatoes grown in the 36 greenhouses were of exceptional quality (RU4), with the highest purchase price reaching CNY 12 per pound, and each greenhouse generated a net income exceeding CNY 140,000 (RU5).

④ The dynamic interaction between actors and resource units led to the rapid expansion of the resource system, the growth of agricultural infrastructure, and sustained improvements in productivity. In 2017, the poverty alleviation industrial park, covering over 260 mu and comprising 36 greenhouses, produced 1.12 million jin (1 jin = 1.1023 pounds) of cherry tomatoes, generating over CNY 5.6 million in sales revenue (RS3, RS4, RS5).

⑤ The rapid growth of the resource system signified a new phase in the development of specialty agriculture. The implementation of the government's poverty alleviation projects not only brought in construction funds but also revitalized the land and labor in six surrounding impoverished villages. The previously fragmented or idle land was consolidated and developed into a poverty alleviation industrial park, generating employment opportunities for hundreds of impoverished farmers in the surrounding areas. Land use became more efficient, agricultural output and economic returns improved, benefiting a greater number of surrounding villages and farmers. The project has generated no less than CNY 50,000 in collective income for the six impoverished villages annually (O1). However, it is somewhat regrettable that the local government, which relied on poverty alleviation projects to promote the development of agriculture, provided only national poverty alleviation funds and land use incentives to facilitate the quick implementation of the project, but failed to offer further support or guidance to the cooperative and farmers. Even in planning specialty agriculture, the focus has been primarily on expanding the scale of specialty crop cultivation in the short term to achieve the goal of economic poverty alleviation for more farmers, rather than considering the ecological balance and sustainable land use (O2). Moreover, after the rural tax and fee reforms and restructuring of township governments (S4b), C Town lost its primary fiscal sources, and its administrative capacity diminished. After using the initial special funds for industrial park construction, the local government struggled to provide continued support. These issues led C Town's specialty agriculture to quickly enter a period of stagnation in sustained development after initial gains from poverty alleviation investments.

3.3. Regional Proliferation (2019–2022), Characterized by Market-Driven Investments

The rapid development and agglomeration effect of C Town's specialty agriculture sector attracted a significant influx of investment from both enterprises and agricultural companies. The infusion of commercial capital, combined with the professional management of agricultural businesses, has propelled the large-scale, intensive, and industrialized expansion of specialty agriculture. However, this growth has also been accompanied by the marginalization of the cooperative and farmers [25,40,76]. Figure 4 illustrates the dynamics of the SES in this phase, with Supplementary Table S5 providing empirical evidence regarding the secondary variables.

① In China, both industrial and commercial capital, as well as agricultural enterprises, are recognized as pivotal forces in driving agricultural reform and modernization [25]. In 2013, the central government issued policies encouraging industrial and commercial capital to invest in modern agricultural practices in rural areas (S4c). By the end of 2017, the country initiated the Rural Revitalization Strategy, which explicitly called for promoting investments by industrial and commercial capital in agricultural projects suitable for industrialization and large-scale operations (S4a). The revision of the "Land Contract Law" in 2018 institutionalized the "separation of three rights" land reform system, reinforcing the legal standing of industrial and commercial enterprises (S4b) [77]. These national policies provided strong institutional support for agricultural enterprises to engage in specialty agriculture. Moreover, the agglomeration effect and notable benefits of C Town's specialty agriculture attracted profit-seeking capital, while the sector's inherent development challenges called for the intervention of enterprises. Around 2019, spurred by robust local government efforts to attract investment, C Town's tomato industry drew over a dozen companies, including large state-owned enterprises like Ping Agricultural Tourism Group and private agricultural firms such as Huanken Agricultural Company (GS2a).

② With the entry of enterprise actors, the governance system of specialty agriculture shifted to a new "Government-Enterprise-Cooperative" model, prompting changes in governance rules. In this framework, new relationships and interactions were established between the government, enterprises, cooperatives, and farmers. The local government assumed the dual responsibility of "poverty alleviation and effective linkage with rural revitalization", attracting industrial and commercial capital, as well as leading enterprises, by offering policy support and land incentives to accelerate the industrialization and scaling of specialty agriculture. Faced with management challenges, the cooperative relinquished management rights to more capable enterprises, focusing instead on serving as intermediaries for land transfer. Enterprises, in turn, assumed responsibility for managing and operating industrial parks, frequently employing farmers for cultivation (GS5, GS6).

③ Under the "Government-Enterprise-Cooperative" governance model, various actors and resource units interacted dynamically with resource systems. Firstly, the government's interaction with the resource system evolved from internal industrial planning to external industrial packaging and brand promotion. Both P City and C Town governments rebranded agricultural industrial parks multiple times. The initial poverty alleviation industrial park was renamed several times, first to "Farmers' Entrepreneurship and Innovation Park" (also known as "Farming Innovation Body"), then to "Modern Characteristic Industrial Park", "Tomato Town", and finally "Xiangyang Common Prosperity Industrial Park". Additionally, the C Town government took over the private Tomato Festival, elevating it to an official event, the "C Town Tomato Exhibition and Trading Fair and Tourism Culture Festival". The government also strengthened media promotion, with the local tomato industry being featured in national-level official media outlets (I11). Through these actions, the local government successfully turned agricultural development into tangible political achievements. However, the emphasis on industrial planning and support considerably diminished. Secondly, due to limited managerial capabilities, the cooperative no longer managed the industrial parks but instead focused on facilitating centralized land transfer services (I10). Thirdly, large state-owned enterprises primarily engaged in investment and construction (I5), erecting numerous intelligent greenhouses and independent modern industrial parks on large-scale contracted farmland (I10). Smaller agricultural companies rented land and greenhouses from these enterprises. Moreover, enterprises took over the daily management of industrial parks, including soil improvement, facility maintenance, technical support, and product sales (I11). Finally, farmers who transferred their land were predominantly employed by enterprises to carry out tasks such as planting, harvesting, and packaging. A few farmers leased greenhouses from enterprises for independent cultivation.

④ The involvement of enterprises introduced higher capital investment, as well as more advanced agricultural technologies and operational management systems. The local government provided significant policy incentives, facilitating access to large, contiguous plots of land, while the cooperative successfully connected farmers to implement large-scale land transfers. Through the combined efforts of all three parties, the number of greenhouses and the planting area saw substantial growth, agricultural facilities and cultivation technologies were upgraded, and the yield and quality of the products improved. Consequently, the asset value and economic returns of the land resources also increased (RU3). As a result, the scale of the specialty agriculture system expanded considerably, with a marked increase in output value, achieving both large-scale operations and commercialization. By 2019, P City had attracted over CNY 500 million in investment, leading to the construction of more than 1000 greenhouses (RU5), each generating an average net income of CNY 200,000 (RU4). By 2022, the number of greenhouses exceeded 2400 (RS4), covering more than 15,000 mu (RS3) of land, with an annual output of 90 million jin and annual sales revenue reaching CNY 1.03 billion (RS5). Furthermore, the proactive industrial packaging and brand promotion efforts of the local government rapidly elevated the profile and reputation of C Town's specialty agriculture. In just a few years, the town's poverty alleviation industrial park was transformed into a national-level rural revitalization demonstration area, becoming a prominent political achievement for the governments at the municipal, district, and town levels.

⑤ After enterprises integrated into the governance system, C Town's specialty agriculture experienced rapid large-scale development and industrialization. The large-scale land transfer had, to some extent, optimized the allocation of land resources, providing essential support for the development and upgrading of specialty agriculture. The enterprises' unified management and efficient utilization of land resources had effectively enhanced the efficiency, asset value, and economic returns of agricultural land. Through its specialized cultivation methods and market-oriented operations, the enterprises had increased the economic value of specialty agriculture, facilitated the return of villagers to employment in their hometowns, boosted both individual and collective incomes, and revitalized rural development, thereby making a substantial contribution to rural poverty alleviation and revitalization (O1). However, driven by short-term economic incentives, the large-scale and intensive land use and cultivation models adopted by the enterprises had weakened the ecological resilience of the specialty agriculture. The greenhouse transformation of large-scale farmland not only encroached on essential arable land but also exacerbated environmental issues, including excessive consumption of water resources and energy, overuse of chemical fertilizers and pesticides, and plastic pollution. Furthermore, the long-term reliance on monocropping had amplified risks such as soil degradation, increased pest outbreaks, loss of biodiversity, and contamination from fertilizers and pesticides (O2). Additionally, the involvement of enterprises had significantly diminished the influence of farmers and their self-governing organizations within the governance subsystem. The farmers' cooperative lost their decision-making, management, and control powers. Farmers who transferred their land to enterprises and became laborers grew increasingly detached from the land, losing their initiative and agency in agricultural development. These outcomes could present potential risks to the long-term sustainability of specialty agriculture.

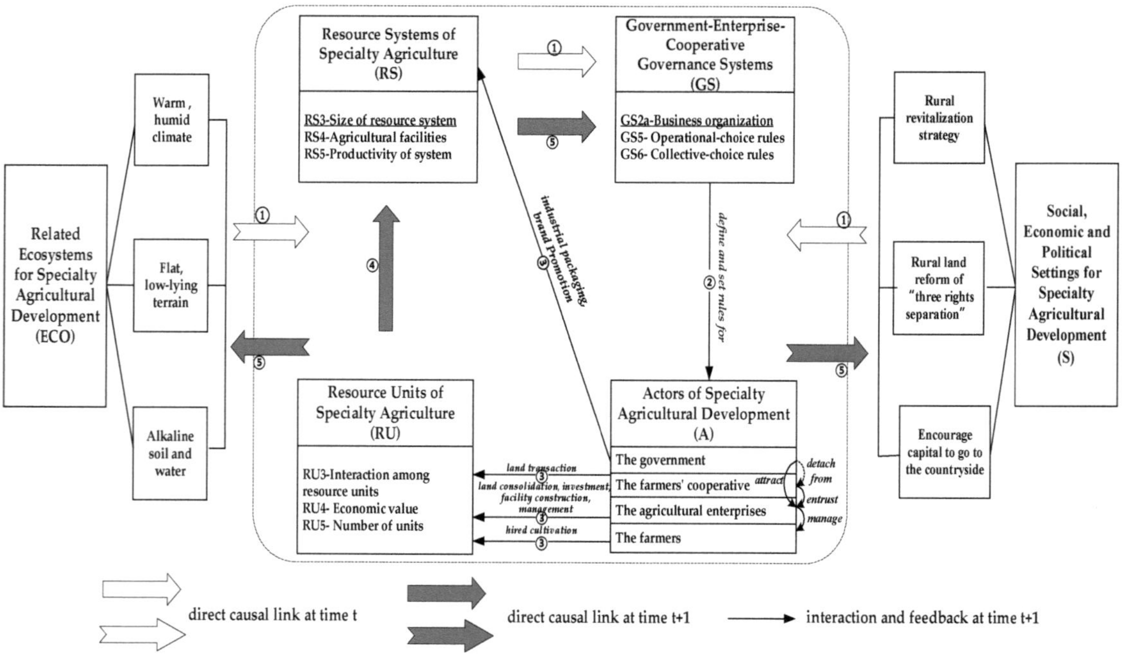

Figure 4. Regional proliferation (2019–2022). With government support, newly introduced enterprises carried out broader land transfers, converting large areas of farmland into greenhouse agricultural land. While this facilitated the scaling up and intensive management of specialty agriculture, it also heightened the vulnerability of the specialty agriculture SES, thereby creating potential risks to the long-term sustainability of both specialty agriculture development and land use.

3.4. Abrupt Decline (2023–2024), Characterized by Systematic Vulnerability

Since 2023, under the combined pressures of social, political, economic, and environmental risks, the cherry tomato industry in C Town has experienced a dramatic decline. Several companies have withdrawn after years of sustained losses, leaving a significant number of greenhouses idle and vast areas of farmland abandoned. Figure 5 illustrates the dynamics of the SES in this phase. Empirical evidence for secondary variables is provided in Supplementary Table S6.

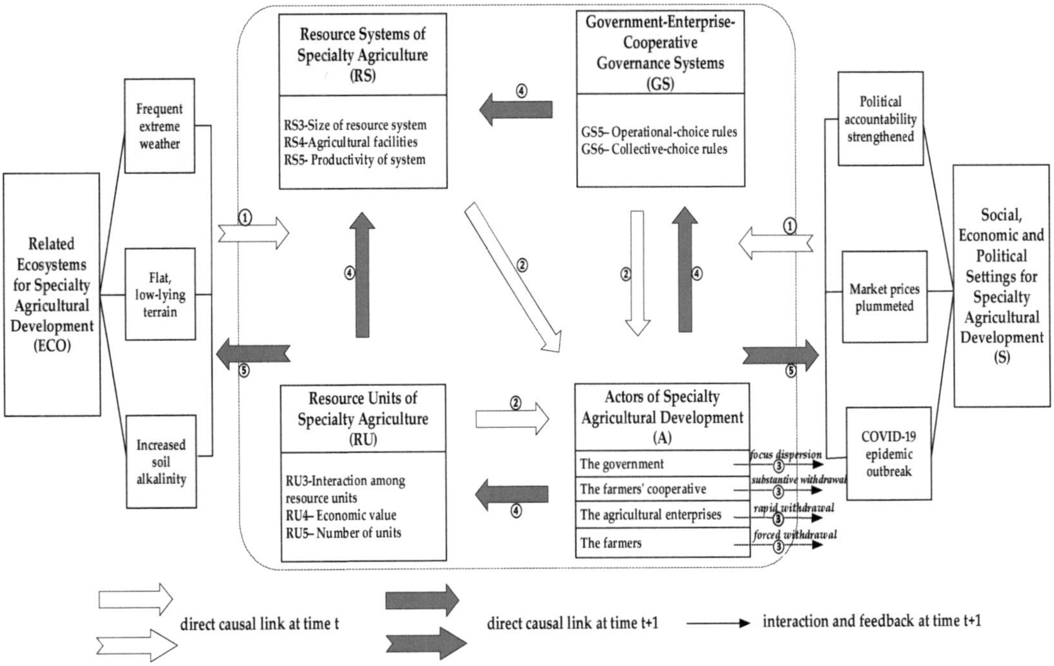

Figure 5. Abrupt decline (2023–2024). The detrimental interactions among the governance systems, farmland resources, and the agricultural ecosystem had amplified the vulnerability of the SES. Under external shocks, the withdrawal of the three key actors triggered the collapse of the governance systems, leading to the abandonment of vast greenhouse land. As a result, local specialty agriculture entered a state of decline.

① Since 2023, the political, economic, and environmental conditions affecting the development of the specialty agricultural sector have undergone a series of significant transformations. Politically, the revision of the "Regulations on Accountability of the Communist Party of China" by the Central Committee has led to a marked increase in political accountability (S4). Economically, the protracted effects of the COVID-19 pandemic and the extended control measures have intensified, resulting in long-term disruptions in social production and stagnation in economic growth (S1). Additionally, with the market's shift in tomato varieties, the price of the single-crop, highly homogeneous "Busan 88" variety plummeted (S5), falling sharply from CNY 14 per jin to a mere CNY 3 (RU4). In terms of environmental factors, in 2023, frequent heavy rainfall occurred (ECO1). Due to the low-lying topography of the specialty agriculture area (ECO4), compounded by the dilapidated state of the greenhouse back walls and drainage systems (RS4) led to flooding inside the greenhouses. As a result, a large number of tomato seedlings perished due to prolonged submersion, and some greenhouses collapsed due to water accumulation

(RU5). At the same time, the long-standing highly intensive and excessive monocropping of greenhouse-based specialty agriculture began to reveal its detrimental effects on land resources and the agricultural ecosystem. Soil nutrient imbalance, severe compaction, salinization, and a marked increase in alkalinity became increasingly evident (ECO4). Furthermore, the continuous monocropping of "Busan 88" over the years had exacerbated pest and disease issues (RU3). These combined factors had resulted in a decline in both the quality and yield of tomatoes (RU5).

② The changes in the external environment and the resulting dynamics of resource units and systems have directly influenced the motivations of key actors. For the government, as the specialty agricultural sector developed and upgraded, the marginal returns from continuous investment and policy support in terms of industry growth and political achievements have diminished [78]. Furthermore, in the previous stage, the three levels of government and their officials had already maximized their political returns. Government departments received accolades from the central government, and many officials were promoted. However, with the increased pressure of political accountability and growing industry risks, local governments' motivations have shifted from seeking political gain to avoiding political fallout. In essence, the drive to continue supporting the tomato industry has markedly weakened. For businesses, driven by the pursuit of high returns and low-risk opportunities, capital investment is particularly sensitive to market demand and price fluctuations [19,24]. The scaling, marketization, and industrialization of operations carry higher costs and greater risks. In 2023, with yields sharply dropping and prices remaining low, agricultural enterprises and growers suffered substantial losses. Farmers who rented greenhouses struggled to meet rental payments, and the laborers saw their earnings sharply reduced. For the farmer cooperative managed by village committees, their focus shifted towards land transfers, village affairs management, and rural construction after enterprises took over operations.

③ This situation has prompted the government, enterprises, cooperative, and farmers to widely adopt exit strategies. While the tomato industry remains a distinctive feature of the region, the local government's attention has shifted towards supporting other specialty agricultural products, such as carrots, in an effort to create new political achievements. Concurrently, industry promotion activities, including media coverage and cultural festivals, have essentially come to a halt. Numerous small and medium-sized agricultural companies, after years of financial losses, have ceased operations and withdrawn from C Town. Large investment companies, still holding land management rights and greenhouse ownership due to unexpired land transfer contracts, have suspended maintenance and operations of facilities due to poor returns. As enterprises gradually withdraw, the cooperative have largely ceased their services and activities. On the individual farmer level, with the exception of a few self-operating greenhouse growers, most farmers who had rented greenhouses have relinquished their leases. Migrant workers, who lost their job opportunities after the enterprises left, have been forced to return to cities to find work.

④ As the government and cooperative effectively withdrew, enterprises followed suit, and most farmers abandoned cultivation, the "Government-Enterprise-Cooperative" governance systems collapsed, losing its rule-setting and governance functions. Due to neglect and lack of maintenance, the actual planting area in greenhouses sharply declined (RU5), and many greenhouses remained idle for extended periods (RS4), no longer generating any economic value (RU4). Since the land transfer contracts had not yet expired, a substantial portion of the greenhouse land cannot be reclaimed or re-cultivated in the short term, resulting in the prolonged abandonment of large tracts of land (RU3). The tomato cultivation area shrank rapidly (RS3), and production dropped dramatically (RS5). These developments clearly reflect the significant decline of the tomato industry in C Town.

⑤ The decline of the tomato industry in C Town has led to a cascade of negative consequences. From a socio-economic perspective, the decline weakened the market competitiveness and regional distinctiveness of agricultural products, potentially leading to a return to traditional, inefficient farming methods. As a key economic pillar for local farmers, the decline in cherry tomato cultivation had directly resulted in a sharp reduction in farmers' incomes, with some facing livelihood difficulties. This shift had also triggered a new wave of rural labor outflow, exacerbating rural depopulation. Furthermore, the industry's decline had led to reduced activities within farmers' cooperative and collective economic organizations, weakening rural social cohesion and governance capacity (O1). At the agricultural ecosystem level, the widespread abandonment of greenhouses and the long-term neglect of large tracts of farmland had not only resulted in the extreme waste of arable land resources but had also intensified the contradiction between peasants and land. The issues caused by the long-term monoculture agricultural, such as soil degradation, water pollution, and biodiversity loss, were also difficult to naturally restore in the short term (O2). In conclusion, the decline of specialty agriculture is a direct consequence and external manifestation of the imbalance in the internal interactions and the misalignment of the human-environment relationship within SES.

4. Discussion

4.1. Theoretical Rationale Underlying the Rise and Fall of Specialty Agriculture

In recent years, the Chinese government has made substantial efforts to promote rural specialty agriculture, viewing it as a critical strategy for achieving targeted poverty alleviation and rural revitalization. Additionally, it is seen as an effective means of agricultural structural adjustment, increasing farmers' incomes, and driving agricultural modernization. However, from the perspective of agricultural practice, some regions have experienced a phenomenon of "short-term prosperity" in the development of specialty agriculture, where industries rapidly shift from rise to fall [78,79]. The cherry tomato industry in C Town stands as a quintessential example of this trend.

The dynamic SESF analysis of the four development stages in C Town vividly illustrates its development trajectory (Figure 6). Driven by the social, economic, and political settings, the participation of farmers, government, and enterprises in the development of C Town's specialty agriculture has shaped its path, transitioning from a livelihood logic to alternating political logic and market logic.

In the early stages, local farmers, seeking to increase their income and resolve poverty and livelihood issues, spontaneously cultivated tomato varieties suitable for the local soil and climate, thereby achieving higher economic returns. To stabilize and organize agricultural production, farmers self-organized into cooperatives, establishing effective self-governance and fostering the endogenous emergence of specialty agriculture. Later, faced with the dual political tasks of poverty alleviation and rural revitalization, the local government, eager to meet performance assessments, create political achievements, and accumulate promotion capital, rushed to rapidly expand the scale and visibility of specialty agriculture. To accelerate this process and streamline governance, the government actively attracted external investments. Enterprises, driven by economic interests and incentivized by government, joined the initiative, following a market-driven approach. These enterprises, oriented towards market demand and profit maximization, implemented large-scale monoculture and highly intensive greenhouse farming, thereby accelerating the commercialization and industrialization of specialty agriculture.

In China, where government intervention is high and social organizations are relatively weak [80], the political logic followed by the government plays a decisive role in shaping the trajectory of specialty agriculture. Under the dominance of political logic, the market-driven logic of enterprises gradually replaced the livelihood logic of farmers. This shift is reflected in the SES, where the involvement of the government and enterprises progressively diminished the role of farmer cooperatives and farmers in the governance system. Their governance function and agency were eroded, eventually marginalizing the farmers and their self-governing organizations. This was the primary driver behind the rapid decline of specialty agriculture in C Town.

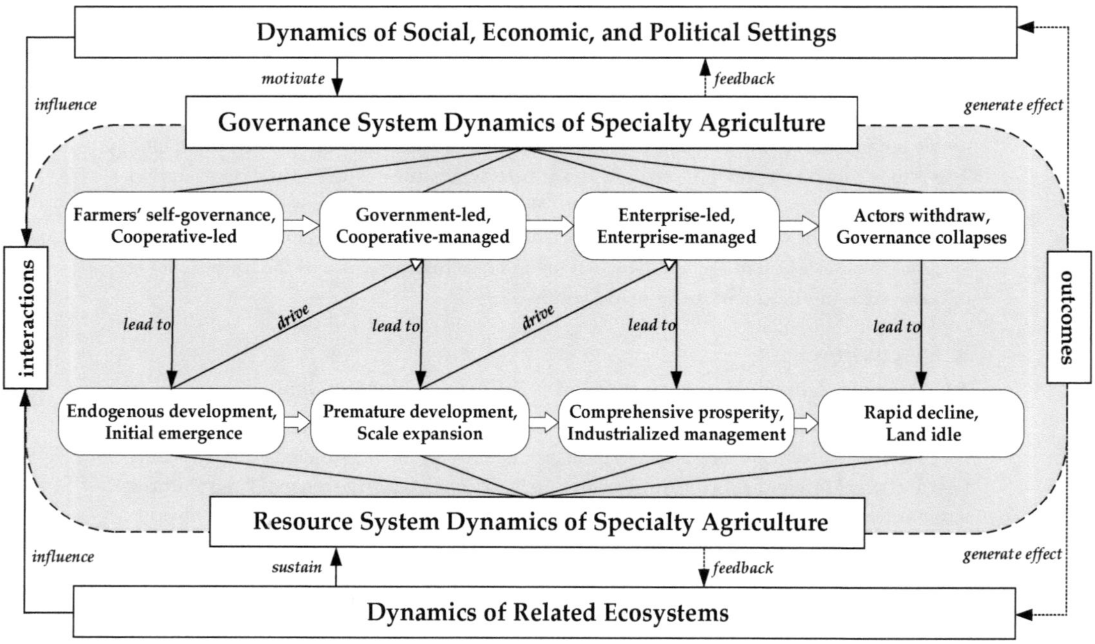

Figure 6. Theoretical rationale underlying the rise and fall of the specialty agriculture in C Town. The dynamic incentives of the social, economic, and political context shaped the distinct action logics of the three main actors. Driven by short-term interests, the government and enterprises led the development of specialty agriculture, achieving rapid scaling, industrialization, and intensive management in the short term. However, this also gradually marginalized farmers and their self-governing organizations, weakening the resilience of the SES and ultimately leading to the rapid decline of specialty agriculture.

As Ostrom's profound insights have pointed out, externally imposed, larger-scale governance systems may destroy the existing governance mechanisms within an SES, thereby hindering the achievement of long-term sustainability [50]. In the case of C Town, under the dominance of political logic, the local government focused on rapid industrial expansion and immediate income increases for farmers, neglecting the protection of land resources, agricultural landscapes, and the agricultural ecosystem. This tendency was reflected not only in the lack of long-term sustainable development planning but also in the absence of specific guidance and effective regulation for farmers and enterprises. Under the influence of market logic, enterprises prioritized profit maximization, which exacerbated resource overconsumption and environmental degradation. More significantly, the marginalization of farmers and their organizations not only led to the gradual erosion of traditional knowledge and local expertise but also greatly reduced the agricultural system's capacity to withstand stress and unexpected shocks [81], making it more vulnerable to external disturbances such as climate change and market fluctuations [82]. The case of C Town deeply illustrates the sustainability dilemma caused by improper external governance interventions and an imbalanced internal governance structure. In other words, the sustainable development of the specialty agriculture SESs must achieve a positive interaction and dynamic balance between the governance systems and the resource systems.

4.2. Governance–Land Interactions and Feedbacks in Transition of Specialty Agriculture SES

Arable land serve as the spatial carrier and core production factor for the development of specialty agriculture, playing an essential role in supporting its growth. At the early stage of the development of specialty agriculture in C Town, the area's abundant arable land and slightly alkaline soil provided the foundation for the cultivation of cherry tomatoes. The farmers, acting voluntarily, integrated fragmented farmlands, built greenhouse facilities, and introduced cherry tomato varieties with higher added value. These activities successfully turned a small portion of the farmland into specialty agricultural land, significantly improving land use efficiency and economic returns. Subsequently, the government took the lead in implementing large-scale land transfer and land consolidation, which not only facilitated the effective integration of arable land but also promoted the scale and intensive management of specialty agriculture. This further enhanced the efficiency, economic returns, and asset value of the farmland, accelerating the transformation of land use patterns and the productive function of the land [83].

However, large-scale land transfer led to a widespread conversion of basic farmland into greenhouse agricultural land for a single economic crop, posing a threat to food security and the continuous provision of ecological services. Land resource management has always required a delicate balance between economic development and farmland protection. Entering the 21st century, the Chinese central government has consistently strengthened farmland protection, with the 2008 policy clearly demarcating 1.8 billion mu bottom-line of basic farmland reserve. However, to meet the needs of modern agriculture, the country has adopted relatively flexible and lenient management of facility agriculture land. For example, the 2019 policy allowed the use of permanent basic farmland for facility agriculture, provided the cultivation layer remains intact [84].

Thanks to the flexibility of national policies and proactive promotion by local governments, C Town was able to use 15,000 mu of farmland to develop greenhouse cherry tomato agriculture. However, the unchecked expansion of greenhouse agricultural land had encroached upon food-growing land, exacerbating the trend of "non-grain conversion of cultivated land". In grassroots practices, the definition of facility agricultural land is ambiguous, regulatory oversight is difficult, and the costs of land reclamation and cultivation are high, making recycling difficult [85]. In the construction of the industrial park and demonstration area in C Town, extensive ground hardening for roads, buildings, and storage areas damaged the farming layers of certain farmlands. After the decline of the industry, large areas of farmland were left fallow for extended periods due to contractual constraints and the high costs of restoration.

Furthermore, the expansion of greenhouse agricultural land has significant ecological negative externalities, including increased greenhouse gas emissions, soil acidification, salinization, nutrient imbalances, severe pollution from fertilizers, pesticides, and plastics, loss of biodiversity, increased crop pests and diseases, and the destruction of agricultural landscapes [86].

In summary, during the development of specialty agriculture, the interactions between the governance systems and farmland resources has led to a transformation in land use and triggered a series of socio–ecological outcomes. These outcomes, through feedback mechanisms, further propel the dynamic evolution of farmland resources and the specialty agriculture system. Unlike small farmers who are deeply connected to the land and possess local knowledge [87,88], the government and enterprises, driven by short-term profit maximization, often neglect the long-term value of farmland, including food security, ecological services, and social security functions. Due to an excessive pursuit of economic profits in the development of specialty agriculture, the government failed to guide the industry toward eco-friendly practices, while farmers and enterprises lacked green and organic farming practices, thereby hindering the sustainable development of specialty agriculture and the sustainable use of land resources.

4.3. Policy Implications for Sustainable Development of Specialty Agriculture

In China's agricultural development blueprint, driven by political motives, the government is committed to rapidly realizing the large-scale, modernized, and commercialized operation of agricultural enterprises [89]. However, this study uncovers a paradox: driven by short-term interests, both the government and enterprises—while advancing the development of specialty agriculture—may marginalize farmers and their self-governed organizations. Furthermore, the excessive monoculture farming and neglect of agricultural ecosystem dynamics could undermine the sustainable of specialty agriculture development and the agricultural land use. From this perspective, the C Town fails to provide a viable path for promoting sustainable agricultural and rural development.

In contrast, successful cases from other countries illustrate a very different approach. Dorsey's early study on the development of specialty coffee cultivation in central Kenya reveals that, the diversification of specialty crops and small-scale production is more conducive to successful commercialization. This not only boosts coffee yields and farmers' income but also optimizes the use of increasingly fragmented land resources. Similarly, Roest and the colleagues, in their comparative study of vegetable and fruit farming development across seven EU member states, argue that the high degree of agricultural specialization and scale diminishes the economic resilience of farms. In contrast, small-scale, diversified farming, along with well-developed farmer networks, plays a crucial role in building a more resilient and sustainable agricultural system. A more comparable example to China's agricultural development context is Japan [90]. In 1961, Oyama Town in Oita Prefecture launched the One Village, One Product (OVOP) movement, encouraging farmers to transition from traditional rice farming to a diverse range of specialty crops, including chestnuts, mushrooms, and herbs. The movement adhered to principles of local wisdom and resource conservation. This initiative achieved remarkable success. Subsequently, the governor of Oita Prefecture promoted the OVOP movement across the entire region, urging each village to specialize in producing one unique product. Unlike C Town, however, the local government's role in OVOP was primarily catalytic—emphasizing community-driven development rather than heavy reliance on government support. It viewed local governments as facilitators, not as direct providers of subsidies for farmers [91]. Some scholars assert that the sustainable development of specialty agriculture in Oita is rooted in three key principles: "local yet global", "self-reliance and creativity", and "human resource development" [92].

By contrasting the failure of C Town with the successes of these international examples, the following policy recommendations can be made for the development of rural specialty agriculture in China:

- Farmers and their self-governing organizations are critical to the sustainable development of specialty agriculture. Local governments should be cautious to protect and nurture the elements of farmers' autonomy, rather than undermine them, when introducing industrial capital and supporting leading enterprises. Especially in regions where development foundations have already been established, targeted policy support and public services should be provided to farmers to stimulate their endogenous development potential.

- The sustainable development of specialty agriculture hinges on the rational and efficient utilization of agricultural land resources. Both the government and enterprises must adopt more prudent and accountable land management strategies, striving to balance the dual goals of agricultural development and farmland resource conservation, thereby achieving positive interactions and feedback within the human–land system [93]. They should collaborate effectively with farmers and cooperatives to jointly explore a sustainable farmland use system, ensuring that the development of specialty agriculture does not come at the expense of the long-term benefits of land resources.
- The pursuit of large-scale, intensive, and specialized approaches is not a viable path for ensuring the sustainable development of specialty agriculture. Instead, a small-scale, diversified, and organic cultivation model, led by farmers, not only strengthens the resilience of the specialty agriculture system but also fosters the preservation of landscapes and the ecological security of the land [94].
- Government intervention in the development of specialty agriculture is driven by performance incentives and political logic, which are deeply rooted in the incentive structures of national macro policies. While the central government has consistently emphasized the "central role of farmers" and the need to "stimulate farmers' intrinsic vitality" in its policies for specialty agriculture, the task assessments of local governments are often focused on scale and efficiency. Economic indicators such as "industry scale, value, and development level [95]", along with park construction, are commonly used to evaluate local government performance [80]. Therefore, the central government should adjust the incentive structure from the top down, ensuring that socioeconomic and ecological benefits such as farmers' participation, sustainable land use, rural social harmony, and ecosystem services are comprehensively considered.

5. Concluding Remarks

As the spatial nexus of social–ecological interactions, land is nonnegligible in understanding the sustainability dynamics of SESs. Unfortunately, the complex roles of land in shaping SES sustainability in various contexts remains underrecognized by some of the traditional research fields in social sciences, which are showing increasing interests in advancing sustainability (e.g., rural sociology, governance studies, and political economy). Our study explicitly takes a social–ecological land system perspective to analyze how the interactions between various actors (e.g., governments, enterprises, and farmers) and their engagement with land modifications have influenced the social–ecological transition of a specialty agriculture SES. By tracing the lifecycle of cherry tomato agriculture in C Town, P City, China—from its gradual rise to its swift decline—we illustrate the dynamic interplay of governance systems, resource systems, and external shocks in driving agricultural land system transformations. Using qualitative data from field surveys and in-depth interviews, we apply an extended SESF to analyze these processes, highlighting lessons for fostering sustainable agricultural practices and resilient land use systems.

Our findings clearly show that land has played critical roles in both the rise and fall of the focal industry—or transitions of SES, more broadly speaking. On one hand, the previous successful development of a specialty agriculture in our case study would be impossible without favorable land endowment and necessary modifications to the sitting land system such as land transfer, land consolidation, and land development. Often the case the difficulty in making such land transformations happen is severely underestimated. Institutional barriers, social frictions, biophysical disadvantages and associated economic costs can rule out at the very beginning the possibility or farmers' right of development. Nevertheless, the flexibility of national policies, coupled with the strong advocacy of local governments, has enabled the transformation of large-scale farmland into specialty agricultural land. On the other hand, transitions originated from the socioeconomic side of SESs, such as the abrupt decline of specialty agriculture in our case due to under the leadership of the government and enterprises, farmers' self-organizations are relegated to the margins within the governance subsystem, would have cascading effects on the land-related ecological subsystem and environmental sustainability implications. Looking ahead, the mission of advancing sustainability through land-related approaches calls for a more explicit social–ecological land system perspective and more inter- and trans-disciplinary research–action practices.

Supplementary Materials: The following supporting information can be downloaded at: https://www.mdpi.com/article/10.3390/land14020254/s1. Figure S1a: SES analysis framework; Figure S1b: The dynamic flows in the SES Framework; Table S1: Second-tier variables in framework for analyzing an SES; Table S2: Interviewee list; Table S3: Second-tier variables analysis in the endogenous emergence stage; Table S4: Second-tier variables analysis in the rapid development stage; Table S5: Second-tier variables analysis in the regional proliferation stage; Table S6: Second-tier variables analysis in the abrupt decline stage.

Author Contributions: Conceptualization, X.W.; methodology, X.W., M.W. and C.Y.; software, M.W. and C.Y.; validation, C.Y. and L.G.; formal analysis, M.W.; investigation, M.W., X.W. and C.Y.; resources, X.W. and L.G.; data curation, M.W. and C.Y.; writing—original draft preparation, X.W. and M.W.; writing—review and editing, X.W., C.Y. and L.G.; visualization, M.W. and C.Y.; supervision, C.Y.; project administration, C.Y.; funding acquisition, X.W. All authors have read and agreed to the published version of the manuscript.

Funding: This work was financially supported by grants from China Postdoctoral Science Foundation (No. 2022M722974o) and Shandong Provincial Natural Science Foundation (No. ZR2023QG090, No. ZR202111170088).

Institutional Review Board Statement: Not applicable.

Informed Consent Statement: Not applicable.

Data Availability Statement: Data will be available upon reasonable request.

Conflicts of Interest: The authors declare no conflicts of interest.

References

1. Cai, Y.; Xia, C. Interpretive Structural Analysis of Interrelationships among the Elements of Characteristic Agriculture Development in Chinese Rural Poverty Alleviation. *Sustainability* **2018**, *10*, 786. [CrossRef]
2. Xu, J.; Wan, J.; Dai, Z. How Does Digital Technology Application Empower Specialty Agricultural Farmers? Evidence from Chinese Litchi Farmers. *Front. Sustain. Food Syst.* **2024**, *8*, 1444192. [CrossRef]
3. Chen, Z.; Sarkar, A.; Hasan, A.; Li, X.; Xia, X.; Luis, J.; Quintas-Soriano, C.; López-Rodríguez, M. Evaluation of Farmers' Ecological Cognition in Responses to Specialty Orchard Fruit Planting Behavior: Evidence in Shaanxi and Ningxi, China. *Agriculture* **2021**, *11*, 1056. [CrossRef]
4. Lv, H. On Specialty Agriculture. *Soc. Sci. Res.* **2002**, *3*, 27–30. (In Chinese)
5. Cao, C. Analysis of the Development Path and Model for Japanese Characteristic Agriculture. *Geogr. Res. Bull.* **2022**, *1*, 85–92. [CrossRef]
6. Gulati, A.; Minot, N.; Delgado, C.; Bora, S. Growth in High-Value Agriculture in Asia and the Emergence of Vertical Links with Farmers. In *Global Supply Chains, Standards and the Poor: How the Globalization of Food Systems and Standards Affects Rural Development and Poverty*; CABI Books: Oxfordshire, UK, 2007; pp. 91–108, ISBN 978-1-84593-185-8.
7. Ozcelik, A.E. Driving Initiatives for Future Improvements of Specialty Agricultural Crops. *Comput. Electron. Agric.* **2016**, *121*, 122–134. [CrossRef]
8. Vuppalapati, C. *Specialty Crops for Climate Change Adaptation: Strategies for Enhanced Food Security by Using Machine Learning and Artificial Intelligence*; STEAM-H: Science, Technology, Engineering, Agriculture, Mathematics & Health; Springer Nature: Cham, Switzerland, 2023; ISBN 978-3-031-38398-4.

9. Balis, L.E.; Shaw, E.; Fung Uy, W.; Nelson, K.; Isack, M.; Flournoy, L.; Vest, D.; Deelo, J.; Yaroch, A.L. Midwestern Specialty Crop Impacts on the Environment and Health: A Scoping Review. *Agric. Food Secur.* **2024**, *13*, 38. [CrossRef]
10. Sibhatu, K.T.; Krishna, V.V.; Qaim, M. Production Diversity and Dietary Diversity in Smallholder Farm Households. *Proc. Natl. Acad. Sci. USA* **2015**, *112*, 10657–10662. [CrossRef]
11. Lisso, L. Examining the Relationships Between Topography and Suitable Agricultural Land for Specialty Crops. Ph.D. Thesis, University of Guelph, Guelph, ON, Canada, 2023.
12. Borlu, Y.; Glenna, L. Environmental Concern in a Capitalist Economy: Climate Change Perception Among U. *S. Specialty-Crop Producers. Organ. Environ.* **2021**, *34*, 198–218.
13. Lee, W.S.; Alchanatis, V.; Yang, C.; Hirafuji, M.; Moshou, D.; Li, C. Sensing Technologies for Precision Specialty Crop Production. *Comput. Electron. Agric.* **2010**, *74*, 2–33. [CrossRef]
14. Aragona, F.B.; Orr, B. Agricultural Intensification, Monocultures, and Economic Failure: The Case of Onion Production in the Tipajara Watershed on the Eastern Slope of the Bolivian Andes. *J. Sustain. Agric.* **2011**, *35*, 467–492. [CrossRef]
15. Rigal, C.; Xu, J.; Hu, G.; Qiu, M.; Vaast, P. Coffee Production during the Transition Period from Monoculture to Agroforestry Systems in near Optimal Growing Conditions, in Yunnan Province. *Agric. Syst.* **2020**, *177*, 102696. [CrossRef]
16. Wan, N.-F.; Dainese, M.; Wang, Y.-Q.; Loreau, M. Cascading Social-Ecological Benefits of Biodiversity for Agriculture. *Curr. Biol.* **2024**, *34*, R587–R603. [CrossRef] [PubMed]
17. Scott, S.; Si, Z.; Schumilas, T.; Chen, A. *Organic Food and Farming in China: Top-Down and Bottom-Up Ecological Initiatives*; Routledge: London, UK, 2018; ISBN 978-0-203-70170-6.
18. Ren, B.; Xiong, K.; Wang, Q. Revitalization Mechanism of Specialty Industries in the Karst Rocky Desertification Areas: From a Perspective of the Actor-Network Theory. *Growth Chang.* **2022**, *53*, 1362–1383. [CrossRef]
19. Qiao, Y.; Martin, F.; He, X.; Zhen, H.; Pan, X. The Changing Role of Local Government in Organic Agriculture Development in Wanzai County, China. *Can. J. Dev. Stud. Can. Détudes Dév.* **2019**, *40*, 64–77. [CrossRef]
20. Qiao, Y.; Martin, F.; Cook, S.; He, X.; Halberg, N.; Scott, S.; Pan, X. Certified Organic Agriculture as an Alternative Livelihood Strategy for Small-Scale Farmers in China: A Case Study in Wanzai County, Jiangxi Province. *Ecol. Econ.* **2018**, *145*, 301–307. [CrossRef]
21. Thiers, P. From Grassroots Movement to State-Coordinated Market Strategy: The Transformation of Organic Agriculture in China. *Environ. Plan. C Gov. Policy* **2002**, *20*, 357–373. [CrossRef]
22. Liu, M.; Feng, X.; Wang, S.; Zhong, Y. Does Poverty-Alleviation-Based Industry Development Improve Farmers' Livelihood Capital? *J. Integr. Agric.* **2021**, *20*, 915–926. [CrossRef]
23. Brown, C.; Waldron, S.; Longworth, J. Specialty Products, Rural Livelihoods and Agricultural Marketing Reforms in China. *China Agric. Econ. Rev.* **2011**, *3*, 224–244. [CrossRef]
24. Scott, S.; Si, Z.; Schumilas, T.; Chen, A. Contradictions in State-and Civil Society-Driven Developments in China's Ecological Agriculture Sector. *Food Policy* **2014**, *45*, 158–166. [CrossRef]
25. Zhang, Q.F.; Donaldson, J.A. The Rise of Agrarian Capitalism with Chinese Characteristics: Agricultural Modernization, Agribusiness and Collective Land Rights. *China J.* **2008**, *60*, 25–47. [CrossRef]
26. Huang, P.C.C. China's New-Age Small Farms and Their Vertical Integration: Agribusiness or Co-Ops? *Mod. China* **2011**, *37*, 107–134. [CrossRef] [PubMed]
27. Song, Y.; Qi, G.; Zhang, Y.; Vernooy, R. Farmer Cooperatives in China: Diverse Pathways to Sustainable Rural Development. *Int. J. Agric. Sustain.* **2014**, *12*, 95–108. [CrossRef]
28. Jia, X.; Hu, Y.; Hendrikse, G. Chapter 9: Centralized versus Individual: Governance of Farmer Professional Cooperatives in China. In *Cooperatives, Economic Democratization and Rural Development*; Edward Elgar: Northampton, MA, USA, 2016; pp. 203–225, ISBN 978-1-78471-938-8.
29. Bijman, J.; Hu, D. The Rise of New Farmer Cooperatives in China; Evidence from Hubei Province. *J. Rural Coop.* **2011**, *39*, 99–113.
30. Li, Y.; Fan, P.; Liu, Y. What Makes Better Village Development in Traditional Agricultural Areas of China? Evidence from Long-Term Observation of Typical Villages. *Habitat Int.* **2019**, *83*, 111–124. [CrossRef]
31. Fischer, E.; Qaim, M. Linking Smallholders to Markets: Determinants and Impacts of Farmer Collective Action in Kenya. *World Dev.* **2012**, *40*, 1255–1268. [CrossRef]
32. Wollni, M.; Zeller, M. Do Farmers Benefit from Participating in Specialty Markets and Cooperatives? The Case of Coffee Marketing in Costa Rica1. *Agric. Econ.* **2007**, *37*, 243–248. [CrossRef]
33. Hisano, S.; Jussaume, R.A., Jr.; Kim, C.-K.; McMichael, P.; Otsuka, S.; Taniguchi, Y.; Lin, Z. *Preliminary Investigations into the Local Impacts of East Asian Agri-Food Restructuring*; Kyoto University: Kyoto, Japan, 2000.
34. Clegg, J. Rural Cooperatives in China: Policy and Practice. *J. Small Bus. Enterp. Dev.* **2006**, *13*, 219–234. [CrossRef]
35. Huang, J.; Wang, X.; Qiu, H. *Small-Scale Farmers in China in the Face of Modernisation and Globalisation*; IIED/HIVOS: London, UK, 2012.

36. Lin, G.C.S.; Ho, S.P.S. The State, Land System, and Land Development Processes in Contemporary China. *Ann. Assoc. Am. Geogr.* **2005**, *95*, 411–436. [CrossRef]
37. Jiang, Y.; Long, H.; Tang, Y.; Deng, W.; Chen, K.; Zheng, Y. The Impact of Land Consolidation on Rural Vitalization at Village Level: A Case Study of a Chinese Village. *J. Rural Stud.* **2021**, *86*, 485–496. [CrossRef]
38. Ye, J. Land Transfer and the Pursuit of Agricultural Modernization in China. *J. Agrar. Change* **2015**, *15*, 314–337. [CrossRef]
39. Zhou, Y.; Guo, Y.; Liu, Y.; Wu, W.; Li, Y. Targeted Poverty Alleviation and Land Policy Innovation: Some Practice and Policy Implications from China. *Land Use Policy* **2018**, *74*, 53–65. [CrossRef]
40. Long, H.; Tu, S.; Ge, D.; Li, T.; Liu, Y. The Allocation and Management of Critical Resources in Rural China under Restructuring: Problems and Prospects. *J. Rural Stud.* **2016**, *47*, 392–412. [CrossRef]
41. Li, X.; Ito, J. An Empirical Study of Land Rental Development in Rural Gansu, China: The Role of Agricultural Cooperatives and Transaction Costs. *Land Use Policy* **2021**, *109*, 105621. [CrossRef]
42. Zhou, Y.; Li, Y.; Xu, C. Land Consolidation and Rural Revitalization in China: Mechanisms and Paths. *Land Use Policy* **2020**, *91*, 104379. [CrossRef]
43. Li, Y.; Wu, W.; Liu, Y. Land Consolidation for Rural Sustainability in China: Practical Reflections and Policy Implications. *Land Use Policy* **2018**, *74*, 137–141. [CrossRef]
44. Long, H.; Zhang, Y.; Tu, S. Rural Vitalization in China: A Perspective of Land Consolidation. *J. Geogr. Sci.* **2019**, *29*, 517–530. [CrossRef]
45. Li, Y.; Wu, H.; Shi, Z. Farmland Productivity and Its Application in Spatial Zoning of Agricultural Production: A Case Study in Zhejiang Province, China. *Environ. Earth Sci.* **2016**, *75*, 159. [CrossRef]
46. Xu, D.; Deng, X.; Guo, S.; Liu, S. Labor Migration and Farmland Abandonment in Rural China: Empirical Results and Policy Implications. *J. Environ. Manag.* **2019**, *232*, 738–750. [CrossRef]
47. Song, B.; Robinson, G.M.; Zhou, Z. Agricultural Transformation and Ecosystem Services: A Case Study from Shaanxi Province, China. *Habitat Int.* **2017**, *69*, 114–125. [CrossRef]
48. Ostrom, E. A Diagnostic Approach for Going beyond Panaceas. *Proc. Natl. Acad. Sci. USA* **2007**, *104*, 15181–15187. [CrossRef] [PubMed]
49. McGinnis, M.D.; Ostrom, E. IAD and SES Dynamic Flows: Introducing the Program in Institutional Analysis of Social-Ecological Systems (PIASES) Framework. In Proceedings of the 13th Economics of Infrastructures Conference, Delft, The Netherlands, 27–28 May 2010. *Workshop Polit. Theory Policy Anal.*
50. Ostrom, E. A General Framework for Analyzing Sustainability of Social-Ecological Systems. *Science* **2009**, *325*, 419–422. [CrossRef] [PubMed]
51. Ostasiewicz, K.; Magnuszewski, P. Agricultural Sustainability in a Simple Social-Ecological System. *Oper. Res. Decis.* **2011**, *21*, 81.
52. Li, W.; Li, Y. Managing Rangeland as a Complex System: How Government Interventions Decouple Social Systems from Ecological Systems. *Ecol. Soc.* **2012**, *17*, 9. [CrossRef]
53. Binder, C.R.; Hinkel, J.; Bots, P.W.; Pahl-Wostl, C. Comparison of Frameworks for Analyzing Social-Ecological Systems. *Ecol. Soc.* **2013**, *18*, art26. [CrossRef]
54. Berkes, F.; Folke, C.; Colding, J. *Linking Social and Ecological Systems: Management Practices and Social Mechanisms for Building Resilience*; Cambridge University Press: Cambridge, UK, 2000.
55. Liu, J.; Dietz, T.; Carpenter, S.R.; Alberti, M.; Folke, C.; Moran, E.; Pell, A.N.; Deadman, P.; Kratz, T.; Lubchenco, J.; et al. Complexity of Coupled Human and Natural Systems. *Science* **2007**, *317*, 1513–1516. [CrossRef]
56. Fischer, J.; Gardner, T.A.; Bennett, E.M.; Balvanera, P.; Biggs, R.; Carpenter, S.; Daw, T.; Folke, C.; Hill, R.; Hughes, T.P.; et al. Advancing Sustainability through Mainstreaming a Social–Ecological Systems Perspective. *Curr. Opin. Environ. Sustain.* **2015**, *14*, 144–149. [CrossRef]
57. Partelow, S. A Review of the Social-Ecological Systems Framework. *Ecol. Soc.* **2018**, *23*, 36. [CrossRef]
58. Colding, J.; Barthel, S. Exploring the Social-Ecological Systems Discourse 20 Years Later. *Ecol. Soc.* **2019**, *24*, 2. [CrossRef]
59. McGinnis, M.D.; Ostrom, E. Social-Ecological System Framework: Initial Changes and Continuing Challenges. *Ecol. Soc.* **2014**, *19*, 30. [CrossRef]
60. Hinkel, J.; Cox, M.E.; Schlüter, M.; Binder, C.R.; Falk, T. A Diagnostic Procedure for Applying the Social-Ecological Systems Framework in Diverse Cases. *Ecol. Soc.* **2015**, *20*, 32. [CrossRef]
61. Anderies, J.M.; Janssen, M.A.; Ostrom, E. A Framework to Analyze the Robustness of Social-Ecological Systems from an Institutional Perspective. *Ecol. Soc.* **2004**, *9*, 18. [CrossRef]
62. Zeng, Z.; Lam, V.W.; Sumaila, U.R.; Cheung, W.W. Climate Change Alters Social-Ecological Trade-Offs in Achieving Ocean Futures' Targets. *Glob. Change Biol.* **2024**, *30*, e17442. [CrossRef] [PubMed]
63. Lyu, D.; Liu, H.; Deng, C.; Wang, X. Promotion of Cooperation in a Structured Population with Environmental Feedbacks. *Chaos Interdiscip. J. Nonlinear Sci.* **2024**, *34*, 123136. [CrossRef]

64. Gou, F.; Liang, W.; Liu, Y.; Fu, B.; Wang, Z.; Zhang, W.; Chen, Z.; Yan, J.; Li, J.; Lv, Y. Large-Scale Ecological Infrastructures Enhance the Productivity of Agro-Socio-Ecological Systems by Reducing Soil Erosion in the Loess Plateau. *J. Environ. Manag.* **2025**, *373*, 123751. [CrossRef]
65. Allen, T.; Prosperi, P.; Cogill, B.; Flichman, G. Agricultural Biodiversity, Social–Ecological Systems and Sustainable Diets. *Proc. Nutr. Soc.* **2014**, *73*, 498–508. [CrossRef]
66. Maleksaeidi, H.; Karami, E. Social-Ecological Resilience and Sustainable Agriculture under Water Scarcity. *Agroecol. Sustain. Food Syst.* **2013**, *37*, 262–290. [CrossRef]
67. Popkova, E.G.; Sergi, B.S.; Bogoviz, A.V.; Semenova, E.I. *Digital Agriculture for Food Security and Sustainable Development of the Agro-Industrial Complex*; Springer Nature: Berlin/Heidelberg, Germany, 2023.
68. Castonguay, A.C.; Burkhard, B.; Müller, F.; Horgan, F.G.; Settele, J. Resilience and Adaptability of Rice Terrace Social-Ecological Systems: A Case Study of a Local Community's Perception in Banaue, Philippines. *Ecol. Soc.* **2016**, *21*, 15. [CrossRef]
69. Wang, S.; Xue, J.; Zhang, Z.; Sun, H.; Li, X.; Chang, J.; Liu, X.; Yao, L. An Adaptive Cycle Resilience Perspective to Understand the Regime Shifts of Social-Ecological System Interactions over the Past Two Millennia in the Tarim River Basin. *Heliyon* **2024**, *10*, e34184. [CrossRef]
70. Kopainsky, B.; Huber, R.; Pedercini, M. Food Provision and Environmental Goals in the Swiss Agri-Food System: System Dynamics and the Social-Ecological Systems Framework. *Syst. Res. Behav. Sci.* **2015**, *32*, 414–432. [CrossRef]
71. Dwyer, J.; Short, C.; Berriet-Solliec, M.; Déprés, C.; Lataste, F.-G.; Hart, K.; Prazan, J. Fostering Resilient Agro-Food Futures through a Social-Ecological Systems Framework: Public–Private Partnerships for Delivering Ecosystem Services in Europe. *Ecosyst. Serv.* **2020**, *45*, 101180. [CrossRef]
72. Zhu, X.; Wang, G. Impact of Agricultural Cooperatives on Farmers' Collective Action: A Study Based on the Socio-Ecological System Framework. *Agriculture* **2024**, *14*, 96. [CrossRef]
73. Guo, X.-X.; Zhao, D.; Zhuang, M.-H.; Wang, C.; Zhang, F.-S. Fertilizer and Pesticide Reduction in Cherry Tomato Production to Achieve Multiple Environmental Benefits in Guangxi, China. *Sci. Total Environ.* **2021**, *793*, 148527. [CrossRef] [PubMed]
74. Kai, T.; Nishimori, S.; Tamaki, M. Effect of Organic and Chemical Fertilizer Application on Growth, Yield, and Quality of Small-Sized Tomatoes. *J. Agric. Chem. Environ.* **2020**, *9*, 121–133. [CrossRef]
75. Wang, G.; Bai, H. Tournament of the Poverty Alleviation: An Analytical Framework for the Local Poverty Governance. *J. Public Manag.* **2018**, *15*, 108–121, 158–159. (In Chinese with English Abstract)
76. Kledal, P.R.; Sulitang, T. The Organization of Organic Vegetable Supply Chains in China—Flexible Property Rights and Different Regimes of Smallholder Inclusion. In Proceedings of the European Association of Agricultural Economists (EAAE), 106th Seminar, Montpellier, France, 25–27 October 2007.
77. Zhu, D. Farmers are Getting Ever More Distant from the Land_Land Transfer and the "Three Rights Separation" System. *Soc. Sci. China* **2020**, *123–144*, 207. (In Chinese with English Abstract)
78. Liu, J.; Lu, Y.; Li, Z. Active Inertia: A Study on the Mechanism of the Restructuring of the Agricultural Industries by the Local Government. *Sociol. Stud.* **2017**, *32*, 140–165, 245. (In Chinese with English Abstract)
79. Liang, C. Operating Mechanisms of the Anti-poverty Policy and the Role of Local Government. *J. Beijing Univ. Technol. Sci. Ed.* **2015**, *15*, 7–15. (In Chinese with English Abstract)
80. Ye, M.; Ma, L.; Luo, X. Expulsion of Small Producers: The Governance Dynamics of Organized Agriculture. *Open Times*, 2012; 6, 130–145.
81. Soriano, B.; Garrido, A.; Bertolozzi-Caredio, D.; Accatino, F.; Antonioli, F.; Krupin, V.; Meuwissen, M.P.; Ollendorf, F.; Rommel, J.; Spiegel, A.; et al. Actors and Their Roles for Improving Resilience of Farming Systems in Europe. *J. Rural Stud.* **2023**, *98*, 134–146. [CrossRef]
82. Adger, W.N. Vulnerability. *Glob. Environ. Chang.* **2006**, *16*, 268–281. [CrossRef]
83. Long, H.; Qu, Y. Land Use Transitions and Land Management: A Mutual Feedback Perspective. *Land Use Policy* **2018**, *74*, 111–120. [CrossRef]
84. Wang, L.; Wang, Z. Research on the "Trinity" Synergistic Supply Path of Facility Agricultural Land under Strict Arable Land Protection. *Rural Econ.* **2024**, *10*, 44–55. (In Chinese with English Abstract)
85. Liu, J.; Liu, H. Study on the Supply and Regulation of Agricultural Land for Facilities in the Development of Industrial Integration. *Agric. Technol.* **2023**, *43*, 132–137. (In Chinese with English Abstract)
86. Chang, J.; Wu, X.; Wang, Y.; Meyerson, L.A.; Gu, B.; Min, Y.; Xue, H.; Peng, C.; Ge, Y. Does Growing Vegetables in Plastic Greenhouses Enhance Regional Ecosystem Services beyond the Food Supply? *Front. Ecol. Environ.* **2013**, *11*, 43–49. [CrossRef]
87. Šūmane, S.; Kunda, I.; Knickel, K.; Strauss, A.; Tisenkopfs, T.; des Ios Rios, I.; Rivera, M.; Chebach, T.; Ashkenazy, A. Local and Farmers' Knowledge Matters! How Integrating Informal and Formal Knowledge Enhances Sustainable and Resilient Agriculture. *J. Rural Stud.* **2018**, *59*, 232–241. [CrossRef]
88. Liu, G.; Yang, L.; Guo, S.; Deng, X.; Song, J.; Xu, D. Land Attachment, Intergenerational Differences and Land Transfer: Evidence from Sichuan Province, China. *Land* **2022**, *11*, 695. [CrossRef]

89. Luo, Q.; Andreas, J.; Li, Y. Grapes of Wrath: Twisting Arms to Get Villagers to Cooperate with Agribusiness in China. *China J.* **2017**, *77*, 27–50. [CrossRef]
90. De Roest, K.; Ferrari, P.; Knickel, K. Specialisation and Economies of Scale or Diversification and Economies of Scope? Assessing Different Agricultural Development Pathways. *J. Rural Stud.* **2018**, *59*, 222–231. [CrossRef]
91. Natsuda, K.; Igusa, K.; Wiboonpongse, A.; Thoburn, J. One Village One Product–Rural Development Strategy in Asia: The Case of OTOP in Thailand. *Can. J. Dev. Stud. Can. Détudes Dév.* **2012**, *33*, 369–385. [CrossRef]
92. Mukai, K.; Fujikura, R. One Village One Product: Evaluations and Lessons Learnt from OVOP Aid Projects. *Dev. Pract.* **2015**, *25*, 389–400. [CrossRef]
93. Hu, S.; Wang, Y.; Cao, Z. Progress and Prospects of Rural Human-Earth System Coupling Research. *Prog. Geogr.* **2023**, *42*, 2439–2452. (In Chinese with English Abstract) [CrossRef]
94. Zhang, X.; Wang, Y.; Yuan, X.; Yang, Y. Regional Land Ecological Security Evaluation and Ecological Poverty Alleviation Practice: A Case Study of Yangxian County in Shaanxi Province, China. *J. Geogr. Sci.* **2022**, *32*, 682–700. [CrossRef]
95. Feng, M. Local Government and the Development of Local Industries: A casestudy based on Sidong county. *Sociol. Stud.* **2014**, *29*, 145–169, 244–245. (In Chinese with English Abstract)

land

MDPI

Review

Smart Growth and Smart Shrinkage: A Comparative Review for Advancing Urban Sustainability

Yang Yang [1,2], Zhe Dong [1], Bing-Bing Zhou [1,*] and Yang Liu [1]

1 School of International Affairs and Public Administration, Ocean University of China, Qingdao 266100, China; yangyang@ouc.edu.cn (Y.Y.); zhe1997@stu.ouc.edu.cn (Z.D.); lyiu0917@stu.ouc.edu.cn (Y.L.)
2 Institute of Marine Development, Ocean University of China, Qingdao 266100, China
* Correspondence: bbz@ouc.edu.cn

Abstract: In the context of ongoing global urbanization, the disparity in urban development, marked by the dual phenomena of urban sprawl and urban shrinkage at the regional level, has become increasingly evident. In this vein, two land-related governance strategies—smart growth (SG) and smart shrinkage (SS)—emerge as potential remedies to these challenges, targeting urban expansion and shrinkage, respectively. This study bridges the gap in the fragmented discourse surrounding SG and SS by conducting a comprehensive comparative review on the respective literatures. Utilizing the Scopus database, our research employs trend analysis, text and topic mining, time node analysis, and regional analysis, augmented by qualitative reviews of seminal papers. The findings reveal a notable shift in research focus, with interest in SS surging around 2010 (the number of SS-related papers published after 2010 accounts for 92.3% of the total number of the entire study period) as attention to SG waned, suggesting an impending paradigm shift in urban sustainability. The analysis indicates that SS research lacks the disciplinary diversity, thematic breadth, and empirical depth of SG studies, underscoring a need for a more robust theoretical foundation to support urban sustainability. Furthermore, while both SG and SS derive from environmental science foundations, SG predominantly addresses the physical and landscape attributes of urban areas, whereas SS focuses more on socio-economic dimensions. Our findings point to an intrinsic link between SG and SS, which could lay the groundwork for their integration into a unified theoretical framework to better advance urban sustainability.

Keywords: smart growth; smart shrinkage; urban sustainability; spatial governance; comparative review

1. Introduction

As the global landscape continues to urbanize, the question of how to navigate this transformation in a manner that advances sustainability has become paramount. This urgency is rooted in the recognition that urbanization is not just a demographic shift but a complex, multifaceted phenomenon with profound implications for sustainability [1–4]. According to the United Nations' forecasts in 2019, 56.15% of the world's population resides in urban areas, a figure that is expected to further rise to 67% by the middle of this century and reach 85% by the end of this century, signifying an unprecedented concentration of humanity in urban spaces [5,6].

However, this long-term global trend of urbanization is accompanied by simultaneous urban shrinkage in some regions during some periods [7]. On the one hand, urban growth dominates most parts of the world, leading to urban sprawl in some areas that is characterized by dispersed, low-density expansion of urban land use [8,9]. Urban sprawl leads to a suite of problems such as farmland depletion, environmental degradation, and the financial burden of infrastructure investments. On the other hand, amid long-term global urbanization, shrinking cities emerged in Germany after the collapse of the Berlin Wall and have occurred in various parts of the world [10], leading to intertwined issues such as spatial decay, economic downturn, and rising crime rates [11]. Generally speaking, urban growth and urban shrinkage are now present as two parallel research discourses in the urbanization-related literature, each targeting distinct challenges and opportunities [12,13].

In response to urban sprawl and shrinkage, two respective paradigms emerged in the academic and professional communities, often known as smart growth (SG) and smart shrinkage (SS). On the one hand, SG can be interpreted as the third wave of land planning rooted in roughly the late 1990s, following growth controls and subsequent comprehensive land planning [14]. The targeting issues of SG are more than mitigating environmental degradation by containing urban sprawl and coordinating infrastructure provision by comprehensive land planning, and also include urban revitalization and placemaking, thus demanding the coordinated planning of multiple land uses and associated spatial governance policies. In this vein, the "smartness" of SG reflects the demarcation of prioritized urban development areas with a package of advantageous policies like tax incentives, grants, and infrastructure investments. At the core of SG are compact urban development and mixed use of urban land, with comprehensive considerations of related issues crucial to urban vitality, such as housing, transportation, land use, infrastructure, urban design, and environmental sanitation [14–17]. In contrast, SS is a relatively new term and one of quite a few similar terms promoting the philosophy of treating cities from no longer a growth-oriented perspective but focusing instead on reduced urban population and associated problems [18]. Popper and Popper first coined "smart decline" to advocate planning for fewer people, fewer buildings, and less land development [19]. Later, the establishment of an EU research network on urban shrinkage, SHRINK SMART, helped popularize the term of SS, with "smart" strategies like land adjustment, land banking, relocation assistance, and urban renewal, although similar terms like "smart decline" and "right-sizing" are still in use [18]. Mirroring the largely parallel literatures on urban growth and shrinkage, the "smartnesses" of SG and SS remain largely unintegrated [13], falling short of their potential in better advancing urban sustainability.

Increasingly, more and more studies have started to recognize the coupled landscape of urban growth and shrinkage at regional and global scales [20,21]. Especially at national and subnational scales, interregional linkages such as economic competition and population migration among various social-environmental systems make up a metacoupled polycentric system of development and residence. From a core–periphery systems perspective [22–24], the growth of a core city might well be linked to the shrinkage of its relatively peripheral areas. In this sense, the full spectrum of urban development challenges should be widely recognized, the metacoupled and evolving nature of urban growth and shrinkage deserves explicit acknowledgement, and further, the "smart" wisdom of SG and SS merits better integration for advancing global and regional urban sustainability, for which timely research efforts are needed.

To better unleash the potential of SG and SS in advancing urban sustainability, our research seeks to address the following three questions: (1) What are the bibliometric statistical characteristics of SG- and SS-related literature? (2) What are the differences between and similarities of SG and SS? (3) How can SG and SS to integrated to better support sustainable urban development? To address these three questions, following the research design and methodological approach in Zhou et al. [25], we present a multidimensional comparative analysis of SG and SS in terms of research trends, thematic evolution, disciplinary structure, global distribution of research areas, and knowledge base. The remaining of the paper is structured as follows: Section 2 outlines the methodology, including data sources and analytical techniques. Section 3 presents comparative findings of SG and SS. Section 4 highlights the caveats for interpreting results and generalizing findings, clarifies the two research domains' conceptual linkages and differences, and proposes directions for integrating SG and SS into a cohesive theoretical framework that can maximize the potential of their "smartnesses" in advancing urban sustainability. Section 5 concludes the paper with key messages.

2. Materials and Methods

Our entire research procedure was mainly divided into three stages (Figure 1). Firstly, we selected a suitable literature database based on the research needs, established literature screening criteria, and conducted literature retrieval and collection. Secondly, we conducted a systematic comparative analysis of SG and SS based on the bibliographic data of the collected papers, including publication and citation trend analyses, topical and thematic analyses, disciplinary structure, and global coverage. Finally, we interpreted the content of the top 10 most-cited publications in SG and SS respectively, in order to cross-validate the conclusions drawn from the data analysis.

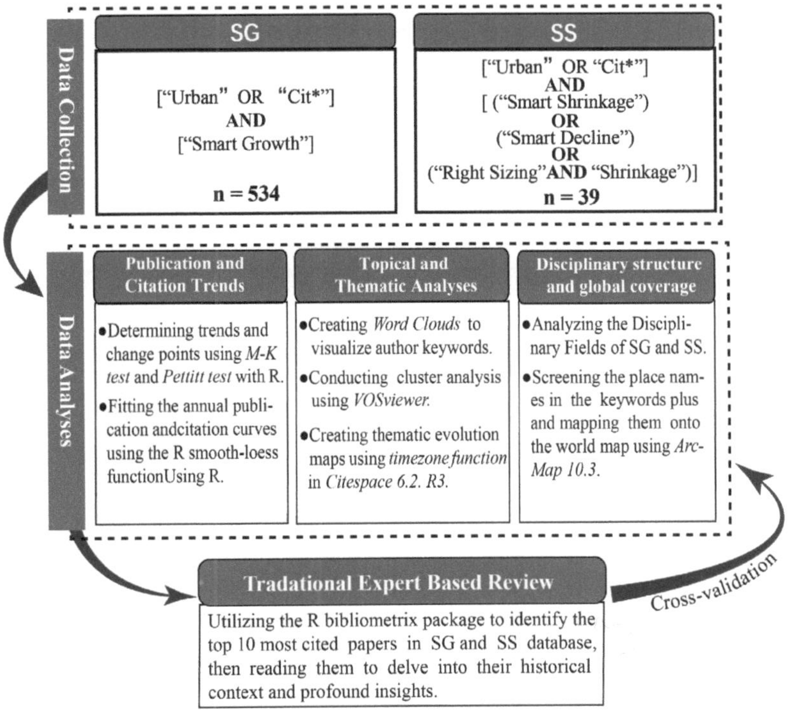

Figure 1. Flowchart of data collection and analyses. The symbol * serves as a wildcard, representing any sequence of characters or suffixes. For instance, "cit*" includes words such as "cities" or "city".

2.1. Data Collection

Referring to the literature of Zhou, Wu, and Anderies [25], we sourced the data from the Scopus database, as of 24 March 2023. Recognized as the most extensive abstract and citation database globally, Scopus boasts broader literature coverage, disciplinary range, and global journal representation across both Northern and Southern Hemispheres, outperforming the Web of Science in these aspects [26]. In addition, Scopus established the International Center for the Study of Research in 2019, composed of bibliometrics experts [27]. This ensures Scopus's high usability in the field of bibliometrics, and as a result, it is extensively used in bibliometric studies across various disciplines [28,29].

Regarding the literature search, direct usage of the term "Smart Growth" as the sole keyword resulted in retrieving a significant number of irrelevant publications outside the urban studies field. To mitigate this, the SG search was narrowed to urban research specifically by employing the terms "Cit*" or "Urban" in conjunction with "Smart Growth," forming the query ["Cit*" OR "Urban"] AND ["Smart Growth"]. Additionally, considering the diverse nomenclature for SS, including terms like Smart Shrinkage, Smart Decline, and

Right Sizing, a comprehensive search query ["Urban" OR "Cit*"] AND [("Smart Shrink*") OR ("Smart Decline") OR ("Right Sizing" AND "Shrink*")] was formulated to encompass the varied terminology associated with SS. The execution of these search strategies was further refined by restricting the publication date to include only articles published up to 2022, and specifying the document types to "Article" and "Review". Considering that English is the most widely used language in the world and also the universal language for international communication in academic fields, we limited the language of search results to English. In the end, a collection of 534 papers on SG and 39 papers on SS was sampled. The bibliographic details of these papers were extracted and served as the foundation for the subsequent bibliometric analysis (Table 1).

Table 1. A comparative overview of sampled papers on SG and SS.

	SG Subset	SS Subset	Ratio of SS to SG
Number of articles	534	39	7.3%
Timespan	1997–2022	2002–2022	80.77%
Number of authors	1078	80	7.42%
Number of sources	224	25	11.16%
Annual growth rate	10.8%	7.18%	66.48%
Number of Keywords Plus	1685	133	7.89%
Number of author's keywords	1127	105	9.3%

2.2. Publication and Citation Trend Detection

To explore the evolution and identify potential inflection points within the research domains of SG and SS, this study employed a time series analysis on yearly publication and citation data. The analysis utilized the Mann–Kendall test (a non-parametric test used to identify trends in time series data) and Pettitt's test (a non-parametric test used to detect change points in time series data), as implemented in the R "trend" package, to assess trends and changes in publication and citation frequencies over time. For visual representation of these trends, the R "ggplot2" package was utilized, specifically leveraging the "loess" method within the "geom_smooth" function (a function for performing smoothing operations in R, used to reduce noise or abrupt changes, thereby revealing underlying patterns or trends within the data) to model and depict the annual variations in publication and citation counts [30]. Furthermore, to examine the relationship between annual publication and citation counts for SG and SS, the correlation function from the R "see" package was applied. The results of this analysis were then illustrated through a correlation heatmap, providing a visual representation of the interconnectivity between publication output and citation impact in these fields.

2.3. Text Mining and Thematic Mapping

The initial step involved recording the quantity of publications within each disciplinary field related to SG and SS as categorized by Scopus. This was followed by employing Excel and Adobe Illustrator to graphically represent the disciplinary spread of these research areas. Subsequent bibliometric analysis was performed using the R Bibliometrix package, which facilitated the identification and consolidation of key terms, the elimination of irrelevant data, and the creation of a keyword-based word cloud via www.wordclouds.com (accessed on 15 April 2023) [25,31,32]. Additionally, VOSviewer was utilized to construct and visualize keyword co-occurrence networks for both SG and SS, allowing for the deduction of primary research themes from the interlinking patterns of keywords [33]. The study also leveraged the timezone function in Citespace 6.2. R3 to generate a timeline graph, marking the chronological emergence of significant terms and thus tracing the historical development of topics within SG and SS research [34]. Moreover, a detailed examination of keywords led to the identification and categorization of geographical names, which were then mapped globally using ArcMap 10.3 to display the spatial concentration and thematic focus of the research locales across various regional levels, offering insights into the geographic dispersion and focal points of studies within the SG and SS frameworks.

2.4. Content Analysis

In addition to the above-noted quantitative assessments, we utilized the R Bibliometrix package to pinpoint the top 10 most-cited articles within both the SG and SS literature groups, resulting in a total of 20 key papers. This approach identified papers that were frequently cited within their respective groups, indicating their foundational importance to the field. These pivotal articles were then meticulously reviewed to ensure the bibliometric findings' accuracy and to gain more profound insights into the research directions and developmental trajectory of SG and SS. Such a quantitative–qualitative combined thorough examination [25] enabled us to deepen our understanding of the core themes and progress within these research domains.

3. Results

3.1. Publication and Citation Trends

SG-related research emerged in 1997, and then increased rapidly in terms of publication number over a span of 26 years until 2022 (Figure 2a), with its annual growth rate averaging 10.8% (Table 1). Notably, the growth rate started to slow down around 2006 and it peaked in 2013 (30 papers), followed by a gradual decline. A change point in SG publications occurred around 2006, which is statistically significant. Overall, SG research activity showed a trend of rapid outbreak, stable growth, and slow decline. In terms of the trend in annual citation counts, SG papers published around 2005 received the highest number of citations (Figure 2b).

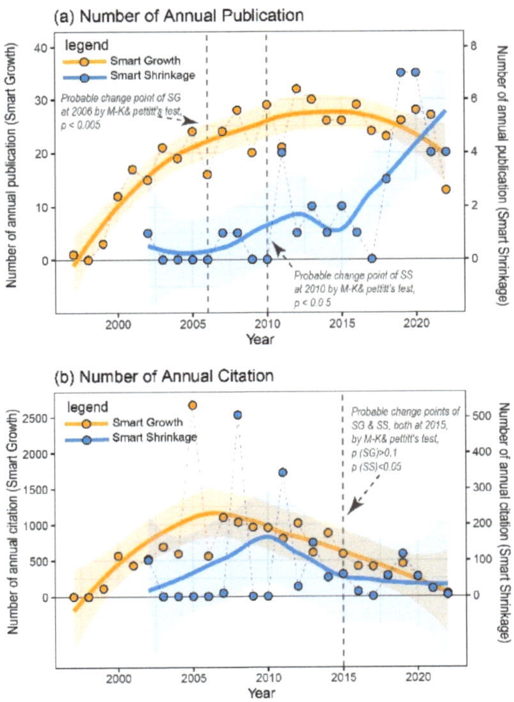

Figure 2. Correlation and trends in annual publications and citations within SG and SS research. (**a**) Yearly publication trends in SG and SS. (**b**) Yearly citation trends per paper in SG and SS datasets. The shaded area illustrates the 95% confidence interval of the smoothing function.

SS-related research began in 2002, and over a span of 21 years until 2022, its publication count grew annually by 7.3% (Table 1). From 2002 to 2010, SS research went through a period of tepid and prolonged incubation, with 6 out of 9 years having no publications (Figure 2a). Notably, a statistically significant change point in SS publication trend occurred in 2010. There was a minor fluctuation in its annual publication counts during 2010-2015, followed which the past few years saw dramatic growth of SS research output. In terms of the trend in annual citation counts, SS papers published around 2010 received the highest number of citations (Figure 2b). From a comparative perspective, SS papers appeared 5 years later than SG, and SG had over 10 times the quantity of SS in terms of the number of source journals, authors, Keywords Plus, and author's keywords. It is safe to say that SS research intensity is much lower than that of SG. However, a trend seems to be emerging: SG research intensity has started to decline in recent years, while SS research intensity keeps rapidly increasing. Additionally, SG and SS studies present similar research impacts as indicated by annual citation counts: they both show an initial increase followed by a decrease, although with SG reaching its peak earlier than SS.

3.2. Multi-Facted Research Topics and Themes

3.2.1. Prominent Author Keywords

Prominent author keywords reflect scholars' conceptual bases of their SG and SS studies (Figure 3). The two research domains have 33 shared author keywords, mainly focusing on urban planning, landscape coverage, and climate environment. Such shared author keywords account for about one-third of the total author keywords for SS and one-thirty-fourth of those for SG. Among them, "Right-sizing", "Urban Shrinkage", "Vacant Land", and "China" appear more than twice in the SS database, while "Urban Planning", "Sustainable Development", "Climate Change", "GIS", "Regional Planning", "Quality of Life", and "Regionalism" appear more than five times in SG papers.

Figure 3. Visual representation of author keywords in SG and SS publications. (**a**) Frequently occurring SG keywords. (**b**) All SS keywords. Font size indicates keyword frequency, with comparative weights set for SG and SS for balanced analysis.

The difference lies in the fact that SG's frequently used keywords are much more diverse compared to those in SS. Compared to 1119 effective author keywords in SG (an average of 2.10 author keywords per paper), SS only has 100 effective author keywords (an average of 2.56 effective author keywords per paper). A total of 97% of the author keywords in SG are not present in SS, and besides topics related to urban growth itself, frequently used keywords with high frequencies include "Sustainability" (21 times), "Built Environment" (17 times), "New Urbanism" (15 times), "Transit-oriented Development" (15 times), and "Urban Form" (15 times). In general, the unique frequently used keywords in SG compared to SS are related to classical theories on one hand and lean towards practical urban planning on the other. On the contrary, the frequently used keywords in SS focus on the phenomenon of urban shrinkage and the concept of SS itself (Figure 3), indicating that SS is still in the stage of conceptual construction and theoretical research.

Notably, "Sustainable Development" appears 16 times in SG, while it only appears once in SS; "Sustainability" appears 21 times in SG but is not present in the author keywords of SS. Although the concepts in SS align with sustainability science, SS focuses more on specific issues in the urban shrinkage process, while the forward-looking and goal-oriented perspective emphasized by sustainability science [35] receives little attention in SS.

3.2.2. Main Research Themes

To reveal the thematic structure of sampled SG and SS papers, the keyword co-occurrence networks of Keywords Plus of the two research domains were visualized [25].

For SG, its 49 Keywords Plus that occur 15 or more times fall into four main themes (Figure 4a): the red cluster emphasizes land use, urban areas, environment, humanities, and the United States (n = 14), mainly reflecting urban landscapes; the green cluster emphasizes planning, mostly related to management measures and policies for urban growth (n = 13), reflecting management strategies; the blue cluster emphasizes SG and keywords related to urban development concepts (n = 11), mainly reflecting theories and concepts; and the yellow cluster emphasizes China and issues related to urban expansion, urban growth, GIS, and spatial analysis methods (n = 11), mainly reflecting empirical research. The morphology of the co-occurrence network, featured by the cohesiveness within each theme, indicates the SG research is relatively mature with four recognizable subfields.

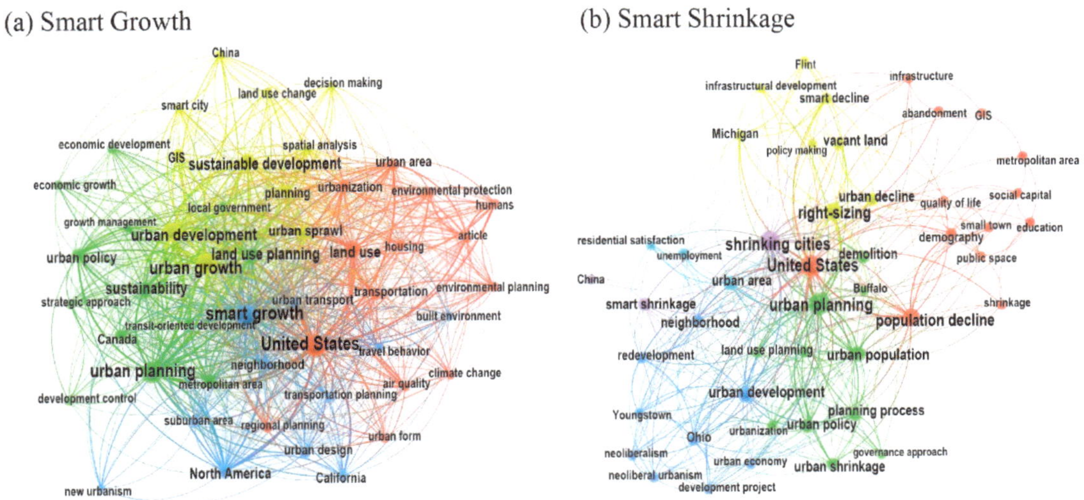

Figure 4. Network analysis of keyword co-occurrence in SG and SS. (**a**) SG with keyword frequencies $\geq$ 15. (**b**) SS with keyword frequencies $\geq$ 2. Frequency thresholds vary to align the networks' keyword density across SG and SS.

For SS, its 47 Keywords Plus that occur two or more times group into six themes (Figure 4b). There are no clear boundaries between these six clusters, and the keywords within each cluster are loosely connected; thus, it is difficult to synthesize their shared theme. This morphology of the SS Keyword Plus co-occurrence network indicates that research related to SS is still in its formative stages.

3.2.3. Thematic Evolution

To further reveal the temporal evolution of research themes and the emergence of new topics [34] in the two research domains, thematic evolution maps were produced to display the years in which author keywords first appeared in the SG and SS paper datasets. On the SG side (Figure 5), its number and frequency of new keywords have decreased over time. Even when the threshold for mapping frequent keywords was reduced from five times to two times, the decreasing trend of new keyword emergence remained consistent. In particular, no keywords have occurred five or more times since 2018, with "smart city" being the last keyword occurring more than five times in 2017. Notably, the chronological appearance of SG keywords clearly shows that SG originated from "growth management" in relation to "urban growth" and "land use planning" as well as "urban planning", and that the research discourse has evolved from SG, to "smart city", and increasingly toward "sustainable city"—although "sustainable development" and "sustainability" had become key terms in the SG literature dating way back to the early 2000s.

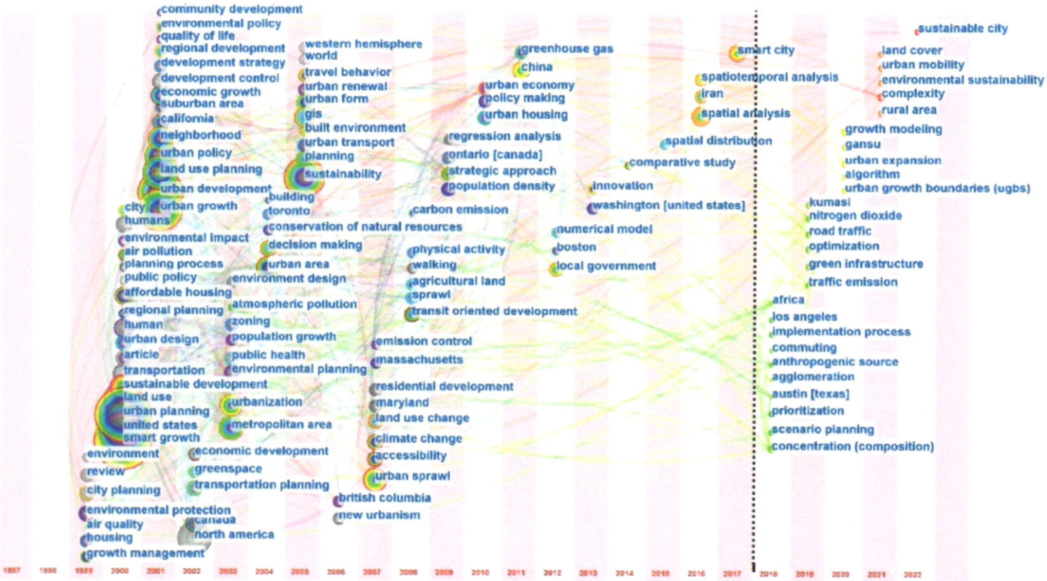

Figure 5. Chronological appearance of keywords in SG research. Created using *Citespace* 6.3.R2, showcasing years of first occurrence and frequency of keyword appearances. The gap in 1997 is noted due to the absence of keywords from Tregoning (1997) [15].

SS is opposite from SG in terms of its overall increasing trend of new keywords' number and frequency (Figure 6). Clearly, SS originated from "land use planning" and "urban planning" as early as 2002, although scholarly interest in researching SG did not regain momentum until 2008 when concerns on "shrinking city" and "right sizing" became elevated. Later in 2011, "smart decline", a term similar to SG, gained popularity, while SS per se did not become a popular term until 2018. Notably, the SS studies before 2018 featured popular terms such as "urban development", "development project", "redevelopment", "residential satisfaction", and "quality of life", indicating more of a growth-oriented, shrinkage-mitigation thinking, whereas SS studies after 2018 present growing concerns on "rightsizing", "small town", and "degrowth coalition", suggesting a shrinkage-adaptation thinking. Another interesting feature of SS studies after 2018 is their attention to "conceptual framework", "governance approach", various methodological issues, and diverse study areas. A more subtle yet important observation is that "sustainable development" and "sustainability" have never been popular terms in the SS literature.

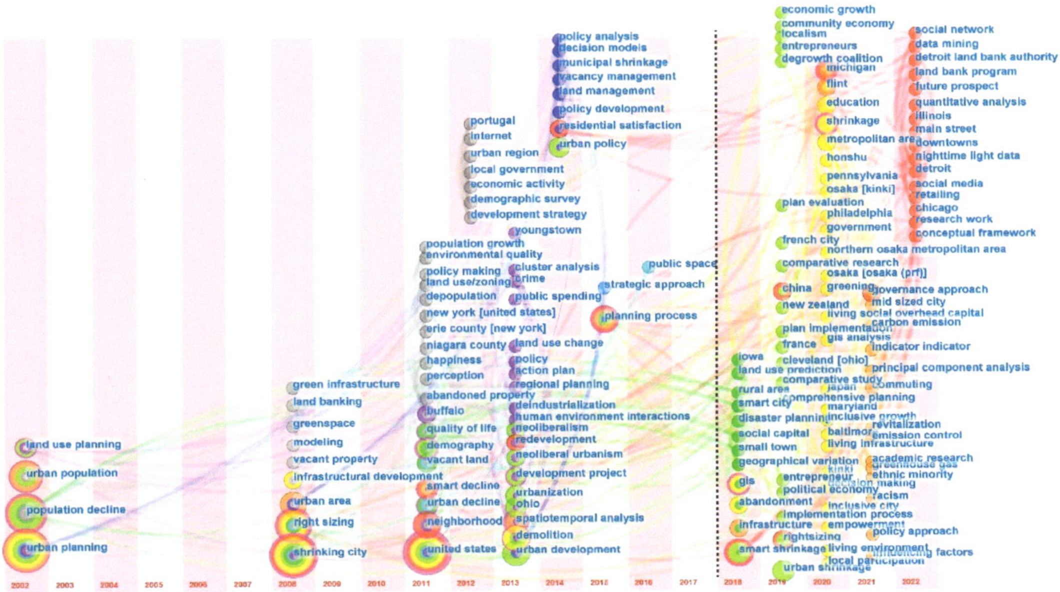

Figure 6. Evolution of SS research keywords over time. Displays keyword frequencies ≥2 up to 2017 and ≥1 from 2018 onwards. Other methodological settings are consistent with those employed in Figure 5.

3.3. Disciplinary Structure and Global Coverage

3.3.1. Contributing Disciplinary Fields

Based on the classification of disciplinary fields in Scopus, the sampled SG papers fall into 18 fields, encompassing the 10 fields that those SS papers belong to (Figure 7). In this sense, SG involves more diverse disciplinary perspectives than SS. Furthermore, in both research domains, the category of social sciences holds an absolute majority (78.09%), followed by environmental science (37.67%). Similarly, 38 out of the 39 SS papers belong to social sciences (97.44%), followed also by environmental science (30.77%). Such similarity reflects the dominance of environmental social science in researching SG and SS. Additionally, the proportion of engineering papers in SG (12.92%) is significantly higher than in SS (5.13%), indicating the slightly greater involvement of natural sciences in researching SG than SS, as exemplified by engineering.

3.3.2. Global Coverage of Study Areas

The study areas covered by SG are significantly broader than those covered by SS, at national, provincial/state, and city scales (Figure 8). SG includes a total of 182 place name keywords, accounting for approximately 10.8% of the total SG Keywords Plus, encompassing 37 countries, 46 provincial/state-level administrative regions, 99 cities, and other locations. SS includes a total of 24 place name keywords, accounting for 18% of the total SS Keywords Plus, including 8 countries, 5 provincial/state-level administrative units, and 9 cities and other locations.

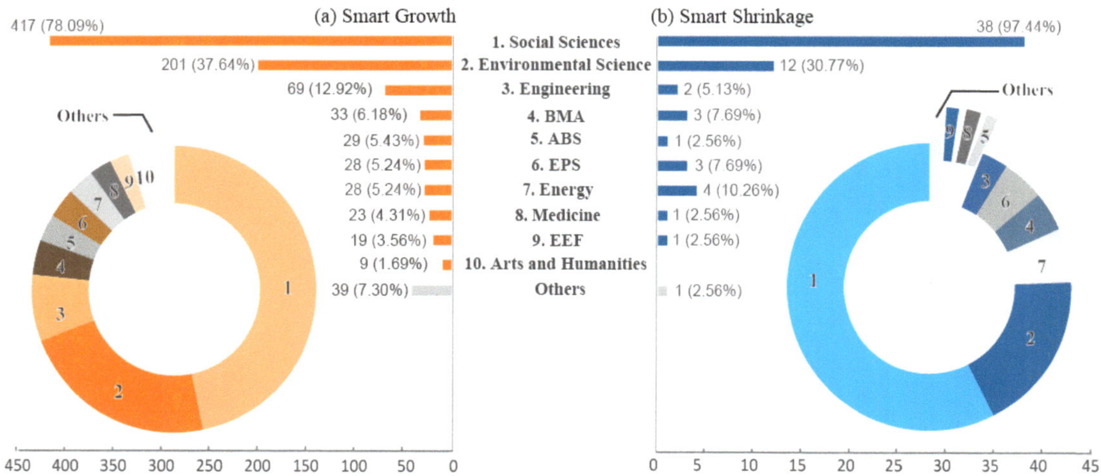

Figure 7. Comparative analysis of disciplinary fields in SG and SS research. (**a**) Distribution in SG. (**b**) Distribution in SS. Bars indicate paper counts with percentages showing field proportions within each dataset. Pie charts reflect field distribution against the total. Notes: (1) Papers may span multiple disciplinary fields; 67.23% of SG and 69.23% of SS papers fall into this category. (2) SG's x-axis scale is tenfold that of SS. (3) Disciplinary abbreviations, such as BMA, ABS, EPS, and EEF are used for clarity, denoting "Business, Management and Accounting", "Agricultural and Biological Sciences", "Earth and Planetary Sciences", and "Economics, Econometrics and Finance", respectively. The "Others" in this graph may be classified under less-fitting disciplines by the Scopus algorithm due to some words in the abstracts. However, their number is small, and their main content does not deviate from the research themes of SG and SS.

At the national scale, the study areas of high interest in SG are the United States (mentioned 195 times), followed by Canada, China, various European countries, and Iran. They also involve numerous economically underdeveloped countries in Asia and Africa, illuminating all continents except Antarctica. Similarly, the most mentioned country in SS is the United States (mentioned 14 times), with other countries like Mexico, Japan, South Korea, France, and Portugal receiving much less attention. The countries involved in SS studies are relatively economically developed, with a knowledge gap remaining for economically underdeveloped regions in Africa and South America (the single point in South America is French Guiana, shown on the map as part of France as a whole) and vast areas of Asia beyond East Asia (white areas in Figure 8).

At the provincial/state scale, the distribution of provincial/state-level SG study areas is quite extensive, with over half of the states in the United States involved. The states/provinces that receive noticeable SG research attention are in China, Australia, Iran, and India. On the other hand, at the provincial/state scale, the study areas of SS are mainly in the Great Lakes region of the United States, including Ohio (three times), Michigan (three times), Pennsylvania (one time), Illinois (one time), Pennsylvania (one time), and Maryland (one time).

Similarly, at the city scale, the contrast in study area distribution is remarkable between SG and SS. In SG, the most attention is given to cities in North America, but areas with concentrated distributions of major cities in East Asia, Western Europe, and elsewhere are also covered. Highly studied cities in SG include Toronto (10 times), Washington (9 times), Seattle and Los Angeles (6 times each), and Austin, Boston, and Seoul (all 5 times). On the other hand, the focal cities of SS, apart from the Japanese city of Ōban [36,37], are in the Rust Belt region near the Great Lakes in the United States [38].

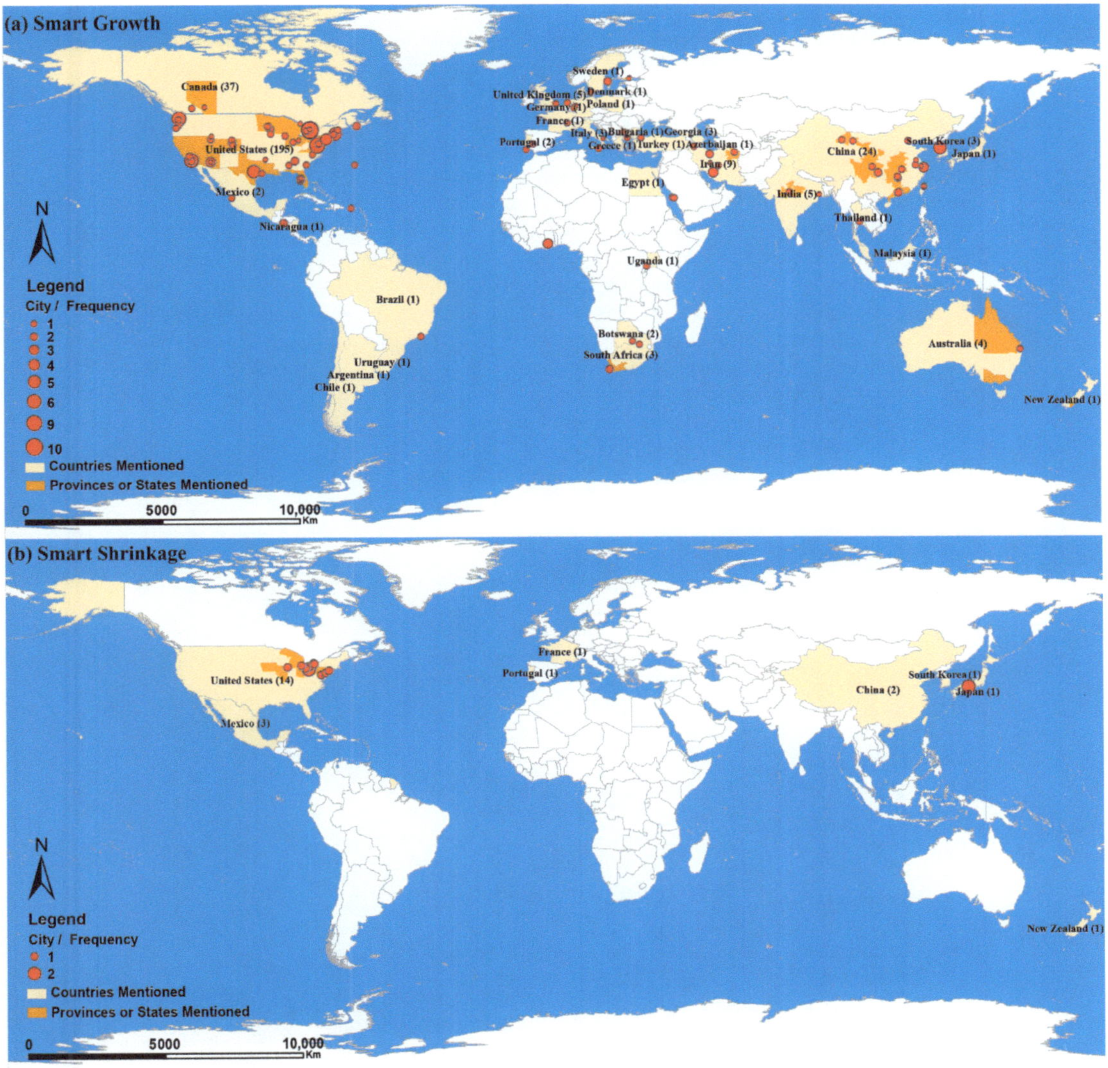

Figure 8. Geographical mapping of SG and SS research focus. Maps illustrate research distribution at national, provincial/state, and city levels, with data derived from Keywords Plus. Base map sourced from Natural Earth (does not represent the author's political views though), with missing place names geocoded via Google Maps for *ArcMap* 10.3 visualization.

3.4. Top Ten Most-Cited Publications of Two Research Domains

Complementary to analysis of contributing disciplinary fields, the most-cited references of a research domain indicate the knowledge base that has nurtured the referring papers. The top 10 most-cited papers in SG span from 1999 to 2007 (Table 2). The first four papers and the seventh paper discuss SG's conceptual and theoretical bases [39–43]. Around the time when the concept of SG emerged (around 1997), the American Planning Association, the Environmental Protection Agency, the Natural Resources Conservation Council, and the Maryland state government introduced policies and regulations related to SG management. At the same time, the Ground Transportation Project Department introduced the

"Smart Growth Toolkit." This top-down promotion quickly sparked widespread academic discussion and led to the formation of an academic community of smart growth—the Smart Growth Network. SG rapidly aroused research interest in the academic community during its emergence and its theoretical foundation was quickly developed in a series of planning practices. The ten SG principles promulgated by the Smart Growth Network have been widely recognized. The themes covered include spatial compactness, housing, transportation, community form, and open space, providing clear and detailed guidance for subsequent academic research and planning practice.

The other five papers are empirical studies. Among them, Song (2005) assessed the extent to which the land development patterns of five American counties conform to SG based on widely accepted SG principles, such as street network connectivity, density, land use mix, accessibility, and walkability [44]. The premise for conducting such assessments is that SG already has clear goals and principles to follow. The remaining articles examined SG in relation to more specific issues including housing [45], small communities [46], legislation [47], and transportation and land use [48]. These empirical focuses are long-lasting concerns of SG research and policymaking.

Table 2. Top ten most-cited publications of SG and SS.

Sampled Papers	Most-Cited Publications	Title	Source	Cited Frequency
Smart Growth	Downs, 2005 [49]	Land preservation: An essential ingredient in smart growth	*Journal of Planning Literature*	8.61%
	Downs, 2001 [40]	What does 'smart growth' really mean?	*Planning*	3.93%
	Burchell et al., 2000 [41]	Smart growth: More than a ghost of urban policy past, less than a bold new horizon	*Housing Policy Debate*	3.75%
	Knaap & Talen, 2005 [42]	New urbanism and smart growth: A few words from the academy	*International Regional Science Review*	3.56%
	Song, 2005 [44]	Smart growth and urban development pattern: A comparative study	*International Regional Science Review*	3.56%
	Danielsen et al., 1999 [45]	Retracting suburbia: Smart growth and the future of housing	*Housing Policy Debate*	3.37%
	Daniels, 2001 [43]	Smart growth: A new American approach to regional planning	*Planning Practice and Research*	3.18%
	Edwards & Haines, 2007 [46]	Evaluating smart growth: Implications for small communities	*Journal of Planning Education and Research*	2.81%
	Talen & Knaap, 2003 [47]	Legalizing smart growth: An empirical study of land use regulation in Illinois	*Journal of Planning Education and Research*	2.43%
	Handy, 2005 [48]	Smart growth and the transportation-land use connection: What does the research tell us?	*International Regional Science Review*	2.06%
Smart Shrinkage	Schilling & Logan, 2008 [50]	Greening the rust belt: A green infrastructure model for right sizing America's shrinking cities	*Journal of the American Planning Association*	56.41%
	Hollander & Németh, 2011 [51]	The bounds of smart decline: A foundational theory for planning shrinking cities	*Housing Policy Debate*	41.03%
	Rhodes & Russo, 2013 [52]	Shrinking smart?: Urban redevelopment and shrinkage in Youngstown, Ohio	*Urban Geography*	20.51%
	Hummel, 2015 [53]	Right-Sizing cities in the United States: Defining its strategies	*Journal of Urban Affairs*	12.82%
	Popper & Popper, 2002 [19]	Small can be beautiful	*Planning*	10.26%
	Hollander & Popper, 2007 [54]	Planning practice and the shrinking city: Reversing the land use allocation model	*Plan Canada*	10.26%

Table 2. *Cont.*

Sampled Papers	Most-Cited Publications	Title	Source	Cited Frequency
Smart Shrinkage	Hollander, 2011 [55]	Can a city successfully shrink? evidence from survey data on neighborhood quality	*Urban Affairs Review*	10.25%
	Hummel, 2015 [56]	Right-Sizing cities: A look at five cities	*Public Budgeting and Finance*	10.25%
	Frazier et al., 2013 [57]	The spatio-temporal impacts of demolition land use policy and crime in a shrinking city	*Applied Geography*	7.70%
	Hollander & Cahill, 2011 [58]	Confronting population decline in the Buffalo, New York, Region: A close reading of the Erie-Niagara framework for regional growth	*Journal of Architectural and Planning Research*	5.13%

The time span of the top 10 most-cited papers in SS ranges from 2002 to 2015. These papers indicate the ambiguity of SS concepts and theories. Popper (2002) proposed the idea of smart shrinkage, a precedent concept of SS, to advocate "planning for fewer people, fewer buildings, less land use" [19]. It is, however, more of a direction instead of actionable guidelines. SS as a revolution compared to traditional growth-oriented planning in the United States has been limited in planning practices, leading to much of the SS research remaining largely theoretical for a long period. During this time, filling and greening of vacant land emerged as an essential SS practice [50]. It was not until 2005 when the city of Youngstown published the renowned "2010 Comprehensive Plan" that SS truly began to be explored in planning practices [52]. Youngstown's planning was phenomenal at that time, and subsequently, cities like Flint, Cleveland, Detroit, and Buffalo also explored planning strategies to address urban shrinkage [56]. Along with exploration in SS planning practices, various policy tools began to emerge, and, in addition to measures such as demolition and reconstruction, land banking became an important approach to address the funding shortage for SS [53]. Notably, the study areas of these seminal SS-cited papers are largely limited to American cities such as Youngstown, Cleveland, Flint, Detroit, and Buffalo, pointing to a potential geographical bias in the knowledge base of SS.

4. Discussion

Urban development is a cyclic process from natural growth to natural decline and then to growth again. As cities around the world develop along different paths, urban sprawl and urban shrinkage often coexist [21,59]. For a long time, people have habitually treated the coexisting issues of urban sprawl and urban shrinkage in isolation. Correspondingly, the SG and SS strategies that have emerged in response to urban sprawl and urban shrinkage are rarely considered within the same context. This easily leads to a "either-or, this but not that" phenomenon in the process of sustainable urban governance. Therefore, based on the systematic comparative analysis of the literature on SG and SS, we analyzed the commonalities and unique characteristics of SG and SS, delineated the logical connection between the two research domains, and attempted to integrate them into a unified framework of "smart development".

4.1. Limitations

Before generalizing findings, three major caveats should be acknowledged. First, our sample size of the SS literature is relatively small, thus making our SS results sensitive to further publications (Figure 2a). Second, the sampled SS papers represent only research published in English, thus making our SS results biased toward English-speaking countries like the United States (e.g., Figures 4b and 8, Table 2). Third, the sampled SS papers explicitly mentioned in their title, abstract, and keywords the terms "smart shrink*", "smart decline", and/or "'right sizing' AND 'shrink*'" in the context of cities or urban areas, yet some studies on shrinking cities that use alternative terms like "urban redevelopment", "urban revitalization", "urban regeneration", and/or "urban resurgence" are insufficiently reflected by our results [60].

Considering these caveats, our interpretation of SS results needs to be very careful to avoid overgeneralization. Essentially, the "smartness" captured in this study is only part of the wisdom that scholars have explored to address urban shrinkage (and urban sprawl); yet it is safe to claim that our results reflect how SS and SG, as two scholarly concepts, have been used in the English literature.

Nonetheless, it is imperative to acknowledge that our research, in its quest to explore the term "SS"—a term birthed in the United States—inadvertently neglects the manifold endeavors undertaken by Europe, situated on the other side of the Atlantic, to counter urban shrinkage. Indeed, the phenomenon of urban shrinkage surfaced earlier and is more pervasive in Europe, with European governments and academicians initiating actions ahead of their American counterparts [61]. In 2001, prior to the introduction of SS in the United States, Germany was grappling with the escalating issue of urban shrinkage in East Germany. In response, they proposed the Stadtumbau Ost (Urban Restructuring East) plan, which aimed at balancing the real estate market in shrinking cities through the demolition of vacant houses and the reinvigoration of urban allure. This initiative can arguably be considered humanity's inaugural attempt at a paradigm shift from growth planning to shrinkage planning [62,63]. Prominent examples emanate from Germany, where numerous cities such as Leipzig, Halle, and Dresden have adopted strategies to adapt to urban shrinkage, underpinned by national financial support. These strategies are predicated on the fiscal transfers and aid spanning regional levels in high-welfare nations [63,64]. In contrast to "SS", scholars within the European narrative framework tend to characterize this series of measures as "urban resurgence/revitalization/regeneration". However, the associated terminologies are seldom reflected in the titles, abstracts, and keywords of academic papers, thereby augmenting the difficulty of retrieving literature pertinent to these policies. We must concede this is a limitation of our research, yet the dispersion and lack of uniformity of these descriptive terms also curtail the international exchange and diffusion of numerous policy innovations. For instance, in China, the term "urban shrinkage" often evokes the concept of "smart decline" among many Chinese scholars. As articulated by Wiechmann, T. there is an urgent need for a transatlantic debate on this matter [3,65].

4.2. Main Findings of Bibliometric Statistics on SG and SS

The study provides the most comprehensive multidimensional quantitative comparative analysis of literature on SG and SS from 1997 to 2022. These results can visually demonstrate the developmental trajectory, current status, research hotspots, and trends in the study of SG and SS. The main findings are as follows: (1) Research on SG and SS emerged successively in the late 20th century and early 21st century. The publication volume related to SG was more than ten times that of SS (Table 1), but the research popularity of SS rapidly increased around 2010 (Figures 2 and 6). (2) Both SG and SS studies are primarily focused on environmental science and social science, with the former being more focused on environmental science and the latter being more oriented towards social science (Figures 3 and 7). (3) The study areas covered by SG are significantly greater and more extensive than those of SS. In addition to developed regions such as the United States and Europe, research on SG also encompasses many underdeveloped economic areas in Asia, Latin America, and Africa (Figure 8).

Our findings point to a seemingly apparent observation that SS, as a relatively new concept signaling good intentions to address urban shrinkage, is still on a path of consolidation (Figures 2, 3b, 4b, 6 and 8b). This is in stark contrast to the maturity of SG as an established terminology with well-defined guidelines widely recognized among scientists [43,44], professionals (e.g., Smart Growth Network), and governments (e.g., Environmental Protection Agency, United States). For SG, its discourse framing to mobilize stakeholders has shifted from growth versus no-growth (i.e., growth control) to "smart" versus "dumb" growth (i.e., sprawl and related problems) (Figure 5), reflecting a mindset change from mitigation to adaptation. Contrastingly, the discourse framing of SS—although intended to transform the traditional growth-oriented mindset of urban planning [19,52,66]—has been largely dominated by growth-oriented mitigation thinking (e.g., "urban development", "development project", and "redevelopment" in Figures 4b and 6), with adaptation thinking emerging only in the past five years (e.g., "degrowth coalition", "right sizing", and "small town" in Figure 6). In comparison with the broader studies on shrinking cities [18], especially since the early 2000s when the German architect Philipp Oswaltand his colleagues popularized the term "shrinking cities" [11], the scholarly interest in "smartly" addressing urban shrinkage and consensus on SS principles are remarkably and regretfully limited.

Although the research domain of SS is still at a nascent stage of conceptual and theoretical development, an encouraging finding of our study is that SS publications have been rapidly growing since 2010 (Figure 2a). This abnormal surge was likely attributable to the funding and implementation of the 7 EU FP project "Shrink Smart—The Governance of Shrinkage within a European Context" (2009–2012; no. 225193) [67]. Such a catalyzing effect of innovative projects, pioneering conferences, and consequent scholar communities on nurturing and advancing new ideas and research fields has been common, e.g., the establishment of the North American School of Landscape Ecology [68], the burgeoning of land system science [69], the birth of sustainability science [70], and the popularization of the concept of shrinking cities [11]. In this regard, funding agencies' bold financial support of such potentially transformative scholarly activities plays a decisive role. Such funding is particularly needed for synthesizing and promoting the existing "smartness" of addressing urban shrinkage in various places across the world; note that the "smart" wisdom of European and China's shrinking cities remains underrepresented [20,21,71] (cf. Figure 8).

In contrast to SS not gaining significant capacity in mobilizing research efforts for advancing "smart" urbanization, the term SG seems to have been losing its mobilizing capacity since 2013 (Figure 2a). As one of the reviewers proposed, this might result from a saturation of research topics, shifts in funding priorities, or changes in academic trends. Actually, in the past decade, there have been some criticisms of the compact urban development advocated by SG [72], including the inequality in urban land [73] and the decrease in residential satisfaction caused by high-density housing [74]. With the increasing influence of the people-centered research paradigm of sustainability science and the continuous expansion of embedded issues, more and more studies are re-examining the compact development advocated by SG from the perspective of the coordinated coupling of humans and the environment [75]. Although the number of new issues related to SG has gradually decreased after 2013, topics related to sustainability science (such as environmental sustainability, complexity, land cover, carbon dioxide, and green infrastructure; see Figure 6) have taken up a high proportion, indicating that SG is shifting towards a paradigm more in line with sustainability science. In fact, as early as 2012, Chapin [14] noticed this trend of shifting from SG to sustainable growth and called for the integration of issues such as industrial decline, employment, environment, and climate into future growth management.

4.3. *Differences and Similarities between SG and SS*

Undoubtedly, there are obvious differences between SG and SS. The former focuses more on spatial compactness, while the latter emphasizes scale simplification. Specifically, the "smartnesses" aspect of SG lies in the compact development of urban elements such as population, land, and industry. On one hand, SG strives to reduce urban sprawl by establishing organic connections between public transportation and land use, designing mixed-use communities, and enhancing the reuse of abandoned land within cities [48,76]. On the other hand, by implementing measures like Urban Growth Boundaries (UGBs), SG aims to maintain compact land use, allocate new urban land demand to existing built-up areas as much as possible, and minimize the impact on agriculture and ecology [40,77].

Unlike SG, the "smartnesses" aspect of SS lies in the intensive development of urban elements. SS mainly focuses on centralizing urban elements such as population, land, and industry to maintain the healthy operation of concentrated areas, especially in the context of a continuous decrease in population [53,65]. The specific approaches taken by different cities generally vary. Currently, the main approach is based on reducing the scale of land development, scientifically and reasonably controlling the size of the city, and strategically arranging the population based on the advantages and functional positioning of urban development. This transformation aims to change the traditional notion of necessary urban growth and maintain the intrinsic drive for sustainable urban development [56,65].

However, the differences mentioned above do not obscure the intrinsic connections between SG and SS. We find that although SG and SS have long been separate in research and practice, there are at least three similarities between them.

First, they share similar value concepts. Due to historical factors like deindustrialization and suburbanization, urban development challenges such as inner-city population decline, industrial stagnation, and low land use efficiency have emerged one after another [78]. Both SG and SS were proposed to address specific problems arising from the urbanization process, and both are committed to sustainable urban development [36,77].

Second, they have similar core objectives. Whether addressing urban sprawl or urban shrinkage, the core goal of both SG and SS is to achieve optimal allocation of urban elements such as population, land, and industry [79,80], resolving issues of unsustainability caused by urban element misallocation (spatial misallocation, scale misallocation, etc.).

Lastly, they share similarities in disciplinary paradigms and research themes. Research related to SG and SS shows an interdisciplinary characteristic, merging social sciences and environmental sciences. Both seek to improve the physical form and landscape pattern of cities [81,82], while also considering policy-making and management issues [40]. For example, they emphasize the collaborative participation of diverse stakeholders like government, market forces, residents, and social groups, and direct research interests toward discussions of social equity [83,84].

4.4. Implications and Future Directions

Currently, both SG and SS fall short in their respective commitment to "urban sustainability" [85], and there is a need to shift the fragmented discourse on SG and SS. Building on the analysis of the similarities in and differences between SG and SS in the previous sections, we abstracted the core value concepts from the numerous principles and initiatives of SG and SS, and attempted to construct a cohesive theoretical framework of "Smart Development" (Figure 9) that can comprehensively consider both SG and SS. This framework, starting from the issues of urban shrinkage and urban sprawl arising from the urbanization process, reexamines the concepts and practical methods of SG and SS from the three dimensions of social, economic, and environmental aspects of urban sustainability. It integrates the core concepts of SG and SS into "Smart Development" to achieve a more comprehensive and coordinated approach to urban sustainable development.

In terms of striving to enhance urban sustainability, SG and SS are not in opposition but rather can complement each other in a coordinated manner. The extensive planning practices of SG can provide experiential and methodological support for SS in shaping urban landscape patterns and optimizing and restructuring public service networks, thus offering valuable insights to avoid spatial fragmentation and systemic dislocation of public services in shrinking cities [39]. The discussions in SS on development rights, equity issues, and unique insights into urban development trends can enrich the value connotations of SG and provide inspiration on how to truly improve human well-being [57]. We hope to achieve the coupling and coordination of SG and SS based on this theoretical complementarity, merging the two into "Smart Development" to comprehensively consider the three dimensions of urban sustainability and promote sustainable urban development.

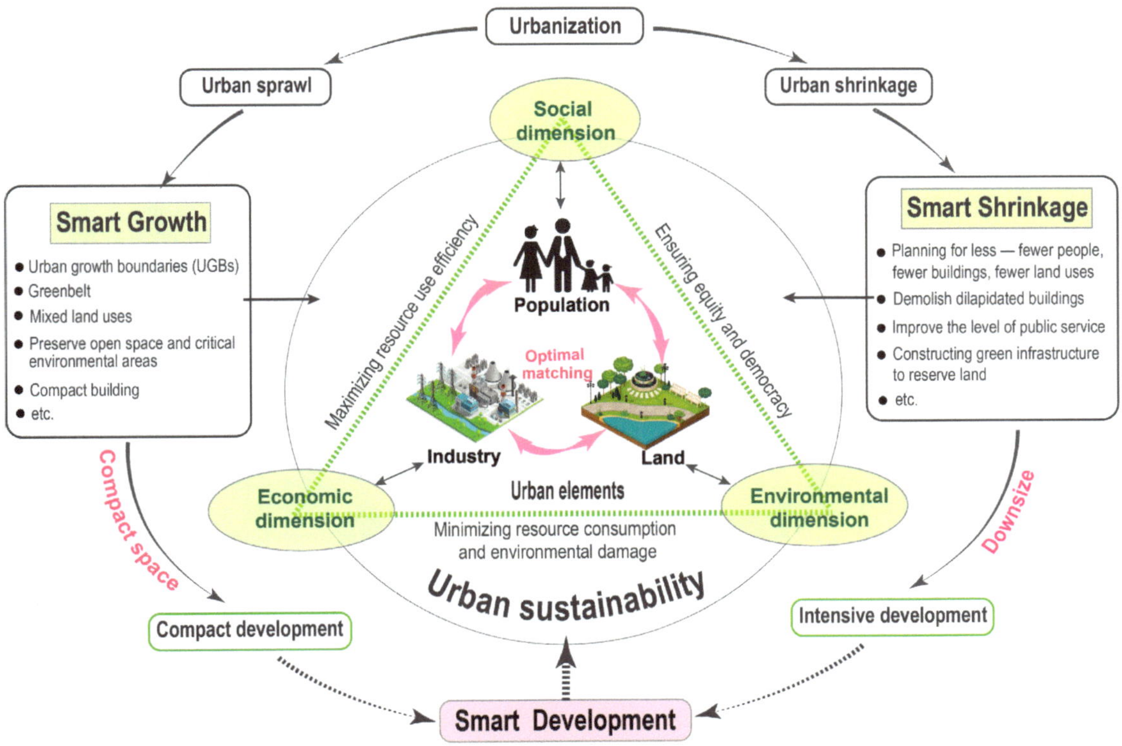

Figure 9. Theoretical framework of "Smart Development" for urban sustainable development.

At the theoretical level, it is necessary to change the existing situation of treating urban growth and urban shrinkage in isolation, and consider the issues of urban growth and shrinkage from a more macro and holistic perspective. Specifically, it includes simultaneously measuring and assessing urban growth and shrinkage, comprehensively analyzing the driving mechanisms of growth and shrinkage from a systemic perspective, and predicting future scenarios of urban growth and shrinkage, as well as potential challenges and opportunities based on current trends and possible influencing factors. It is encouraging that there is an increasing amount of research simultaneously measuring urban growth and shrinkage [86,87]. Moreover, with the advancement in artificial intelligence and machine learning technologies, it has become increasingly possible to more accurately predict various trends and scenarios for the future of cities [88–90].

At the practical level, efforts can be strengthened in the following aspects. Firstly, the selection of strategies must be based on scenarios that match the urban development, and a comprehensive action plan should be formulated according to the scientific prediction of the temporal and spatial dynamics of the urban population, so as to avoid the mismatch in the scale and space of urban elements. Secondly, it is essential to approach the coordination of interests among different groups with greater caution, establishing mechanisms for dialogue and interest communication, and involving a more diverse set of actors in responding to urban sprawl and urban shrinkage [91]. Finally, for the coordinated governance of urban sprawl and urban shrinkage, positive interactions across multiple levels are necessary. On one hand, emphasis should be placed on the coordination among various levels within the city, from individuals and families to neighborhoods, communities, functional areas, and the city. On the other hand, the city itself should be positioned within a larger urban system and network. In this way, the rational flow of resource elements can be ensured among small towns, medium-sized cities, large cities, mega-cities, metropolitan areas, and even national and global cities, ultimately achieving urban sustainable development in which the overall functionality exceeds the sum of its parts.

5. Conclusions

Land-related approaches to advancing sustainability are increasingly recognized for their critical role in shaping the future of urban environments. This paper explores the intertwined roles of SG and SS as complementary strategies within this context, emphasizing their importance in guiding urbanization towards sustainability. By employing bibliometric analysis, we systematically explored the evolution and thematic focus of SG and SS research, highlighting the distinct yet interconnected pathways they offer for sustainable urban development. Our findings reveal a divergence in the scope and clarity of research themes between SG and SS, with SG commanding a broader and more defined research landscape. Despite this, SS has experienced a notable surge in interest since 2010, signaling a shift in the focus of sustainability discussions towards embracing urban shrinkage as a viable and necessary approach to sustainability in post-industrial contexts. This shift underscores the growing recognition of the need to balance urban expansion with shrinkage, reflecting a holistic view of urban dynamics that incorporates both growth and shrinkage as integral to sustainable development.

SG, rooted in New Urbanism, emphasizes the optimization of urban physical form and the mitigation of environmental impacts, thereby addressing the spatial and ecological aspects of sustainability. On the other hand, SS, emerging from critical reflections on urban decline, prioritizes economic revitalization and the enhancement of resident well-being, thus tackling the economic and social dimensions of sustainable urban living. These dual focuses underscore the necessity of a balanced approach that considers both environmental integrity and human well-being. The integration of SG and SS into a unified framework for urban sustainability is crucial for navigating the complexities of urban development. The unified theoretical framework of "Smart Development" (Figure 9) proposed in this paper organically integrates the core concepts of SG and SS, and can promote more comprehensive and coordinated urban sustainable development. As cities evolve into more dynamic and complex systems, the recognition of SG and SS as normal processes of urbanization highlights the need for adaptable and comprehensive strategies that address the full spectrum of sustainability challenges. By fostering more research on SS across various contexts and integrating it with SG principles, we can better understand and implement land-related approaches that advance sustainability in an increasingly urbanized world. This holistic approach not only addresses the immediate challenges of urban growth and decline but also lays the groundwork for a sustainable and resilient urban future.

Author Contributions: Conceptualization, Y.Y.; methodology, Z.D. and B.-B.Z.; software, Z.D. and B.-B.Z.; validation, B.-B.Z., Y.Y. and Y.L.; formal analysis, Z.D.; investigation, Z.D.; resources, Y.Y.; data curation, Z.D. and B.-B.Z.; writing—original draft preparation, Z.D. and Y.Y.; writing—review and editing, Y.Y., B.-B.Z. and Y.L.; visualization, Z.D. and B.-B.Z.; supervision, B.-B.Z. and Y.Y.; project administration, Y.Y.; funding acquisition, Y.Y. All authors have read and agreed to the published version of the manuscript.

Funding: This work was supported by the National Social Science Fund of China (19CJL025).

Data Availability Statement: Data will be available upon reasonable request.

Conflicts of Interest: The authors declare no conflicts of interest.

References

1. McCormick, K.; Anderberg, S.; Coenen, L.; Neij, L. Advancing sustainable urban transformation. *J. Clean. Prod.* **2013**, *50*, 1–11. [CrossRef]
2. Maranghi, S.; Parisi, M.L.; Facchini, A.; Rubino, A.; Kordas, O.; Basosi, R. Integrating urban metabolism and life cycle assessment to analyse urban sustainability. *Ecol. Indic.* **2020**, *112*, 106074. [CrossRef]

3. Wiechmann, T.; Pallagst, K.M. Urban shrinkage in Germany and the USA: A comparison of transformation patterns and local strategies. *Int. J. Urban Reg. Res.* **2012**, *36*, 261–280. [CrossRef]
4. Patias, N.; Rowe, F.; Cavazzi, S.; Arribas-Bel, D. Sustainable urban development indicators in Great Britain from 2001 to 2016. *Landsc. Urban Plan.* **2021**, *214*, 104148. [CrossRef]
5. Acuto, M.; Parnell, S.; Seto, K.C. Building a global urban science. *Nat. Sustain.* **2018**, *1*, 2–4. [CrossRef]
6. Aslam, A.; Rana, I.A. The use of local climate zones in the urban environment: A systematic review of data sources, methods, and themes. *Urban Clim.* **2022**, *42*, 101120. [CrossRef]
7. Oswalt, P.; Rieniets, T. Global Context. Shrinking Cities. Available online: http://www.shrinkingcities.com/globaler_kontext.0.html?&L=1 (accessed on 10 May 2024).
8. Hassan, M.O.; Ling, G.H.T.; Rusli, N.; Mokhtar, S.; Wider, W.; Leng, P.C. Urban Sprawl Patterns, Drivers, and Impacts: The Case of Mogadishu, Somalia Using Geo-Spatial and SEM Analyses. *Land* **2023**, *12*, 783. [CrossRef]
9. Lv, Z.Q.; Dai, F.Q.; Sun, C. Evaluation of urban sprawl and urban landscape pattern in a rapidly developing region. *Environ. Monit. Assess.* **2012**, *184*, 6437–6448. [CrossRef] [PubMed]
10. Adhya, A. Introduction: The shrinking city phenomenon. In *Shrinking Cities and First Suburbs: The Case of Detroit and Warren, Michigan*; Adhya, A., Ed.; Springer International Publishing: Cham, Switzerland, 2017; pp. 1–10.
11. Oswalt, P. *Shrinking Cities: International Research*; Hatje Cantz: Berlin, Germany, 2005; Volume 1.
12. Sun, L.; Chen, J.; Li, Q.; Huang, D. Dramatic uneven urbanization of large cities throughout the world in recent decades. *Nat. Commun.* **2020**, *11*, 5366. [CrossRef]
13. Ma, Z.; Li, C.; Zhang, J.; Liu, D.; Tong, Y.; Liu, W. Understanding Urban Growth and Shrinkage From a Multiscale Perspective: Case Study of Jilin Province, China. *J. Urban Plan. Dev.* **2023**, *149*, 05023006. [CrossRef]
14. Chapin, T.S. Introduction: From growth controls, to comprehensive planning, to smart growth: Planning's emerging fourth wave. *J. Am. Plan. Assoc.* **2012**, *78*, 5–15. [CrossRef]
15. Tregoning, H. Smart growth: A challenge for urban infrastructure development. *Public Work. Manag. Policy* **1997**, *2*, 3–8. [CrossRef]
16. Benfield, F.K.; Terris, J.; Vorsanger, N. *Solving Sprawl: Models of Smart Growth in Communities across America*; Island Press: Washington, DC, USA, 2001; pp. 137–138.
17. de la Cruz, E.E.R. Local political institutions and smart growth: An empirical study of the politics of compact development. *Urban Aff. Rev.* **2009**, *45*, 218–246. [CrossRef]
18. Lee, J.-E.; Park, Y.; Newman, G.D. Twenty years of research on shrinking cities: A focus on keywords and authors. *Landsc. Res.* **2023**, *48*, 884–899. [CrossRef] [PubMed]
19. Popper, D.E.; Popper, F.J. Small can be beautiful. *Planning* **2002**, *68*, 20–23.
20. Wolff, M.; Wiechmann, T. Urban growth and decline: Europe's shrinking cities in a comparative perspective 1990–2010. *Eur. Urban Reg. Stud.* **2017**, *25*, 122–139. [CrossRef]
21. Sun, Y.; Jiao, L.; Guo, Y.; Xu, Z. Recognizing urban shrinkage and growth patterns from a global perspective. *Appl. Geogr.* **2024**, *166*, 103247. [CrossRef]
22. Krugman, P. *Increasing Returns and Economic Geography*; National Bureau of Economic Research: Cambridge, MA, USA, 1991.
23. Krugman, P. Increasing returns, monopolistic competition, and international trade. *J. Int. Econ.* **1979**, *9*, 469–479. [CrossRef]
24. Zhou, B.-B.; Aggarwal, R.; Wu, J.; Lv, L. Urbanization-associated farmland loss: A macro-micro comparative study in China. *Land Use Policy* **2021**, *101*, 105228. [CrossRef]
25. Zhou, B.-B.; Wu, J.; Anderies, J.M. Sustainable landscapes and landscape sustainability: A tale of two concepts. *Landsc. Urban Plan.* **2019**, *189*, 274–284. [CrossRef]
26. Pranckutė, R. Web of Science (WoS) and Scopus: The titans of bibliographic information in today's academic world. *Publications* **2021**, *9*, 12. [CrossRef]
27. Baas, J.; Schotten, M.; Plume, A.; Côté, G.; Karimi, R. Scopus as a curated, high-quality bibliometric data source for academic research in quantitative science studies. *Quant. Sci. Stud.* **2020**, *1*, 377–386. [CrossRef]
28. Zyoud, S.H. Global landscape of COVID-19 research: A visualization analysis of randomized clinical trials. *Clin. Exp. Med.* **2024**, *24*, 14. [CrossRef] [PubMed]
29. Nobre, G.C.; Tavares, E. Scientific literature analysis on big data and internet of things applications on circular economy: A bibliometric study. *Scientometrics* **2017**, *111*, 463–492. [CrossRef]
30. Wickham, H.; Çetinkaya-Rundel, M.; Grolemund, G. *R for Data Science*; O'Reilly Media, Inc.: Sebastopol, CA, USA, 2023.
31. Jalal, S.K. Co-authorship and co-occurrences analysis using bibliometrix r-package: A case study of india and bangladesh. *Ann. Libr. Inf. Stud.* **2019**, *66*, 57–64.
32. Aria, M.; Cuccurullo, C. bibliometrix: An R-tool for comprehensive science mapping analysis. *J. Informetr.* **2017**, *11*, 959–975. [CrossRef]
33. van Eck, N.J.; Waltman, L. Software survey: VOSviewer, a computer program for bibliometric mapping. *Scientometrics* **2010**, *84*, 523–538. [CrossRef] [PubMed]
34. Lin, M.; Chen, Y.; Chen, R. Bibliometric analysis on Pythagorean fuzzy sets during 2013–2020. *Int. J. Intell. Comput. Cybern.* **2020**, *14*, 104–121. [CrossRef]

35. Griggs, D.; Stafford-Smith, M.; Gaffney, O.; Rockström, J.; Öhman, M.C.; Shyamsundar, P.; Steffen, W.; Glaser, G.; Kanie, N.; Noble, I. Policy: Sustainable development goals for people and planet. *Nature* **2013**, *495*, 305–307. [CrossRef]
36. Newman, G.; Hollander, J.B.; Lee, J.; Gu, D.; Kim, B.; Lee, R.J.; Horney, J.A.; Bearfield, D.; Li, Y. Smarter Shrinkage: A Neighborhood-Scaled Rightsizing Strategy Based on Land Use Dynamics. *J. Geov. Spat. Anal.* **2018**, *2*, 11. [CrossRef]
37. Kato, H.; Kanki, K. Development of walkability indicator for smart shrinking: Case study of sprawl areas in Northern Osaka metropolitan. *Int. Rev. Spat. Plan. Sustain. Dev.* **2020**, *8*, 39–58. [CrossRef] [PubMed]
38. Hollander, J.B. *Sunburnt Cities: The Great Recession, Depopulation and Urban Planning in the American Sunbelt*; Routledge: London, UK, 2011; pp. 1–165.
39. Daniels, T.; Lapping, M. Land preservation: An essential ingredient in smart growth. *J. Plan. Lit.* **2005**, *19*, 316–329. [CrossRef]
40. Downs, A. What does 'smart growth' really mean? *Planning* **2001**, *67*, 20–25.
41. Burchell, R.W.; Listokin, D.; Galley, C.C. Smart growth: More than a ghost of urban policy past, less than a bold new horizon. *Hous. Policy Debate* **2000**, *11*, 821–879. [CrossRef]
42. Knaap, G.; Talen, E. New urbanism and smart growth: A few words from the academy. *Int. Reg. Sci. Rev.* **2005**, *28*, 107–118. [CrossRef]
43. Daniels, T. Smart growth: A new American approach to regional planning. *Plan. Pract. Res.* **2001**, *16*, 271–279. [CrossRef]
44. Song, Y. Smart growth and urban development pattern: A comparative study. *Int. Reg. Sci. Rev.* **2005**, *28*, 239–265. [CrossRef]
45. Danielsen, K.A.; Lang, R.E.; Fulton, W. Retracting suburbia: Smart growth and the future of housing. *Hous. Policy Debate* **1999**, *10*, 513–540. [CrossRef]
46. Edwards, M.M.; Haines, A. Evaluating smart growth: Implications for small communities. *J. Plan. Educ. Res.* **2007**, *27*, 49–64. [CrossRef]
47. Talen, E.; Knaap, G. Legalizing smart growth: An empirical study of land use regulation in Illinois. *J. Plan. Educ. Res.* **2003**, *22*, 345–359. [CrossRef]
48. Handy, S. Smart growth and the transportation-land use connection: What does the research tell us? *Int. Reg. Sci. Rev.* **2005**, *28*, 146–167. [CrossRef]
49. Downs, A. Smart growth: Why we discuss it more than we do it. *J. Am. Plan. Assoc.* **2005**, *71*, 367–378. [CrossRef]
50. Schilling, J.; Logan, J. Greening the rust belt: A green infrastructure model for right sizing America's shrinking cities. *J. Am. Plan. Assoc.* **2008**, *74*, 451–466. [CrossRef]
51. Hollander, J.B.; Németh, J. The bounds of smart decline: A foundational theory for planning shrinking cities. *Hous. Policy Debate* **2011**, *21*, 349–367. [CrossRef]
52. Rhodes, J.; Russo, J. Shrinking smart?: Urban redevelopment and shrinkage in Youngstown, Ohio. *Urban Geogr.* **2013**, *34*, 305–326. [CrossRef]
53. Hummel, D. Right-Sizing Cities In The United States: Defining Its Strategies. *J. Urban Aff.* **2015**, *37*, 397–409. [CrossRef]
54. Hollander, J.B.; Popper, F. Planning practice and the shrinking city: Reversing the land use allocation model. *Plan. Can.* **2007**, *47*, 38–40.
55. Hollander, J.B. Can a city successfully shrink? evidence from survey data on neighborhood quality. *Urban Aff. Rev.* **2011**, *47*, 129–141. [CrossRef]
56. Hummel, D. Right-Sizing Cities: A Look at Five Cities. *Public Budg. Financ.* **2015**, *35*, 1–18. [CrossRef]
57. Frazier, A.E.; Bagchi-Sen, S.; Knight, J. The spatio-temporal impacts of demolition land use policy and crime in a shrinking city. *Appl. Geogr.* **2013**, *41*, 55–64. [CrossRef]
58. Hollander, J.B.; Cahill, B. Confronting population decline in the Buffalo, New York, Region: A close reading of the Erie-Niagara framework for regional growth. *J. Archit. Plan. Res.* **2011**, *28*, 252–267.
59. Kim, Y.E.; Lee, J.S.; Kim, S. Proposing the classification matrix for growing and shrinking cities: A case study of 228 districts in South Korea. *Habitat. Int.* **2022**, *127*, 102644. [CrossRef]
60. Li, X.; Hui Eddie, C.M.; Long, Y.; Chen, T.; Lang, W. Shrinking cities and towns: Policy responses to the challenges, urban planning, and development strategies. *J. Urban Plan. Dev.* **2021**, *147*, 02021002. [CrossRef]
61. Haase, A.; Athanasopoulou, A.; Rink, D. Urban shrinkage as an emerging concern for European policymaking. *Eur. Urban Reg. Stud.* **2016**, *23*, 103–107. [CrossRef]
62. Bernt, M. Partnerships for demolition: The governance of urban renewal in East Germany's shrinking cities. *Int. J. Urban Reg. Res.* **2009**, *33*, 754–769. [CrossRef]
63. Haase, A.; Rink, D.; Grossmann, K.; Bernt, M.; Mykhnenko, V. Conceptualizing urban shrinkage. *Environ. Plan. A* **2014**, *46*, 1519–1534. [CrossRef]
64. Bartholomae, F.; Woon Nam, C.; Schoenberg, A. Urban shrinkage and resurgence in Germany. *Urban Stud.* **2017**, *54*, 2701–2718. [CrossRef]
65. He, H. Smart decline: Planning measures for urban decay and its practice in USA. *J. Urban Reg. Plan.* **2011**, *3*, 157–168.
66. Schmidt-Thomé, K.; Lilius, J. Smart shrinkage and multi-locality—The appeal of hope, illustrated through Puolanka, a rural municipality in Finland. *Urban Plan. Transp. Res.* **2023**, *11*, 2165140. [CrossRef]
67. Bernt, M.; Cocks, M.; Couch, C.; Grossmann, K.; Haase, A.; Rink, D. *Policy Response, Governance and Future Directions, Shrink Smart Research Brief No. 2*; Helmholtz Centre for Environmental Research—UFZ: Leipzig, Germany, 2012.

68. Risser, P.G.; Karr, J.R.; Forman, R.T.T. *Landscape Ecology: Directions and Approaches, a Workshop Held at Allerton Park, Piatt County, Illinois, April 1983*; Illinois Natural History Survey: Champaign, IL, USA, 1984; p. 18.
69. Turner II, B.L.; Skole, D.; Sanderson, S.; Fischer, G.; Fresco, L.; Leemans, R. *Land-Use and Land-Cover Change: Science/Research Plan*; International Geosphere-Biosphere Programme: Stockholm, Sweden, 1995.
70. NRC. *Our Common Journey: A Transition Toward Sustainability*; National Academies Press: Washington, DC, USA, 1999.
71. Yang, Y.; Wu, J.G.; Wang, Y.; Huang, Q.X.; He, C.Y. Quantifying spatiotemporal patterns of shrinking cities in urbanizing China: A novel approach based on time-series nighttime light data. *Cities* **2021**, *118*, 103346. [CrossRef]
72. Leigh, N.G.; Hoelzel, N.Z. Smart growth's blind side: Sustainable cities need productive urban industrial land. *J. Am. Plan. Assoc.* **2012**, *78*, 87–103. [CrossRef]
73. McDonald, R.I.; Forman, R.T.T.; Kareiva, P. Open space loss and land inequality in United States' cities, 1990–2000. *PLoS ONE* **2010**, *5*, e9509. [CrossRef]
74. Howley, P. 'Sustainability versus liveability': An exploration of central city housing satisfaction. *Int. J. Hous. Policy* **2010**, *10*, 173–189. [CrossRef]
75. Zhou, B.-B.; Qim, M.; Wu, J.-G.; Hu, G.-H.; Mao, D.-H.; Zeng, X.-J.; Guo, J.; Fang, X.-N.; Liu, Y.-P.; Lyu, L.-G. Sustainability science revisited: Recent advances and new opportunities. *Yingyong Shengtai Xuebao* **2019**, *30*, 325–336.
76. Rahimi, A. A methodological approach to urban landuse change modeling using infill development pattern—A case study in Tabriz, Iran. *Ecol. Process.* **2016**, *5*, 1. [CrossRef]
77. Jang, G.; Kim, S. Are decline-oriented strategies thermally sustainable in shrinking cities? *Urban Clim.* **2021**, *39*, 100924. [CrossRef]
78. Rieniets, T. Shrinking cities: Causes and effects of urban population losses in the twentieth century. *Nat. Cult.* **2009**, *4*, 231–254. [CrossRef]
79. Chen, X.; Wei, L.; Zhang, H. Spatial and Temporal Pattern of Urban Smart Development in China and Its Driving Mechanism. *Chin. Geogr. Sci.* **2018**, *28*, 584–599. [CrossRef]
80. Echenique, M.H.; Hargreaves, A.J.; Mitchell, G.; Namdeo, A. Growing cities sustainably. *J. Am. Plan. Assoc.* **2012**, *78*, 121–137. [CrossRef]
81. Gospodini, A.; Manika, S. Conceptualising 'smart' and 'green' public open spaces; Investigating redesign patternsfor Greek cities. *Civ. Eng. Archit.* **2020**, *8*, 371–378. [CrossRef]
82. Artmann, M.; Kohler, M.; Meinel, G.; Gan, J.; Ioja, I.C. How smart growth and green infrastructure can mutually support each other—A conceptual framework for compact and green cities. *Ecol. Indic.* **2019**, *96*, 10–22. [CrossRef]
83. Ewing, R.; Lyons, T.; Siddiq, F.; Sabouri, S.; Kiani, F.; Hamidi, S.; Choi, D.A.; Ameli, H. Growth Management Effectiveness: A Literature Review. *J. Plan. Lit.* **2022**, *37*, 433–451. [CrossRef]
84. Németh, J.; Hollander, J.B.; Whiteman, E.D.; Johnson, M.P. Planning with justice in mind in a shrinking Baltimore. *J. Urban Aff.* **2020**, *42*, 351–370. [CrossRef]
85. Huang, L.; Wu, J.; Yan, L. Defining and measuring urban sustainability: A review of indicators. *Landsc. Ecol.* **2015**, *30*, 1175–1193. [CrossRef]
86. Zhou, Y.; Zheng, W.; Wang, X.; Xiong, Y.; Wang, X. The Mechanism Behind Urban Population Growth and Shrinkage from the Perspective of Urban Network Externalities. *Chin. Geogr. Sci.* **2023**, *33*, 189–204. [CrossRef]
87. Jaoude, G.A.; Mumm, O.; Karch, A.; Carlow, V.M. Understanding land take in small and medium-sized cities through scenarios of shrinkage and growth using autoregressive models. *Front. Built Environ.* **2022**, *8*, 908698. [CrossRef]
88. Newton, D. Generative deep learning in architectural design. *Technol. Archit. Des.* **2019**, *3*, 176–189. [CrossRef]
89. Ibrahim, M.R.; Haworth, J.; Cheng, T. Understanding cities with machine eyes: A review of deep computer vision in urban analytics. *Cities* **2020**, *96*, 102481. [CrossRef]
90. Li, F.; Yigitcanlar, T.; Nepal, M.; Nguyen, K.; Dur, F. Machine learning and remote sensing integration for leveraging urban sustainability: A review and framework. *Sustain. Cities Soc.* **2023**, *96*, 104653. [CrossRef]
91. Yang, W. Pro-growth urban policy implementation vs urban shrinkage: How do actors shift policy implementation in Shrinking Cities in China? *Cities* **2023**, *134*, 104157. [CrossRef]

land

MDPI

Review

History, Progress, and Prospects of Urban Fringe Research in China: An Idiosyncratic Synthesis from a Spatial Perspective

Zhi Li [1], Lixin Zhang [1], Tian Hu [2,*] and Yifei Wu [1]

[1] College of City Construction, Jiangxi Normal University, Nanchang 330022, China; gmorning1985@jxnu.edu.cn (Z.L.); zzzzz@jxnu.edu.cn (L.Z.); yifei_wu12@163.com (Y.W.)

[2] Beijing Key Laboratory of Megaregions Sustainable Development Modelling, School of Urban Economics and Public Administration, Capital University of Economics and Business, Beijing 100070, China

* Correspondence: tianahu@cueb.edu.cn

Abstract: Urban fringes, as transitional zones between urban and rural systems, have played a pivotal role in China's rapid urban expansion. Though there have been very rich urban fringe studies in China, little efforts have been made to introduce this research progress to global peers. To fill this gap, this paper presents an idiosyncratic synthesis based on the relevant Chinese literature, from a spatial perspective—with urban fringes as a land system type of its own kind. We first recap the historical trajectory of urban fringe research in China, revealing its progression from early international influences to context-specific investigations shaped by China's unique governance and socio-economic conditions. We then introduce, in more detail, the main research progresses on revealing the spatial patterns, underlying mechanisms, and governance practices of China's urban fringe areas. We argue for studying urban fringes as part of and the critical zone for coupling urban–rural systems, and call for a holistic, inclusive approach in future studies. Specifically, we identify three research prospects: (1) understanding the new patterns of social-environmental dynamics in China's New Era; (2) promoting spatial governance from a coupled urban–rural system perspective; and (3) leveraging artificial intelligence for producing spatially actionable knowledge.

Keywords: urban fringe area; land system; urban-rural system; review; spatial governance

1. Introduction

With over four decades of economic reform and rapid development, China has expanded its urban footprint dramatically, with an estimated 50,000 square kilometers of urbanized land. The majority of such remarkable urban growth has occurred in China's urban fringe, which has largely shaped the quality and sustainability of China's urbanization [1,2]. The term "urban fringe" is also seen in the scholarly literature and governmental documents as urban fringe areas [3], metropolitan fringe areas [4], the urban fringe belt [5], megacity fringe areas [6], the urban–rural interlocking zone [7], and the urban–rural transition zone [8], which have been variously defined. Essentially, the urban fringe describes the transitional area between urban and rural systems—featuring complex physical, economic, and social transformations from the rural to the urban as well as the dynamic landscape mosaic resulting from those intertwined urban–rural interactions [9–11].

The dynamic nature of the urban fringe adds significance to its role in shaping the sustainability of China's urbanization [12]. Prior to 1960, the urban fringe was an integral component of China's agricultural economy, providing sustenance to China's small yet growing urban populations [13]. After the 1980s, China's urban fringe area underwent a fundamental transformation in terms of policy orientation, gradually turning into a region primarily for industrial development across the country and indirectly booming China's urbanization [14]. In the 1990s, particularly, a vast number of China's urban fringe areas had developed into known hubs with concentration effects in specific industries like clothing and toys [15]. Contrastingly, since the year 2000, China's urban fringe has increasingly become a multifaceted and multifunctional zone for developing a more diverse economy and better serving its urbanizing residents [16]. However, in the recent years, there have emerged new and sometimes daunting sustainability challenges across China's urban fringe areas [10], such as land use conflicts due to urban encroachment [17], social conflicts among diverse residents [18], poor living conditions because of infrastructure underinvestment [19], and growing crimes overwhelming weak governance capacity [20].

Due to the significance of the urban fringe as the frontier of China's urbanization and the intricacy of the urban fringe as influenced by complex urban–rural interactions, there have been long-lasting scholarly efforts in researching China's urban fringe since the 1980s. In the Chinese literature, a few reviews of varying depths have been published about the urban fringe, contributing by introducing the relevant international research insights [21], calling for enhancing China's urban fringe research [22], comparing the international and Chinese literature on the urban fringe [23,24], and synthesizing concepts on the urban fringe [25]. Contrastingly, in the international literature, reviews that systematically introduce the urban fringe research in China are still rare. One exceptional review was published recently, focusing on synthesizing the characteristics of China's urban fringe areas though [26].

In the above context, this study aims primarily to provide the global peers with the big picture of urban fringe research in China by synthesizing the relevant literature from, particularly, a spatial perspective. Our spatial perspective positions the urban fringe as a land system type with characteristics of its own in sustainable development and contributes to land-based approaches to advancing sustainability through spatial governance and the related human interventions. In this sense, our synthesis focuses on the spatial patterns and processes across China's urban fringe areas and underlying social-environmental mechanisms, as well as China's spatial governance practices concerning the urban fringe. The remainder of the paper is organized as follows: Sections 2 and 3 elaborate on the history and progress of urban fringe research in China, respectively; Section 4 presents some idiosyncratic thoughts on moving the existing research forward to better serve China's sustainable urbanization; and finally, Section 5 concludes with three notes in relation to the theme of this Special Issue, i.e., "Advancing Sustainability through Land-Related Approaches: Theoretical Foundations and Empirical Investigations."

2. History of Urban Fringe Research in China

Any long-lasting scientific field must echo the themes of its time. In retrospect, the urban fringe research in China has been influenced internally by China's rapid and widespread urbanization since the 1980s and externally by pioneering studies in mostly western countries that have a much longer urbanization history. In this vein, meaningful assessment of the urban fringe research in China needs to consider both aspects. We highlight representative terminology and scholars of urban fringe areas. The determination of the representative term is premised upon its first appearance in the published literature. By pinpointing the earliest instances of each term's use, we are able to trace the historical development of concepts within the field. The selection criteria of representative scholars are centered around their standing and acclaim within the academic community dedicated to this field, as well as the far-reaching citation impact their research works have exerted over time.

2.1. International Context of Urban Fringe Research: From the 1930s to Present

The emergence of the international research on urban fringes dated back to the 1930s when the German geographer Herbert Louis first the concept of urban fringe in 1936. His German studies used the term "Stadtrandzonen" to describe those areas on the outskirts of urban settlements [27]. However, the early root of urban fringe research could trace back to the "garden city" studies by the English urban planner Ebenezer Howard, who made perhaps the first attempt in the late 1890s to plan the landscape transitions from rural to urban [28]. By the late 1990s, more than 500 scholarly articles containing the term "fringe area" had been published in English. These studies covered a wide range of disciplines and yielded substantial findings (Table 1) [2,29]. At the emergence stage, a majority of the scholars focusing on the study of the urban fringe had their roots in geography and landscape ecology [30]. As deeply influenced by the "Age of Exploration", those emerging research efforts mainly concentrated on revealing the phenomena of land and landscape changes in the urban–rural interface zones, while largely neglecting the underlying processes and mechanisms that drive such changes [14].

The western research on the urban fringe experienced a rapid growth during the 1950s and the 1980s when the world recovered from WWII (World War II) and saw a relatively flourishing economy and associated urbanization. By the late 1960s, scholars had enhanced their investigations on the spatial differentiation of social-environmental elements in urban fringe areas. Such research led to a series of derivative concepts such as "inner fringe areas" and "outer fringe areas". Meanwhile, studies identified some common features of urban fringe areas, like periodicity, mobility, and relative stability [21,31,32]. The early 1970s witnessed urban fringes receiving new research interests from scholars with background in economics, sociology, and political science. Along with these multi-disciplinary research efforts involving land economics, sociology, demography, and political economy came a new emphasis beyond biophysical space (e.g., economic, social, and power spaces) and associated social-environmental processes within urban fringes [9]. Efforts were then made to explore the spatial patterns, implicit characteristics, and driving mechanisms of growing urban fringes. Relatedly, a series of theories were proposed, such as core–periphery theory, metropolitan structure theory, urban agglomeration structure theory, cyclical mechanism theory, community-driven theory, and commodification-driven theory [33–38].

Table 1. The emergence, development, and maturing of international urban fringe research.

Stage	Representative Scholars and Terms	Contents
Emergence stage	Redfield (urban–rural continuum); Wehrwein (rural–urban fringe); R. B. Andrews (rural–urban fringe)	In 1941, Redfield introduced the concept of the urban–rural continuum [39]. Subsequently, in 1942, Wehrwein proposed that fringe areas are transitional zones between clearly defined urban land and agricultural regions [40]. In the same year, R.B. Andrews suggested that the concept of the rural–urban fringe more comprehensively encapsulates the transition zone between urban and rural areas [41].
Development stage	Queen and Thomas (metropolitan region); W. C. Mckain and R. O. Bumight (the inner fringe, the outer fringe); R. G. Golledge (no-man's region); M. R. G. Conzen (inner fringe, intermediate fringe, outer fringe); E. J. Taaffe and B. J. Garner and M. H. Yeates (suburb); J. W. R. Whitehand (urban fringe belts); R. J. Pyror (rural–urban fringe); L. H. Russwurm (urban fringe)	From the early 1950s to the late 1970s, the formation of fringe areas developed in tandem with the phenomenon of suburbanization. In 1953, Queen and Thomas named this region the metropolitan region and divided this regional structure into three parts, the inner city, the urban fringe, and the urban hinterland, thus pioneering the research on the urban fringe area [42]. W. C. Mckain and R. O. Bumight further refined this by dividing the fringe area into the inner fringe and outer fringe [43].

Table 1. *Cont.*

Stage	Representative Scholars and Terms	Contents
Development stage	Queen and Thomas (metropolitan region); W. C. Mckain and R. O. Bumight (the inner fringe, the outer fringe); R. G. Golledge (no-man's region); M. R. G. Conzen (inner fringe, intermediate fringe, outer fringe); E. J. Taaffe and B. J. Garner and M. H. Yeates (suburb); J. W. R. Whitehand (urban fringe belts); R. J. Pyror (rural–urban fringe); L. H. Russwurm (urban fringe)	In 1960, Golledge referred to fringe areas as no-man's region [44]. That same year, British geographer M. R. G. Conzen introduced the concepts of the inner fringe, intermediate fringe, and outer fringe. He also proposed a cyclical theory, suggesting that the expansion of urban fringe areas undergoes cyclical changes consisting of three phases: acceleration, stasis, and deceleration [45].
		In 1963, E.J. Taaffe, B.J. Garner, and M.H. Yeates proposed an ideal city model, suggesting that an ideal city consists of a concentric structure made up of five parts: the central business district (CBD), the CBD fringe, the intermediate zone, the outer fringe, and the near suburb. The near suburb, composed of industrial, agricultural, and residential areas, is similar to what we refer to as the fringe area [46]. Weisick, on the other hand, defined this area as a "zone of great variation," due to its random development and mixed land-use patterns [47]. In 1967, J. W. R. Whitehand conducted research on the issue of locational competition in fringe areas. He emphasized that the planning and architectural forms in fringe zones are just as important as land use patterns, and he referred to these areas as the urban fringe belts [5]. In 1968, R.J. Pryor proposed that fringe areas are zones of land-use transition between urban and rural areas, serving as transitional zones for various elements and characteristics between the city and the countryside, which he called the rural–urban fringe [3]. In 1975, L. H. Russwurm proposed that urban structure should be composed of the core area, fringe area, impact zone, and rural region. He specifically pointed out that the urban fringe area is a continuous continuum between urban and rural areas, heavily influenced by urban radiation. It is a zone where urban development factors penetrate, and land-use transitions occur [48].
Maturing stage	Countryside Commission (Urban fringe)	In the 1970s, the Countryside Commission, through its research reports, emphasized the urban fringe as an important area characterized by the intermingling of urban development, agriculture, and open space [49].

Since the late 1970s, with the Countryside Commission's thematic symposium on urban fringe research in 1977 as sort of a landmark, global urban fringe research has seen more organized research activities and more cohesive research communities, entering more of a maturing stage [49]. Particularly, since the turn of the 21st century, with sustainability as the new theme of the time, this research field has seemed to be evolving toward a more use-inspired orientation, emphasizing increasingly practical solutions and policy implications. For example, some studies now examine urban fringes in the context of restructuring and transforming urbanization trajectories from monocentric to polycentric patterns [16].

2.2. Two Stages of Urban Fringe Research in China: From Late 1980s to Present

The urban fringe research in China is closely linked to China's rapid development and urbanization since the reform and opening up in 1978. Nonetheless, it was in the late 1980s that weakly managed urbanization started to bring about various issues associated with urban fringe areas, thus catching governmental and scholarly attention. Back then, the initial urban fringe research in China, like many other research fields in China, was more at an inheritance stage, largely focusing on introducing and applying the ideas, theories, and methods of the western pioneering studies to the context of China [50] (Table 2). However, Chinese studies were featured by their demarcation of urban fringe areas based on usually the official administrative boundaries adopted by the Chinese government instead of the urbanization pattern itself. This, consequently, led to a multiplicity of conceptualizations regarding urban fringe.

Table 2. The inheritance and expansion stages regarding to urban fringe research in China.

Stage	Representative Scholars and Terms	Contents
Inheritance stage	Ministry of Land (urban–rural integration); Chaolin, Gu and Gonghao, Cui (urban fringe); Youqi, Chen (urban–rural interlocking zone); Xueqiang, Xu (urban fringe belt)	In 1987, China's land and planning authorities first used the concept of the "urban–rural integration" in a series of guiding documents on urban land management, referring to a sub-urban zone emerging in rural areas surrounding cities [51]. From 1989 to 1993, Chaolin, Gu, Gonghao, Cui, Fulong, Wu, and Jin, Wu introduced the term "urban fringe zone" and translated it as "urban fringe area" while studying the fringe spaces of major cities [13,14,52,53]. In 1995, Youqi, Chen, in his published paper, proposed defining the transition zone between urban and rural areas as the "urban–rural interlocking zone" [7]. In 1997, Jianming, Zhang, Xueqiang, Xu, and others, after reviewing over a decade of domestic research on fringe areas, suggested that it should be referred to as the "urban fringe belt" [21].
Expansion stage	Lei, Fan (urban–rural fringe); Xiuqi, Fang (urban–rural transition zone); Wangming, Li, Qifeng, Yuan, and Shidong, Chen (metropolitan fringe); Qingyuan, Yang (metropolitan fringe area)	In 1998, Lei, Fan, based on an analysis of fringe area theory and the development context of the concept, first defined the term "urban–rural fringe" as areas where interactions and mutual influences between urban and rural spaces result in land use, population, and social characteristics displaying both urban and rural features [54]. In 2002, Xiuqi, Fang, based on his study of images of Beijing's urban area over the past century, proposed the concept of the "urban–rural transition zone," and argued that this zone is an indispensable part of urban areas [8]. In 2008, Wangming, Li, using the urban fringe area of Hangzhou as a case study, explored the mechanisms behind the decline in ecological carrying capacity from the perspective of ecological carrying capacity and referred to this area as the "metropolitan fringe" [55]. In 2015, Qifeng, Yuan, Shidong, Chen, and others conducted research on the issues of urban–rural integration in the fringe areas, focusing on the peasants and related relationships under rapid urbanization in Guangzhou, and continued to use the term "metropolitan fringe" [56]. In 2015, Qingyuan, Yang introduced the concept of the "metropolitan fringe area" in her study on land use impact in Chongqing's fringe areas. Metropolitan fringe areas are dynamic transitional zones on the outskirts of large cities, balancing urban expansion and rural transformation with high complexity and diversity [57].

After joining in the WTO (World Trade Organization) in 2001, China has experienced unprecedented development and urbanization, along with its rapid growth in population, thus leading to remarkably growing urbanization-associated challenges in, particularly, urban fringes. The insufficiency of China's governance system that was originated from a 5000-year rural history became increasingly prominent, given that China officially turned into an urban society in 2010. Issues at urban fringe areas have gained elevated momentum in the recent years, e.g., inequalities in land utilization, population mobility, and resource allocation [57,58]. These issues are largely rooted in China's characteristic dual-governance system, with its urban part more of a market economy and the rural part retaining features of a planned economy in many aspects (e.g., rural land use rights cannot be directly traded in the national market except in the few policy pilot areas). That said, these Chinese-featured issues call for more place-based and actionable research, giving the urban fringe research in China more emphases different from those of the western counterpart.

3. Research Progress of Urban Fringe Areas in China

In line with the general pattern of epistemology, existing research on China's urban fringe areas can be roughly classified into three categories, from the spatial patterns of China's urban fringe dynamics (i.e., empirical regularities), to their underlying mechanisms (i.e., theoretical rationales), and then, to practical implications (i.e., informed actions). Roughly speaking, the most research output is on the spatial patterns, followed by the practical studies, and last are those mechanistic studies. Increasingly, however, the usefulness of the practical studies relies on a deeper understanding of the social-environmental mechanisms underlying various spatial patterns within urban fringes.

3.1. *Spatial Patterns of China's Urban Fringe Dynamics*

From a spatial perspective, urban fringe development is characterized by a transformation of predominantly rural landscapes into a mosaic of urban–rural land uses [9]. To date, the spatial patterns in urban fringe areas remain a topic of substantial research interest. Such research encompasses not only the area's physical morphology dynamics but extends to its economic, social, and environmental aspects, integrating elements such as land use, industrial activities, social structure, and ecological functioning in a comprehensive manner.

3.1.1. Physical Morphology of Urban Fringe Areas Shaped by Land Use

Urban fringes in China are afflicted with a multitude of land-use conflicts, mainly stemming from the conversion of agricultural land to accommodate urban development [59]. The urban–rural dualistic governance regime of the housing, land, and population registration systems following China's reform and opening up in 1978 set a starting point for the initial layout of land use in urban fringe areas [60]. During the early 1990s, with the intensification of economic activities, urban fringe areas gradually witnessed land value rises due to spillover from the expanding urban centers. The expansion of built-up areas, accompanied by their encroachment on agricultural lands and the emergence of new mixed-use developments, has shaped the physical morphology of urban fringe areas [61]. In this vein, a considerable number of studies emerged with a focus on the spatial structure and characteristics of the urban fringes around major Chinese cities. Scholars classified the patterns of urban expansion morphology in urban fringe areas into five types, i.e., finger-like, fan-shaped, strip-shaped, scattered-bead, and network, revealing the diversity of empirical regularities of urban development within China's urban fringes [52,53].

3.1.2. Industrial Evolution in Urban Fringe Areas

Meanwhile, the urban fringe areas in China have been the sites of a sustained and ongoing process of industrial evolution. Following the pervasive implementation of local tax-sharing systems and economic zones in the 1990s, a considerable number of labor-intensive industries were attracted to China's urban fringe areas due to their relatively lower labor and land costs compared to their urban counterparts [62]. This resulted in a fragmented pattern of land use in urban fringe areas. In accordance with the historical context of industrial expansion and clusters in the 1980s and 1990s, the spatial pattern of the urban fringe underwent a process of differentiation, with the emergence of distinct industrial zones [63]. The subsequent construction of knowledge cities and university towns in the 2000s gave rise to more complex spatial patterns in urban fringe areas. It was accompanied by the conversion of agricultural and rural land into more lucrative residential, commercial, and industrial developments. In recent years, there has been a growing recognition that the spatial layout of the urban fringe has evolved from a concentration of similar industries to a more diverse and mixed-use pattern [64].

3.1.3. Social-Spatial Differentiation Resulting from Social Stratum Segmentation

The social stratum segmentation has been pivotal in shaping the spatial pattern of urban fringe landscapes, which in western countries is known for urban gentrification. In China, a lesser degree of spatial segregation was observed between high-income and low-income populations in urban fringes [65], which impacted the configuration of transportation infrastructures, access to public services, social interaction, and community cohesion [19,66]. Those high-income communities frequently benefited from priority access to public services, housing, and other resources, whereas the low-income communities were constrained to striving for mere survival and development with limited surplus resources. Still, the unequal distribution of resources has resulted in disparate spatial developments. Urban fringe areas were characterized by the clear demarcation of high-end residential neighborhoods from urban villages, giving rise to the formation of distinct social-spatial landscapes. Nonetheless, with the acceleration of urban development and the intensification of social exchanges, there emerged an increasing tendency towards spatial integration in China's urban fringe areas. The interaction of diverse social groups has contributed to the social-spatial evolution of the urban fringe, from an initially simple urban–rural transition structure to a complex space system. The spatial distribution of residences, consumption patterns, and employment situations among various social classes has been continuously adjusted over time. This has propelled the spatial processes within urban fringe areas to develop in a more diversified and inclusive manner [67,68].

3.1.4. Green Space and Degraded Ecosystem Functioning in Urban Fringe Areas

Large-scale and spatially continuous green spaces have long been a prominent feature of China's urban fringe areas. However, during the last few decades, the rapid urban expansion has led to substantial changes in ecological patterns and the degradation of ecosystem functions in urban fringes [69]. In this regard, the conversion of agricultural land into built-up areas has severely disrupted food production. Moreover, the encroachment onto wetlands and forested areas has given rise to the weakening of water yield, the loss of flooding control, and the destruction of wildlife habitats. Given the increasing recognition of the complex interplay between urban growth and environmental integrity, the focus of urban fringe studies has shifted towards various examinations of urbanization-associated eco-environmental impacts, particularly through landscape ecological perspectives [70,71]. Studies have identified the most typical landscape ecological patterns observed in China's urban fringes, i.e., concentric, scatter, network, and radial structures. Thereinto, the concentric layout depends on the configuration of forest and water bodies that encircle urban land, thereby forming a separation belt between urban and rural landscapes. The network layout is primarily composed of interconnected ecological nodes ad corridors. Such elements contribute to safeguarding the integrity of ecological landscapes and guaranteeing long-term benefits. The radial pattern is determined by a complex combination of natural conditions and transportation routes, while the scatter layout is shaped by the distribution of ecological patches [72,73].

3.2. Spatial Mechanisms Underlying China's Urban Fringe Dynamics

As documented in most studies, the drivers of urban fringe dynamics in China are a blend of economic growth, industrial development, transportation infrastructure optimization, rural–urban migration, and government policies [74]. Among these multifaceted drivers, rapid economic development is a prerequisite for the expansion of urban fringes. Since the late 20th century, extensive research has been dedicated to exploring the relationship between urbanization processes and urban expansion. This discourse has emerged in the context of urban centers confronting the challenges of economic structure adjustment and industrial upgrading. Manufacturing sectors have been compelled to relocate to urban fringe areas due to the urban hinterland's demand for large residential, commercial, and public service spaces [75]. Consequently, there was a noticeable rise in the conversion of agricultural land to industrial land in urban fringes. Meanwhile, the construction of the transport network has improved connectivity between the core areas and the surrounding suburbs [14]. Certain factories that were heavily reliant on transportation had a concentrated and spatially continuous development pattern along the main roads and/or rivers in the inner parts of urban fringes. Shanghai is among the most striking instances of urban fringe expansion primarily due to economic drivers. The establishment of industrial zones in urban fringe areas was proposed in the Shanghai Master Plan in 1959 [76,77]. It described how the exploding population would be navigated to the urban fringe during different periods and presented the layout of the "inner edge belt" (Figure 1a). From 1959 to the 1970s, industrial clusters in the urban fringe areas of Shanghai started to form, integrating industrial parks with residential zones. Following this period, satellite towns including Minhang, Wujing, and Songjiang progressively emerged and experienced development. These strategic plans played a crucial role in maintaining a well-organized industrial layout while effectively curbing the uncontrolled expansion of the urban boundary. Moving into the 1980s and thereafter, the overcrowding phenomenon in central Shanghai became increasingly pronounced. Consequently, portions of the population shifted away from the central urban area. This shift necessitated ongoing investments in the infrastructure and public services of the urban fringe areas. Such efforts have significantly contributed to the development of the city's multi-center spatial structure. Especially with the expansion of the industrial activities in the planned areas of Jinshanwei and Wusong-Baoshan, a large amount of agricultural land was transformed into industrial land. As an integral part of the environmental enhancement project initiated in recent years, these regions underwent a further transformation into land dedicated to the development of eco-friendly chemical industries, business operations, tourism-related undertakings, and information technology-based activities. Over the past several decades, the proportion of developed land in the urban fringe of Shanghai has reached 66% [78]. In a sense, Shanghai pioneered in the decentralization of industrial activities from the urban core to the fringe.

The phenomenon of rural–urban migration has also been widely acknowledged as a crucial factor in the evolution of urban fringes. On one hand, the expansion of industrial activities in the urban fringe promoted a significant influx of migrants from rural regions [80]. The employment opportunities to rural migrants and the possibility of their turning into urban household registration were especially attractive [81]. On the other hand, urban residents in pursuit of spacious living conditions and exhibiting sensitivity to housing prices congregated in urban fringe areas [58]. The migration of populations to urban fringe areas further stimulated the development of residential settlements, public infrastructure, and traffic networks.

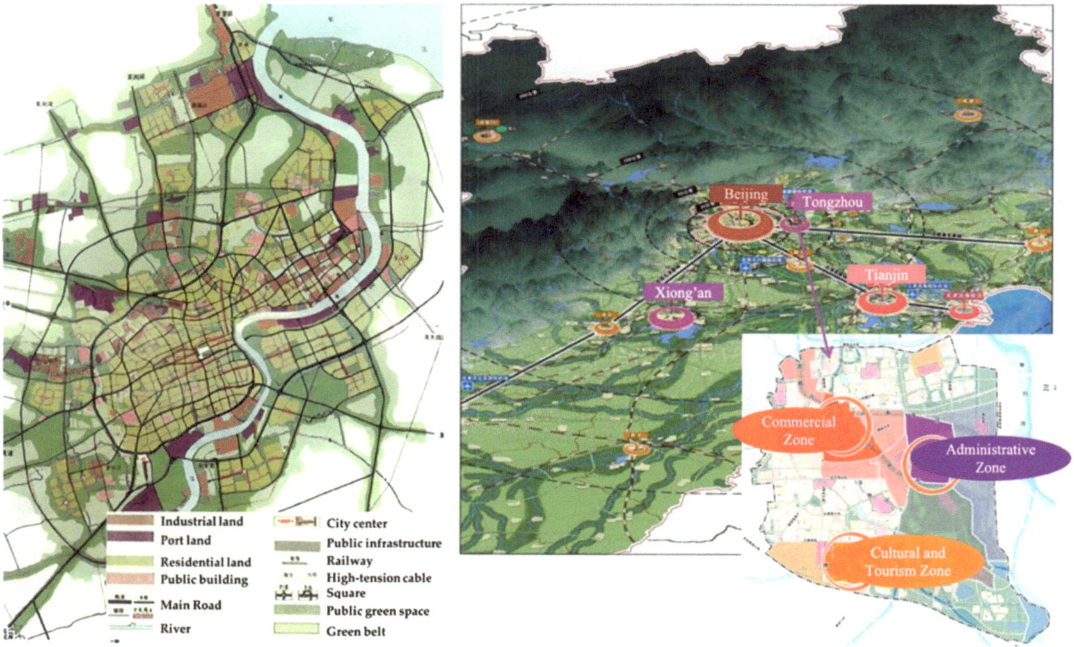

(a) (b)

Figure 1. Urban planning in Shanghai and Beijing City. (**a**) Shanghai Master Plan in 1959. (**b**) The Planning of Beijing and Tongzhou districts from 2016 to 2035. (**a**) from *The Evolution of Urban Planning in Shanghai* edited by Shanghai Urban Planning and Design Institute, p47 [76], translated by the author. (**b**) from *Regulatory Detailed Planning of Beijing Sub-Center (Block Level) (2016–2035)* [79], edited by the author.

Moreover, governmental interventions at various levels regarding land use, urban planning, and infrastructure investments exert a decisive influence on the trajectory of urban fringes [82]. In contrast to the largely bottom-up processes in urban fringes of western countries, the expansion of the urban fringe in China represents a strong top-down feature driven by diverse policy interventions [83]. The development of transport and infrastructure networks is a central policy tool for governing urban fringes. Local governments offer tax incentives, subsidies, and other financial benefits within industrial zones targeting the urban fringe as a way to bridge economic disparities. In addition, the establishment of new urban centers in urban fringe areas has been regarded as an avenue for facilitating technological innovation. One of the primary examples of how governance and policy drive the transformation of an urban fringe area into a pivotal node of urban expansion is the Tongzhou district of Beijng City (Figure 1b) [79]. The strategy of relocating government functions and constructing a national capital sub-center in Tongzhou district prompted the outward migration of the original inhabitants from the older urban areas to the originally fringe areas. Supported by the planning of residential, commercial, and industrial spaces, Tongzhou has now become a hot spot for immigration and investment. The population growth facilitated industrial upgrading, with policies encouraging the development of high-tech industries replacing traditional manufacturing. At the same time, rapid urban expansion led to significant changes in land use, particularly, the reduction in farmland and green spaces. These shifts, while boosting economic vitality, also intensified ecological challenges, such as increased air pollution and strain on water resources, showcasing the trade-offs of policy-driven urban transformation.

3.3. Spatial Governance of China's Urban Fringe Areas

It is of paramount importance to ensure the orderly utilization of land resources in urban fringes. The land use within the urban fringe is characterized by the expansion of developed areas, the accelerated non-agricultural transformation, and alterations in the agricultural structure [84]. Along with China's remarkable economic success in the past few decades, the Chinese government has also made notable progress in advancing its governance capacity, which also applies to their management of urban fringes. The urban–rural dualistic land tenure system enables the Chinese government to intervene in regional land use through systematic planning and management endeavors. In this regard, farmland protection is a long-standing policy objective, which is achieved through the demarcation of conservation areas (i.e., "farmland redline") and the execution of requisition–compensation balance measures [85]. In addition, the location of industrial zones has been planned by taking into account both the convenience of transportation and the potential environmental repercussions. Moreover, to enhance land-use efficiency and expand the scale of agricultural operations, a series of land reform initiatives have been initiated in rural areas, such as the intensified management of land resources and the transfer of land contractual rights [86,87]. These policies have exerted recognized impacts on China's urban fringe dynamics.

In the recent years, the Chinese government has made more spatial governance efforts for improving environmental sustainability, which used to receive insufficient attention. In this regard, the eco-environment in urban fringes is highly vulnerable to human activities, and thus, both central and local governments have formulated various policy interventions. These measures range from the imposition of strict land-use regulations to the execution of ecological restoration projects. They also involve the establishment of green belts and wildlife corridors for the protection of biodiversity in urban fringes. Specifically, the Ecological Redline policy represents a sustainability bottom line (similar to the Farmland Redline) aimed at curbing unregulated urban sprawl and strengthening the protection of natural elements [88]. Additionally, public participation initiatives have been increasingly launched to deal with environmental issues [89]. For example, the establishment of public-friendly platforms allows the public to report environmental violations, while also cultivating environmental awareness and promoting green lifestyles among residents [90].

Last but not the least, the effective planning and management of public open spaces in urban fringes is considered as a way of enhancing the well-being of local residents. However, public open spaces in urban fringe areas face challenges of relatively chaotic development and limited functionality. With the rapid expansion of the built-up environment in urban fringes, illegal occupations of public open spaces occur frequently. Unfortunately, the governance responsibilities concerning public open spaces are currently scattered among relevant departments. It has been posited that the governance of public open spaces should be based on effective coordination and regular supervision [91]. Another proposal is to promote a community-based participatory approach given that public open spaces are not just physical entities but integral components of community life [92]. This approach is beneficial for sustaining the natural landscape of public open spaces considering the diverse interests of the various stakeholders in the urban fringe. Overall, a balance between protection and utilization is needed regarding public open spaces in urban fringe areas.

4. Research Prospects Toward Sustainable Urbanization in China's New Era

4.1. Differences in the Urban Fringe Research Between China and the West

Although the central concerns within the realm of urban fringe research in China and the West share commonalities, there are distinct variances in the objects of investigation as well as the methodologies. In China, research on urban fringe areas typically focuses on the land-use changes during the urbanization process, the ambiguity of the urban–rural demarcation, and the resultant spatial governance challenges. In recent years, scholars have explored the coupling relationship between land development and tenure in urban fringe areas, as well as the role of administrative spillover effects in the spatial governance of these areas, through methods such as structural equation modeling [93,94]. In contrast, urban fringe research in Western countries lays more emphasis on the impacts of urban sprawl on society, the economy, and the environment [95]. Especially under the backdrop of globalization and urbanization, issues such as the "leapfrog" expansion of cities and the livelihoods of residents in fringe areas are of great concern [96]. In some developed countries, research regarding urban fringe areas also involves the construction of diversified ecological networks, aiming to mitigate the negative impacts of urban expansion on ecosystems.

Yet another significant difference in the urban fringe research between China and the West lies precisely in the choice of research methods. Chinese scholars tend to employ quantitative methods, such as remote sensing technology and spatial analysis, to explore the evolutionary process of urban fringes and changes in land use [97]. In contrast, western research more frequently uses the case-study method, by combining multidisciplinary perspectives from sociology, geography, and ecology to analyze the interrelationships among social dynamics, environmental changes, and land use in urban fringe areas [98].

Overall, with the advancement of globalization, there are likely to be more intersections in future research on urban fringe areas between China and the West, especially in terms of how to achieve sustainable development and coordinated urban–rural development.

4.2. Understanding New Patterns of Social-Environmental Dynamics Across China's Urban Fringe Areas in the New Era

The phenomenon of urban fringes and its characteristics evolve with time. The dynamism and complexity of urban fringes will intensify or diminish depending on societal development needs. It is imperative to reexamine the dominant development challenge within China's urban fringes in the post-2012 New Era with new norms of economic development, urbanization trends, and political tendency, among many other aspects. For a long period of time after China's reform and opening up, the officially identified dominant development challenge to China is "the contradiction between Chinese people's ever-growing material and cultural demands and the backward social production [99]." As China officially declared its entering the New Era of socialism with Chinese characteristics in 2012, the dominant development challenge was re-identified as "the contradiction between Chinese people's ever-growing needs for a better life and inadequate, unbalanced development [100]." In such a context, it is expected that China's urban fringes would present, in the upcoming few decades, new patterns different from those that have been studied.

For example, environmental sustainability is currently one of the inadequately developed dimensions across China at large and, similarly, in its urban fringes [101]. Unlike in foreign countries, China's urban fringe development has been predominantly driven by economic factors and industrial activities [25], thereby resulting in more adverse impacts on the environmental system. Luckily, in the past few years, there has been a significant push towards ecological restoration initiatives in China's urban fringes. These initiatives, such as the greening of wastelands and the enhancement of water systems, are designed to rehabilitate damaged or degraded habitats that have been affected by previous urban fringes expansion and development activities [102]. Consequently, future policies for developing urban fringes in China may see more integration of President Xi Jinping's idea of ecological civilization with economic, social, political, and cultural development.

For another instance, to address the unbalanced development issue, which will likely remain a long-lasting challenge facing China's urban fringe areas, future policies may prioritize developing areas currently underdeveloped, e.g., fringe areas of the small and medium sized cities, fringe areas of the shrinking cities [103], and unique cultural landscapes within a fringe area. Particularly, there are rich cultures deeply rooted in urban fringe areas with promising prospects for the development of tourism and cultural industries [13,104]. Recent studies have shown that through the introduction of diverse commercial activities, cultural exhibitions, and comprehensive services, some of those traditional villages are undergoing revitalization. Meanwhile, this gives rise to a wide range of employment opportunities and prompts the return of the indigenous population to the urban fringe villages [105]. Therefore, the new policy norms of urban fringe development in the New Era calls for timely research on potentially new patterns of environment- and balance-oriented development activities.

4.3. Promoting Spatial Governance of China's Urban Fringe Areas from a Coupled Urban–Rural System Perspective

The urban fringe is initially defined as the transitional zone between urban and rural areas, long studied as a "side" product of studying the urban core [106]. After many years, the urban fringe has now been recognized in China as a research unit of its own kind for advancing sustainability. Yet, from a meta-coupling sustainability perspective, it is still insufficient. Urban fringe areas, while presenting unique social-environmental patterns and facing unique development challenges, are closely coupled with the rural and urban subsystems. It is the coupled urban–rural system instead, rather than the urban fringe (or the rural/urban), that makes a meaningful governance unit of sustainability. For example, the planning and building of an ecological network of green spaces for an urban–rural system would be more appropriate than for, merely, the urban fringe. In other words, sustainability governance at the landscape/regional or above scale makes more sense.

The spatial governance of China's urban fringe areas from a coupled urban–rural system perspective, as elaborated by Tan et al. (2024) in this Special Issue, involves not only the various stocks of social-environmental elements (e.g., labor, resources, capital), but also the uni-/bi-directional flows of social-environmental interactions (e.g., migration, ecosystem services, trade), as well as the overall functional complementarity between heterogeneous spatial units (e.g., agriculture-oriented, environment-prioritized, urban-oriented) [107]. Urban fringe areas are the bridging zones between the rural and the urban, and thus, at the core of diverse elements, flows, and functions [108]. In the sustainability community, it is often advocated to "think globally, plan regionally, and act locally." Yet, for the spatial governance of China's urban fringe areas in the future, the most challenging may be the transboundary governance as a coupled urban–rural system, which involves different administrative units, and each may have its own socio-political architecture and policy agenda.

The good news is that there has been more and more transboundary spatial planning explored in China, e.g., metropolitan area planning and watershed planning [109–111]. However, an inclusive sustainability blueprint benefiting the whole urban–rural system is far from sufficient. The "dirty" work for realizing such cooperation interests lies in effective coordination between vertical and horizontal governments and among the government, society, and market. Sustainability is not an end, but a journey. Such muti-agent coordination in various action arenas and situations for the spatial governance of coupled urban–rural systems remains an underexplored task. In this regard, insights from political sciences, public administration, and public policy, such as polycentric governance, collaborative governance, and adaptive governance, may provide complementary actionable knowledge in addition to the existing planning-centered literature and practices [112,113].

4.4. Leveraging Artificial Intelligence for Producing Spatially Actionable Sustainability Knowledge

Another missing point in the existing Chinese literature on the urban fringe is how to leverage artificial intelligence for producing actionable sustainability knowledge. We are now in an era with more and more freely available geospatial big data, which are improving toward better spatiotemporal resolutions [114]. Such improved data in combination with better big-data models has shown encouraging promises in areas including but not limited to high-resolution monitoring, predictive modeling, and real-time decision making while studying various land system types or landscapes in general, though barely regarding urban fringe areas.

In terms of high-resolution monitoring, the landscape, population, and other social-environmental dynamics across various urban fringe areas or even nature can now be better mapped based on integrating big earth data via platforms like Google Earth Engine. Better data and better models are more inclined to produce better maps. Similar studies have been published on mapping global urban land, national wetland, and forests, and even on monitoring progress toward the United Nation's Sustainable Development Goals [115–117].

In terms of predictive modeling, the capability of artificial intelligence to fuse socioeconomic data (e.g., census records, infrastructure databases, transportation network information) with environmental data (e.g., climate variables, soil quality, biodiversity hotspots) would better capture the often non-linear and complex social-environmental interactions, thus producing more accurate modeling predictions. Similar studies are very rich now, for instance, the GeoSOS-FLUS model (http://www.geosimulation.cn/FLUS.html, accessed on 20 October 2024) [118,119]. Following these studies, thematic predictive models may be developed with tailored model modules for capturing urban fringe processes.

In terms of real-time decision making, it is not methodologically challenging to employ artificial intelligence algorithms on continuously updated streams of sensor data to create interactive, real-time dashboards in areas like transportation and disaster risk management [120,121]. However, for applications in urban fringe research, an additional issue is how to integrate the existing methodological practices with citizen science and participatory decision making. In this regard, ethical and transparent digital governance seems especially understudied.

5. Concluding Notes

So, how does our study contribute to advancing sustainability through land-related approaches? The above synthesis reveals a somewhat "narrow-minded" research landscape on the urban fringe in the Chinese literature—one that focuses on the spatial patterns and underlying mechanisms of urban land expansion across China's urban fringe areas and that relates to sustainability mostly concerning the various unsustainable outcomes of such urban land expansion. We argue that the urban fringe is a land system type featuring a complex adaptive social-environmental system of its own kind, distinctive from the urban and rural counterparts that it couples with. Such an up-to-date understanding of the urban fringe per se should serve the purpose of better understanding and governing the multidimensional spatial tradeoffs and synergies of the coupled urban–rural system with land/space as an entry point and intervention instrument.

It is a prerequisite to explore the new patterns or long-term trends of land-use changes and associated social-environmental dynamics across China's urban fringe areas. We identify that the issue of environmental sustainability has become particularly prominent in urban fringe areas. Integrating the collaborative strategies of ecological restoration and land development in the urban fringe areas could be an innovative direction in the future.

The spatial governance of China's urban fringe areas from a perspective of coupled urban–rural systems remains an underexplored task. An inclusive sustainability mindset calls for producing spatially actionable knowledge to adaptively balance the multidimensional outcomes of urban expansion and associated landscape dynamics in the urban fringe and its coupled urban and rural systems.

Moreover, the blossoming artificial intelligence technologies have introduced growing opportunities that could enhance data analysis, predictive modeling, and decision-making processes toward more effective spatial governance efforts, which have remained underrecognized.

To this end, we urge scholars, practitioners, and policymakers to collaboratively reconceptualize China's urban fringe areas—not as peripheral zones of unchecked growth, but as pivotal landscapes where strategic research, inclusive governance, and cutting-edge technologies converge—to transform these areas into catalysts for sustainable urbanization. We would like to propose embracing a more inclusive mindset in future research on China's urban fringe areas, with our sincere hope to advance multi-scale spatial governance of the urban fringe toward sustainable urbanization in China's New Era.

Author Contributions: Conceptualization, Z.L., T.H., and L.Z.; methodology, Z.L. and T.H.; software, L.Z. and Y.W.; writing—original draft preparation, Z.L., T.H., and L.Z.; writing—review and editing, Z.L., T.H., L.Z., and Y.W.; funding acquisition, Z.L. All authors have read and agreed to the published version of the manuscript.

Funding: This work was supported by the National Natural Science Foundation of China (42261046).

Data Availability Statement: No new data were created or analyzed in this study. Data sharing is not applicable to this article.

Conflicts of Interest: The authors declare no conflicts of interest.

References

1. National Bureau of Statistics of China. Significant Improvement in Urbanization Level and a New Look of Cities—Economic and Social Development Achievements Series Report on the 40th Anniversary of Reform and Opening Up, Chapter Eleven. Available online: https://www.stats.gov.cn/zt_18555/ztfx/ggkf40n/202302/t20230209_1902591.html (accessed on 5 April 2024).
2. Ban, M.; Fang, C. Progress in Research on Urban Fringe and Basic Frame of Research in the Future. *Urban Plan. Forum* **2007**, *3*, 49–54.
3. Pryor, R.J. Defining the Rural-Urban Fringe. *Soc. Forces* **1968**, *47*, 202–215. [CrossRef]
4. Browder, J.; Bohland, J.; Scarpaci, J. Patterns of Development on the Metropolitan Fringe: Urban Fringe Expansion in Bangkok, Jakarta, and Santiago. *J. Am. Plan. Assoc.* **1995**, *61*, 310–327. [CrossRef]
5. Whitehand, J.W.R. Urban fringe belts: Development of an idea. *Plan. Perspect.* **1988**, *3*, 47–58. [CrossRef]
6. Guo, X.; Li, H.; Mu, M.; Mai, F. Spatial evolution, influencing factors, and governance paths of megacity fringe areas: Taking Baiyun District of Guangzhou City as an example. *Prog. Geogr.* **2024**, *43*, 1744–1755.
7. Chen, Y. Definition of urban-rural interlocking zone. *Geogr. Geoinf. Sci.* **1995**, *11*, 47–52.
8. Fang, X.; Zhang, W.; Zhang, L.; Luo, H.; Li, Z. The Urban Expansion and the Evolution of Urban Fringe in Beijing in the 20th Century. *City Plan. Rev.* **2002**, *26*, 56–60.
9. Yang, S. Study on the Spatial Dynamics and Mechanisms of Urban Fringe Areas. *Geogr. Geoinf. Sci.* **1998**, *14*, 19–23.
10. Liang, C.; Zeng, J.; Xin, R. Evolution of Ecological Adaptability in Urban Fringe Areas: Reflections and Paradigm Shifts. *Urban Probl.* **2023**, *7*, 13–20.
11. Allen, A. Environmental planning and management of the peri-urban interface: Perspectives on an emerging field. *Environ. Urban.* **2003**, *15*, 135–148. [CrossRef]
12. Deng, W.; Jiang, Z.; Zhang, S.; Ren, P.; Zhang, H.; Wang, Z. Dynamic Identification and Evolution of Urban-suburban-rural Transition Zones Based on the Blender of Natural and Humanistic Factors: A Case Study of Chengdu, China. *Chin. Geogr. Sci.* **2024**, *34*, 791–809. [CrossRef]
13. Cui, G.; Wu, J. The Spatial Structure and Development of Chinese Urban Fringe. *Acta Geogr. Sin.* **1990**, *45*, 399–411.
14. Jin, W.; Ma, Q. Analysis of the Mechanisms of Spatial Structure Evolution in Urban Fringe Areas. *City Plan. Rev.* **1990**, *2*, 38–42, 64.
15. Zhang, T. The Urban Restructuring of Chinese Cities in the 1990s and Its Dynamic Mechanism. *City Plan. Rev.* **2001**, *25*, 7–14.
16. Rong, Y.; Rong, S.; Zhang, Y. A Review of Researches on the Urban Fringe. *Urban Plan. Forum* **2011**, *4*, 93–100.
17. Yang, L.; Fang, C.; Chen, W.; Zeng, J. Urban-rural land structural conflicts in China: A land use transition perspective. *Habitat Int.* **2023**, *138*, 102877. [CrossRef]
18. Liu, D. Governance Issues and Strategies in Urban-Rural Fringe Areas. *Peoples Trib.* **2020**, *23*, 72–73.

19. Liu, J.; Zhang, Y.; Liu, J. The Movement in the Urban Fringe: A Case Study of Zhenru Town in Shanghai. *City Plan. Rev.* **2000**, *4*, 41–43, 64.
20. Huo, X. Research on Social Environment Problems and Optimization Approaches in Urban Fringe: Based on the Perspective of Social Governance Innovation. *Econ. Probl.* **2017**, *11*, 125–128.
21. Zhang, J.; Xu, X. Review and Perspective of the Studies on Rural-Urban Fringe. *Human Geogr.* **1997**, *12*, 9–12, 37.
22. Luo, Y.; Zhou, C. The Review and Outlook on Urban-rural Fringe Studies in China. *Urban Dev. Res.* **2005**, *12*, 25–30.
23. Ma, X. A Review of the Driving Mechanism of Spatial Reconstruction in Urban Fringe Areas. *Spec. Zone Econ.* **2012**, *10*, 251–254.
24. Mao, Y.; Zhang, Y. Research Overview of Urban Fringe Domestic and Overseas. *Dev. Small Cities Towns* **2019**, *37*, 5–11.
25. Xue, B.; Fu, B. Conceptual evolution and classification system reconstruction of urban fringe. *Arid Land Geogr.* **2023**, *46*, 1903–1914.
26. Wang, K.; Mell, I.; Carter, J. Characterising the urban-rural fringe area (URFA) in China: A review of global and local literature on urban-rural fringe areas. *Town Plan. Rev.* **2024**, *95*, 617–642. [CrossRef]
27. Louis, H. *Die Geographische Gliederung von Gross-Berlin*; Engelhorn: Berlin, Germany, 1936.
28. Ebenezer, H. *Garden Cities of To-Morrow*; Swan Sonnenschein: London, UK, 1902.
29. Zhang, Y.; Wang, Y.; Cui, L. Ecology Oriented City Fringe Area Planning: Luoyang Longmen Mountain Ecological Preservation Planning. *Planners* **2016**, *32*, 132–139.
30. Ye, L.; Xing, Z.; Yan, W. Reflection on Eco-oriented Planning Strategies of Peri-urban Area. *Urban Plan. Forum* **2011**, *55*, 68–76.
31. Anas, A.; Arnott, R.; Small, K.A. Urban Spatial Structure. *J. Econ. Lit.* **1998**, *36*, 1426–1464.
32. Tu, R. Preliminary Study on Urban Fringe Areas: A Case Study of Wuhan City. *Geogr. Geoinf. Sci.* **1990**, *11*, 36–39.
33. Friedmann, J. *Regional Development Policy: A Case Study of Venezuela*; MIT Press: Cambridge, UK, 1966.
34. He, N. Research on Contemporary Urbanization Theories in China (1979–2005). Ph.D. Thesis, Fudan University, Shanghai, China, 2007.
35. Tang, Z. Descriptions and Explanations of Urban Spatial Structure: A Review of Research Developments. *Urban Plan. Forum* **1997**, *6*, 1–11, 63.
36. Zhang, X. Foreign Urban-fringe Studies and Its Revelation. *Urban Plan. Int.* **2005**, *20*, 72–75.
37. Guo, S.; Rong, Y.; Zhang, Y.; Wu, G. City Periphery Development Research. *Planners* **2012**, *28*, 57–62.
38. Soko, Y. *Urban Geography*; Hubei Education Press: Hubei, China, 1986.
39. Redfield, R. *The Folk Culture of Yucatan*; University of Chicago Press: Chicago, IL, USA, 1941.
40. Wehrwein, G.S. The Rural-Urban Fringe. *Econ. Geogr.* **1942**, *18*, 217–228. [CrossRef]
41. Andrews, R.B. Elements in the Urban-Fringe Pattern. *J. Land Public Util. Econ.* **1942**, *18*, 169–183. [CrossRef]
42. Queen, S.A. *The American City*; McGraw-Hill: New York, NY, USA, 1953.
43. McKain, W.C.; Burnight, R.G. The Sociological Significance of the Rural-Urban Fringe from the Rural Point of View. *Rural Sociol.* **1953**, *18*, 109–116.
44. Golledge, R.G. Sydney's Metropolitan Fringe: A Study in Urban-Rural Relations. *Aust. Geogr.* **1960**, *7*, 243–255. [CrossRef]
45. Conzen, M.R.G. *Alnwick, Northumberland: A Study in Town-Plan Analysis*; George Philip: London, UK, 1960.
46. Taaffe, E.J.; Garner, B.J.; Yeates, M.H. *The Peripheral Journey to Work: A Geographic Consideration*; Published for the Transportation Center at Northwestern University; Northwestern University Press: Evanston, IL, USA, 1963.
47. Wissink, G.A. *American Cities in Perspective: With Special Reference to the Development of Their Fringe Areas*; Royal Vangorcum: Assen, The Netherlands, 1962.
48. Russwurm, L.H. *Urban Fringe and Urban Shadow*; Holt, Rinehart and Winston: Toronto, ON, Canada, 1975.
49. Harrison, C. Countryside Recreation and London's Urban Fringe. *Trans. Inst. Br. Geogr.* **1983**, *8*, 295–313. [CrossRef]
50. Ren, B.; Xing, Y. Visual Analysis of Domestic Urban Fringe Research Based on the CiteSpace. *Urban Archit.* **2022**, *19*, 112–117.
51. Yu, Q.; Chen, Y.; Xu, X. Concept Discrimination on "Rural-Urban Fringe" in Land Use Science. *China Land Sci.* **2010**, *24*, 46–51.
52. Gu, C.; Chen, T.; Ding, J.; Yu, W. The Study of the Urban Fringes in Chinese Megalopolises. *Acta Geogr. Sin.* **1993**, *48*, 317–328.
53. Wu, F. Types and Characteristics of Urban Communities in China. *Urban Probl.* **1992**, *5*, 24–27.
54. Fan, L. Study of the Concept and Theory of the Rural-Urban Fringe. *Tianjin Business Univ. J.* **1998**, *3*, 29–34.
55. Li, W.; Wang, J.; Ge, D. Analysis of ecological carrying capacity change and driving factors at urban fringe: A case of Hangzhou city. *J. Zhejiang Univ. (Eng. Sci.)* **2008**, *42*, 39–43.
56. Chen, S.; Yuan, Q. Effects of Space Control Policy to Different Types of Property Rights and Their Influences in Urban Fringe of Guangzhou, China: A Perspective of Chinese Basic Farmland Conservation Area Policy. *Sci. Geogr. Sin.* **2015**, *35*, 852–859.
57. Mu, W.; Yang, Q. Effects of Land Use Planning on the Landscape Pattern in Metropolitan Fringe Area. *Hubei Agric. Sci.* **2015**, *54*, 4199–4204.
58. Chen, C.; Liu, A.; Zhou, W. Reconstruction of Population Spatial Structure in City Fringe under the Background of Rapid Urbanization: A Case Study of Tianjin. *City Plan. Rev.* **2016**, *40*, 17–22.
59. Ma, W.; Jiang, G.; Chen, Y.; Qu, Y.; Zhou, T.; Li, W. How feasible is regional integration for reconciling land use conflicts across the urban–rural interface? Evidence from Beijing–Tianjin–Hebei metropolitan region in China. *Land Use Policy* **2020**, *92*, 104433. [CrossRef]

60. Yeh, A.G.-O.; Wu, F. The New Land Development Process and Urban Development in Chinese Cities. *Int. J. Urban Reg. Res.* **1996**, *20*, 330–353. [CrossRef]
61. Zhang, C.; Miao, C.; Zhang, W.; Chen, X. Spatiotemporal patterns of urban sprawl and its relationship with economic development in China during 1990–2010. *Habitat Int.* **2018**, *79*, 51–60. [CrossRef]
62. Yeh, A.G.-O.; Chen, Z. From cities to super mega city regions in China in a new wave of urbanisation and economic transition: Issues and challenges. *Urban Stud.* **2019**, *57*, 636–654. [CrossRef]
63. Zhang, L.; Yue, W.; Liu, Y.; Fan, P.; Wei, Y.D. Suburban industrial land development in transitional China: Spatial restructuring and determinants. *Cities* **2018**, *78*, 96–107. [CrossRef]
64. Qian, Z.; Meng, Q.; Chen, X. Development Models of Urban Fringe Areas in Large Cities. *Urban Probl.* **2005**, *6*, 13–17.
65. Yin, S.; Ma, Z.; Song, W.; Liu, C. Spatial Justice of a Chinese Metropolis: A Perspective on Housing Price-to-Income Ratios in Nanjing, China. *Sustainability* **2019**, *11*, 1808. [CrossRef]
66. Li, H.; Li, J. Idea Management System and Planning Method of Urban Fringe Development. *J. Chongqing Jianzhu Univ.* **2004**, *3*, 1–5.
67. Dai, J.; Dong, J.; Yang, S.; Sun, Y. Identification Method of Urban Fringe Area based on Spatial Mutation Characteristics. *J. Geoinf. Sci.* **2021**, *23*, 1401–1421.
68. Zou, Y. Research on the Spatial Relations and Its Influencing Factors of Residents in Nanchang. Master's Thesis, Jiangxi Normal University, Nanchang, China, 2020.
69. Fang, G.; Sun, X.; Liao, C.; Xiao, Y.; Yang, P.; Liu, Q. How do ecosystem services evolve across urban–rural transitional landscapes of Beijing–Tianjin–Hebei region in China: Patterns, trade-offs, and drivers. *Landsc. Ecol.* **2023**, *38*, 1125–1145. [CrossRef]
70. Ma, T.; Yang, F.; Li, B.; Chen, J. City—Country Interlocking Belt—Special Ecology Zone. *Urban Environ. Urban Ecol.* **2004**, *17*, 37–39.
71. Chen, C.; Hu, F.; Li, H. Landscape Ecological Problems and the Countermeasures of Nanjing Urban-rural Fringe. *J. Nanjing For. Univ. (Nat. Sci.)* **2000**, *43* (Suppl. S1), 17–23.
72. Liu, L.; Li, Z.; Ma, J. Study on Landscape Ecological Features and Landscape Ecological Construction of Rural Landscape in Rurban Fringe. *China Popul. Resour. Environ.* **2006**, *16*, 76–81.
73. Hao, R.; Yang, J.; Li, S.; Lei, J.; Yang, B. Dynamic Changes of Landscape Patterns and Problems in Urban Fringe of the Semi-Arid Area: A Case Study from Huhhot City. *Resour. Sci.* **2005**, *27*, 154–160.
74. Gao, C.; Cheng, L. Tourism-driven rural spatial restructuring in the metropolitan fringe: An empirical observation. *Land Use Policy* **2020**, *95*, 104609. [CrossRef]
75. Shi, S.; Ju, Z. House Price Effect of "Vacating Cage to Change Birds": Using the Relocation and Transformation of Old Industrial Areas as an Example. *J. Quant. Tech. Econ.* **2023**, *40*, 113–132.
76. Shanghai Urban Planning Design Research Institute. *Evolution of Urban Planning in Shanghai*; Tongji University Press: Shanghai, China, 2007.
77. Huang, F.-X. Planning in Shanghai. *Habitat Int.* **1991**, *15*, 87–98.
78. Zhou, L. A Case Study on Mechanism of Spatial Evolution and Spatial Strategies in the Megalopolis Fring. *Urban Plan. Forum* **2017**, *3*, 85–94.
79. Tongzhou District People's Government of Bejing Municipality. Regulatory Detailed Planning of Beijing Sub-Center (Block Level) (2016–2035). Available online: https://www.bjtzh.gov.cn/bjtz/xxfb/202201/1506725.shtml (accessed on 15 May 2024).
80. Liu, Y.; Liu, Y.; Chen, Y.; Long, H. The process and driving forces of rural hollowing in China under rapid urbanization. *J. Geogr. Sci.* **2010**, *20*, 876–888. [CrossRef]
81. Wu, A.; Lu, C. Will the Reform of the Household Registration System Promote the Rural Migrant Population to Settle in the City?—Evidence From Quasi-natural Experiments. *Popul. Dev.* **2020**, *26*, 49–61.
82. Zhou, L.; Wei, L.; López-Carr, D.; Dang, X.; Yuan, B.; Yuan, Z. Identification of irregular extension features and fragmented spatial governance within urban fringe areas. *Appl. Geogr.* **2024**, *162*, 103172. [CrossRef]
83. Feng, C.; Zhang, H.; Xiao, L.; Guo, Y. Land Use Change and Its Driving Factors in the Rural–Urban Fringe of Beijing: A Production–Living–Ecological Perspective. *Land* **2022**, *11*, 314. [CrossRef]
84. Tan, M.; Ran, S.; Ma, S. The Environmental Problems in Urban Fringes of the Metropolitan Area and the Countermeasures: A Case Study of Fangshan District of Beijing. *Prog. Geogr.* **2010**, *29*, 422–426.
85. Yan, Z.; Huang, Z. Research on Land Use Regulation Measures in Rapid Urbanization Process: A Case Study of Nanning. *Acda. Forum* **2007**, *4*, 92–96.
86. Wang, Q.; Zhang, X. Three rights separation: China's proposed rural land rights reform and four types of local trials. *Land Use Policy* **2017**, *63*, 111–121. [CrossRef]
87. Zhou, C.; Liang, Y.; Fuller, A. Tracing Agricultural Land Transfer in China: Some Legal and Policy Issues. *Land* **2021**, *10*, 58. [CrossRef]
88. Luo, Q.; Zhang, X.; Li, Z.; Yang, M.; Lin, Y. The effects of China's Ecological Control Line policy on ecosystem services: The case of Wuhan City. *Ecol. Indic.* **2018**, *93*, 292–301. [CrossRef]
89. Ge, T.; Hao, X.; Li, J. Effects of public participation on environmental governance in China: A spatial Durbin econometric analysis. *J. Clean. Prod.* **2021**, *321*, 129042. [CrossRef]

90. Jiang, X. Study of Eco-environmental Problems and Countermeasures in the Process of Urbanization: By the Case of Chang-Zhu-Tan Agglomeration. *Ecol. Econ.* **2009**, *9*, 167–170.
91. Song, L.; Zhou, C.; OuYang, L. Study on Value, Dilemma and Countermeasures of Public Open Space in Urban Fringe Areas. *Mod. Urban Res.* **2012**, *27*, 24–30.
92. Xue, H. Analysis of Space Environment Problems in Urban Fringes and the Tactics of Urban Design. *Mod. Urban Res.* **2009**, *24*, 67–69.
93. Tong, D.; Yuan, Y.; Wang, X. The coupled relationships between land development and land ownership at China's urban fringe: A structural equation modeling approach. *Land Use Policy* **2021**, *100*, 104925. [CrossRef]
94. Zeng, C.; Yin, Y.; Guo, L.; Liu, C.; Zhang, Y.; Huang, Z. Integrating the administrative spillover effect into the spatial governance system to revisit land development: A study in urban-rural fringe areas of Wuhan and neighboring cities, China. *Land Use Policy* **2024**, *139*, 107060. [CrossRef]
95. Magina, F.B.; Paul, B.N.; Lupala, J.M. Impacts of urban sprawl on people's livelihoods: Analysis of urban fringe neighbourhoods in Dar es Salaam, Tanzania. *Int. Plan. Stud.* **2024**, *29*, 198–212. [CrossRef]
96. Fulman, N.; Grinblat, Y.; Benenson, I. A project-based view of urban dynamics: Analyzing 'leapfrogging' and fringe development in Israel. *Cities* **2024**, *148*, 104908. [CrossRef]
97. Yang, J.; Dong, J.; Sun, Y.; Zhu, J.; Huang, Y.; Yang, S. A constraint-based approach for identifying the urban–rural fringe of polycentric cities using multi-sourced data. *Int. J. Geogr. Inf. Sci.* **2022**, *36*, 114–136. [CrossRef]
98. Marques, A.L.; Alvim, A.T.B. Metropolitan fringes as strategic areas for urban resilience and sustainable transitions: Insights from Barcelona Metropolitan Area. *Cities* **2024**, *150*, 105018. [CrossRef]
99. Liu, G. Deepen the Understanding of the Main Contradiction in the Primary Stage of Socialism. *China Rev. Political Econ.* **2013**, *12*, 4–5.
100. Zhao, Z. Essential Attribute and Morphological Characteristics of the Principal Social Contradiction Facing Chinese Society in the New Era. *CASS J. Political Sci.* **2018**, *34*, 55–65, 126.
101. Lee, C.-C.; Zhang, J.; Hou, S. The impact of regional renewable energy development on environmental sustainability in China. *Resour. Policy* **2023**, *80*, 103245. [CrossRef]
102. Guo, C.; Zhang, Y.; Han, J.; Liu, S.; Wang, L. A practice study of land and water comprehensive remediation in urban–rural fringe—With Sichuan River Shaanxi Province, Northwest China, as an example. *Environ. Sci. Pollut. Res.* **2023**, *30*, 74006–74020. [CrossRef]
103. Yang, Y.; Dong, Z.; Zhou, B.-B.; Liu, Y. Smart Growth and Smart Shrinkage: A Comparative Review for Advancing Urban Sustainability. *Land* **2024**, *13*, 660. [CrossRef]
104. Li, Z.; Zhang, W. Landscape Ecological Problems and Countermeasures of the Famous Scenic Area in Urban Fringe. *Chin. Landsc. Archit.* **2008**, *24*, 71–75.
105. Gao, J.; Wu, B. Revitalizing traditional villages through rural tourism: A case study of Yuanjia Village, Shaanxi Province, China. *Tour. Manag.* **2017**, *63*, 223–233. [CrossRef]
106. Fu, B.; Xue, B. Temporal and Spatial Evolution Analysis and Correlation Measurement of Urban–Rural Fringes Based on Nighttime Light Data. *Remote Sens.* **2024**, *16*, 88. [CrossRef]
107. Tan, L.; Cui, Q.; Chen, L.; Wang, L. An Exploratory Study on Spatial Governance Toward Urban–Rural Integration: Theoretical Analysis with Case Demonstration. *Land* **2024**, *13*, 2035. [CrossRef]
108. Han, W.; Zhao, Y. Rural spatial governance mechanism and model in metropolitan fringe based on the background of rural revitalization. *Sci. Geogr. Sin.* **2023**, *43*, 1340–1349.
109. The People's Government of Sichuan Province. Notice of the People's Government of Sichuan Province on Printing and Issuing the Development Plan of Chengdu Metropolitan Area. Available online: https://www.sc.gov.cn/zfgbdb/detail/e316b5073fff4ac59bfc0d181d1c5b33.shtml (accessed on 5 October 2024).
110. Shanghai Municipal People's Government. Spatial Cooperative Plan of Greater Shanghai Metropolitan Area. Available online: https://www.shanghai.gov.cn/nw12344/20220928/43ffc9be207448c4b2c910c17ac02aaa.html (accessed on 5 October 2024).
111. The People's Government of Hunan Province. Yangtze River Economic Belt-Yangtze River Basin Territorial Spatial Planning (2021–2035). Available online: https://www.hunan.gov.cn/zqt/zcsd/202409/t20240911_33452502.html (accessed on 5 October 2024).
112. Bousema, I.; Busscher, T.; Rauws, W.; Leendertse, W. How do polycentric governance systems adapt? The role of forums explored in Dutch metropolitan areas. *Public Adm. Rev.* **2024**, *85*, 1–15. [CrossRef]
113. Nolte, I.M.; Lindenmeier, J. Creeping crises and public administration: A time for adaptive governance strategies and cross-sectoral collaboration? *Public Manag. Rev.* **2023**, *26*, 3104–3125. [CrossRef]
114. Goldberg, D.; Olivares, M.; Li, Z.; Klein, A.G. Maps & GIS Data Libraries in the Era of Big Data and Cloud Computing. *J. Map Geogr. Libr.* **2014**, *10*, 100–122.
115. Mahdianpari, M.; Granger, J.E.; Mohammadimanesh, F.; Warren, S.; Puestow, T.; Salehi, B.; Brisco, B. Smart solutions for smart cities: Urban wetland mapping using very-high resolution satellite imagery and airborne LiDAR data in the City of St. John's, NL, Canada. *J. Environ. Manag.* **2021**, *280*, 111676. [CrossRef]

116. Liu, X.; Hu, G.; Chen, Y.; Li, X.; Xu, X.; Li, S.; Pei, F.; Wang, S. High-resolution multi-temporal mapping of global urban land using Landsat images based on the Google Earth Engine Platform. *Remote Sens. Environ.* **2018**, *209*, 227–239. [CrossRef]
117. Schepaschenko, D.; See, L.; Lesiv, M.; Bastin, J.-F.; Mollicone, D.; Tsendbazar, N.-E.; Bastin, L.; McCallum, I.; Laso Bayas, J.C.; Baklanov, A.; et al. Recent Advances in Forest Observation with Visual Interpretation of Very High-Resolution Imagery. *Surv. Geophys.* **2019**, *40*, 839–862. [CrossRef]
118. Han, L.; Qu, Y.; Liang, S.; Shi, L.; Zhang, M.; Jia, H. Spatiotemporal Differentiation of Land Ecological Security and Optimization Based on GeoSOS-FLUS Model: A Case Study of the Yellow River Delta in China Toward Sustainability. *Land* **2024**, *13*, 1870. [CrossRef]
119. He, Y.; Ma, J.; Zhang, C.; Yang, H. Spatio-Temporal Evolution and Prediction of Carbon Storage in Guilin Based on FLUS and InVEST Models. *Remote Sens.* **2023**, *15*, 1445. [CrossRef]
120. Cirianni, F.M.M.; Comi, A.; Quattrone, A. Mobility Control Centre and Artificial Intelligence for Sustainable Urban Districts. *Information* **2023**, *14*, 581. [CrossRef]
121. Elvas, L.B.; Mataloto, B.M.; Martins, A.L.; Ferreira, J.C. Disaster Management in Smart Cities. *Smart Cities* **2021**, *4*, 819–839. [CrossRef]

land

MDPI

Article

An Exploratory Study on Spatial Governance Toward Urban–Rural Integration: Theoretical Analysis with Case Demonstration

Lin Tan [1,2], Qinyu Cui [3], Lan Chen [1,2,*] and Lan Wang [4]

1 College of Architecture and Environment, Sichuan University, Chengdu 610065, China; tanlin1211@scu.edu.cn
2 Observation and Research Station of Land Ecology and Land Use in Chengdu Plain, Ministry of Natural Resources, Chengdu 610045, China
3 Department of Civil and Transportation Engineering, South China University of Technology, Guangzhou 510641, China; ct_cuiqinyu@mail.scut.edu.cn
4 Sichuan University Engineering Design and Research Institute Co., Ltd., Chengdu 610065, China; scu_lintan@163.com
* Correspondence: cl_jhxy@scu.edu.cn

Abstract: Establishing a territorial space governance system that supports URI is essential for modernizing national governance capacity and systems. It also serves as a critical strategic measure to promote sustainable development in both urban and rural areas. The current research has not adequately explored the theoretical mechanisms and pathways of territorial spatial governance, resulting in challenges in providing the necessary theoretical foundations and practical guidance for advancing URI. The methods used in this study include theoretical analysis, logical deduction, and field research. The findings are as follows: (1) The key characteristics of TSG are the synergy of governance entities, the comprehensiveness of governance objects, the diversification of governance methods, and the systematization of governance content. Territorial spatial governance, driven by multiple stakeholders, influences the structure and functional arrangement of urban and rural territorial systems through means such as spatial planning, rights allocation, and top-level institutional design. (2) The essential process and core driving path to promote URI involve the interaction of urban and rural elements, the complementarity of their functions, and the reciprocity of their values, all guided by territorial space governance. (3) In the case study area, TSG facilitates the construction of an orderly element circulation channel, coordinating land circulation and capitalization, promoting local urbanization, and aligning residential with industrial development. This results in a complementary urban–rural functional structure, multifunctional rural development, and an urban—rural model characterized by stable interest relationships and balanced service facility layouts.

Keywords: urban and rural integration; territorial spatial governance; territorial space planning; rural transformation; China

1. Introduction

Urban–rural integration (URI) is an unavoidable trend in the development of urban–rural relations, representing a global phenomenon. Nations encounter shared challenges in reality, including the urban–rural income gap and the equalization of public services [1,2]. The primary objective of URI is to enhance the overall optimization of urban and rural regional system functions. This progression is vital for China's comprehensive modernization and realization of "the Two Centenary Goals", rendering it a substantial and extensively debated subject in academic research.

During specific historical periods, China adopted various strategies and governance models for urban and rural development. Before the reform and opening-up era, the state uniformly allocated resources to rapidly accumulate industrial capital and meet modernization needs. This was achieved by exploiting the price differential between industrial and agricultural products. Concurrently, a system of separate management for urban and rural areas emerged in economic, political, and social spheres. The household registration system was not fully optimized, resulting in disparities between urban and rural areas in terms of employment, healthcare, pensions, and other welfare systems. Consequently, urban and rural residents did not enjoy equal rights and interests, leading to the formation of two relatively independent governance systems and dual structures [3,4]. This unbalanced development and urban–rural separation have become significant obstacles to integration, hindering the efficient allocation of factors between urban and rural areas in the market [5]. Studies focusing on rural perspectives highlight that rural revitalization and URI are mutually supportive, with the dynamic transformation process of rural revitalization being integral to URI [6]. Therefore, promoting URI requires starting from the rural regional system, developing scientifically tailored plans adapted to rural transformation through classification and zoning, and shaping a new urban–rural pattern [7].

Currently, rural areas in China continue to lag behind urban regions due to the complex and prominent contradictions in the relationship between people and land. Consequently, in the implementation of the rural revitalization strategy, the interaction among people, land, capital, and industry—as core elements linking urban and rural systems—directly affects the achievement of URI [8–11]. Furthermore, some scholars have approached the issue of rural spatial governance by establishing a comprehensive governance system focused on rural areas, utilizing a "material–ownership–organization" framework [12]. This framework offers solutions for improving urban–rural relations through aspects such as rural spatial reconstruction and reconfiguration of development rights. Building on this foundation, a multi-scale governance system defined by "spatial integration–spatial zoning–spatial mobility" has been proposed [13,14]. This approach is crucial for overcoming the spatial differentiation barriers that impede URI and addressing the structural contradictions in the distribution of urban and rural resources, as along with the provision of public services.

On the other hand, the comprehensive functions of strategic guidance and regional regulation of territorial spatial planning are gradually being strengthened, and it is anticipated that these will become indispensable tools for promoting the symbiotic development of urban and rural areas. Previous studies have revealed that territorial spatial planning creates a symbiotic environment of "form–flow coupling" facilitated by "definite zoning", "definite systems", and "definite forms" [14,15]. This approach plays a vital role in the management and supervision of the efficient allocation and collaborative development of urban and rural resources. Simultaneously, territorial spatial planning takes "development" as a fundamental guarantee, integrating the overarching concepts of coordinated development, equity, and ecology throughout the entire life cycle of URI. By innovating the control system of land use for urban and rural areas and promoting land use transformation, it provides essential support for urban–rural symbiotic development [5,16–18].

Territorial space serves as spatial support for the integrated development of urban and rural areas. Optimizing spatial patterns and enhancing the efficiency of resource allocation in urban and rural territorial spaces are intrinsic requirements of URI. However, challenges persist, such as resource allocation barriers and value biases between urban and rural areas, with asymmetric land resource distribution and unequal functional values being the most significant obstacles [19,20]. As China enters a new stage of development, it is imperative to form an intervention force through land system reform, planning controls, and other means to effectively govern urban and rural territorial spaces. As a component of China's national governance system, territorial space governance (TSG) is a vital tool for coordinating the layout, development, construction patterns, and ownership systems of these spaces. Naturally, it also represents an important approach to land management. Sustainable TSG fosters conditions for optimizing urban and rural regional systems and serves as the necessary foundation for the stable advancement of URI. Current research highlights the importance of sustainable land management and planned utilization, emphasizing that optimizing landscape structures and strengthening farmland protection are critical for sustainable land use. These findings provide a scientific basis for formulating reasonable land management policies. The loss of land resources, especially farmland, threatens global food security and challenges sustainable urbanization [21,22].

Some studies emphasized that TSG needs to balance the guidance for macro-regional development with the refined management and control of micro-land elements, linking "regional" and "element" governance to strengthen baseline control and resource management. This helps enhance governmental organizational and coordination capabilities for urban and rural elements, laying the groundwork for the transformation and reconstruction of rural regions and the allocation of urban–rural spatial development rights during URI [23,24]. Comprehensive enhancement of the service capacity of TSG involves optimizing spatial patterns, coordinating system structures, and improving functional systems within certain urban and rural areas [25–27]. However, this requires empowerment at the local government level and the establishment of a territorial governance mechanism [28]. Although TSG is anticipated to effectively address URI challenges, few studies have established a systematic framework for it or elucidated the combined effects and internal mechanisms of the associated governance measures. As URI progresses, it will also reshape the existing governance model, imposing new demands on current territorial spatial governance. However, existing studies mostly explore concepts and implementation strategies from either a single perspective or a dualistic urban–rural approach, leaving the theoretical discourse on TSG and URI fragmented. This highlights a gap in research that integrates these aspects into a cohesive framework for coordinated urban–rural development and effective governance.

How can TSG drive URI? Addressing this question is crucial not only for understanding spatial governance practices within the modernization of the national governance system but also for exploring strategies for URI and urbanization. Therefore, it is essential to revisit the theoretical foundations to clarify the potential pathways through which TSG can influence URI and development. This exploration aims to uncover the mechanisms that facilitate URI, offering valuable insights for its advancement.

It is noteworthy that this paper seeks to analyze the internal mechanisms and pathways of URI from a theoretical perspective at the regional level. The focus is on a broader, macro-level analysis rather than on individual settlements or administrative scales. This study begins by defining the concept of TSG through theories of human–land relationships and governance frameworks. It then establishes a governance system that aligns with the characteristics of URI development. Utilizing an "elements–functions–values" analytical framework, this paper reveals the mechanisms through which TSG facilitates URI. Finally, empirical studies of typical cases in China are conducted to refine the understanding and practical approaches of TSG in promoting URI. This study aims to provide new insights into the effectiveness of TSG in China and the mechanisms driving URI.

The remainder of this paper is organized into five parts. In Section 2, the authors construct the analysis framework of TSG for URI from the theoretical level, and reveal the specific mechanism. Section 3 summarizes the research field and research methods. Section 4 introduces the implementation path of rural–urban integration in case studies. Section 5 provides the discussion, and Section 6 concludes the article.

2. Theoretical Foundations of Spatial Governance for Urban–Rural Integration

2.1. Spatial Governance

Governance refers to a decision-making framework designed to achieve specific goals through the formulation and implementation of public policies or the management of public resources. Unlike traditional top-down administratively driven approaches, governance seeks to establish a collaborative model among government, market, social organizations, residents, and other stakeholders [29,30]. TSG is a crucial element of the national governance system, embodying and practicing governance within the spatial dimension. It encompasses a comprehensive process from planning and management to supervision, aiming for the rational allocation and efficient utilization of land resources [31,32]. In the context of modernizing governance capacity and systems, TSG can be defined as the management of urban and rural land spaces as the field of governance. It involves collective action and cooperation among various stakeholders, leading to external interventions in ownership, use, development intensity, and functional layout of national land space. This is achieved through tools such as territorial spatial planning, spatial rights distribution, and top-level institutional design, optimizing the comprehensive process of element pattern, structural organization, and overall function of the urban and rural land space system. From a geographical analysis perspective, the essence of TSG lies in coordinating the relationship between the people and the land. It focuses on the study of process patterns, external characteristics, and mechanisms of TSG within different spatial scales. Essentially, it seeks to reconstruct the spatial form and human–land relationship patterns through multi-scale and multi-dimensional governance of urban and rural spaces [33].

The concepts of territorial spatial governance, rural spatial governance, comprehensive land consolidation, and territorial spatial planning are both interrelated and distinct. Rural spatial governance addresses challenges related to spatial constraints and disorganized systems during rural transformation, offering solutions to structural problems in rural revitalization [32]. Comprehensive land consolidation involves systematic regulation of elements within urban and rural geographical systems through ecological engineering, land consolidation projects (such as agricultural land regulation and rural construction land regulation, etc.), and cultural restoration. This approach is crucial for promoting the reconstruction of rural spaces and the optimal allocation of urban and rural land resources [6,34,35].

TSG primarily addresses issues such as inconsistent planning, chaotic zoning, and ineffective transmission mechanisms in land use. It establishes an integrated urban–rural governance system with strategic and controlling features, following the top-level design of "five levels, three categories, and four systems", serving as a key initiative for governments at all levels to implement spatial governance actions [36,37]. The connotation of TSG is more comprehensive, emphasizing the integration of governance objects, diversification of governance means, and the systematization of governance content. Both territorial spatial planning and rural spatial governance are important components of the TSG system [31,38]. In summary, TSG integrates time, space, and human elements within a unified governance framework, employing refined strategies to regulate the territorial spatial system and optimize urban and rural spatial patterns, functional structures, and rights configurations, thereby broadening and deepening our theoretical understanding of territorial spatial governance.

2.2. A Tentative Framework for Understanding Spatial Governance System

The primary aim is to establish a multi-dimensional governance approach to address the challenges of land use between urban and rural areas and among different regions, ultimately supporting the URI strategy and rural revitalization.

TSG involves various aspects such as objects, methods, and targeted entities, and is inherently a complex and systematic activity. Considering the core characteristics and overall goals of territorial spatial governance, we have developed a system encompassing governance entities, content, methods, and objectives, following the logic progression of who governs, what is governed, how to govern, and the outcomes of governance. This system is designed to emphasize the developmental aspirations of the urban–rural system by optimizing the material spatial pattern of the national territory, while also focusing on achieving spatial justice and managing social relations [39,40]. The fundamental purpose is to establish a multi-dimensional governance approach to handle the challenges of land use between urban and rural areas and among different regions, ultimately supporting the URI strategy and rural revitalization (Figure 1).

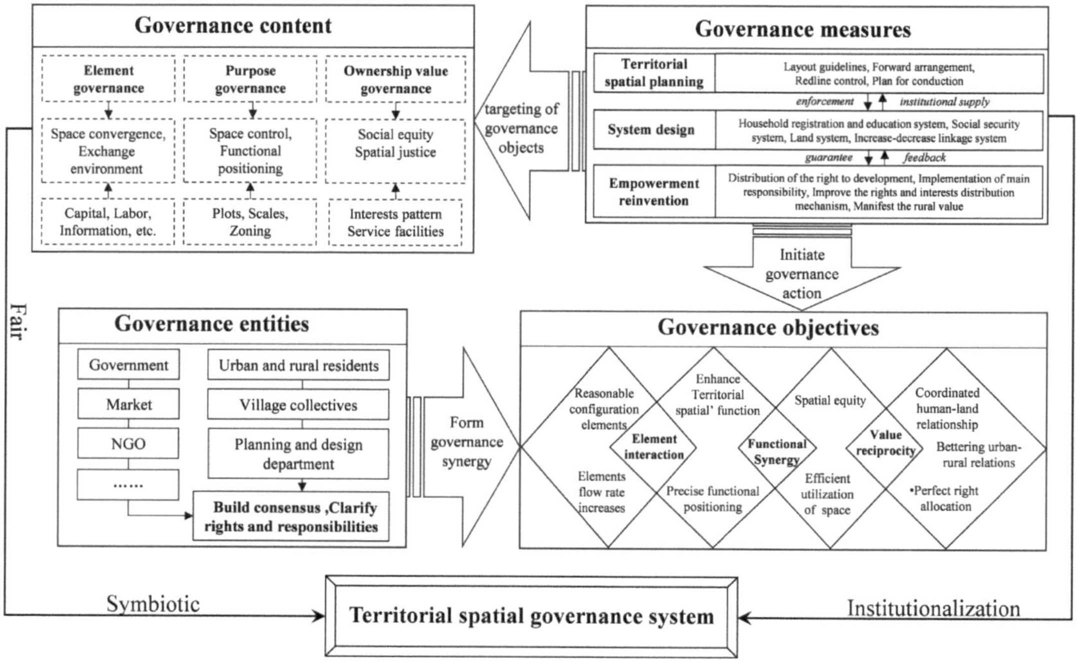

Figure 1. Theoretical illustration on the spatial governance system.

Specifically, in terms of governance subjects, the full-cycle approach to TSG highlights the market's dominant role in the allocation of spatial resources. Reforms aimed at streamlining government, delegating power, and improving services have led to the decentralization of governance authority [41,42]. A redefinition of the boundaries of rights and responsibilities among the government, the market, and social actors fosters social consensus and creates synergistic collaborations among governance subjects. These developments contribute to the establishment of a more scientific governance decision-making system.

In terms of governance content, TSG encompasses three dimensions: element governance, use governance, and tenure value governance. Concretely speaking, the TSG system starts with a fundamental institutional design, attempting to dismantle the dual-track system barriers through reforms such as the homestead system, household registration system, and innovations in land market transaction systems [43–45]. The aim is to address the institutional constraints on the flow of factors between rural and urban areas and to mitigate the inequities caused by unidirectional flows, thereby establishing a smooth interactive pattern for factors like population, capital, and technology between rural and urban areas [46]. Use governance focuses on clarifying the functional roles, zoning layouts, and control strategies of urban and rural land space across various scales, such as parcel, unit, and region, to ensure a balanced mix and differentiation of land space functions. Guided by China's "Strategic Plan for Rural Revitalization (2018–2022)" [47], it explores tailored functional design schemes that consider local conditions, including each village's resource profile, location, geographical environment, and development needs. For instance, in villages designated for special conservation, TSG emphasizes leveraging local cultural resources to create unique living environments, enhancing attractiveness and distinctiveness. In villages undergoing suburban integration, TSG focuses on undertaking industrial transfers from urban areas, attracting capital, promoting multifunctional rural space use, and cultivating integrated units centered around small towns, rationalizing the spatial functional structure between urban and rural areas. The existing differentiated territorial spatial use control and property rights system in urban and rural areas has shaped an initial rural value distribution pattern [5,48], inhibiting value spillover potential and exacerbating spatial development rights inequity between urban and rural areas. TSG tackles this by addressing ownership value governance, establishing an integrated urban–rural use control system, and creating an equivalent market where land enjoys equal rights and value. This improves the value distribution mechanism and expands rural value transformation pathways [49]. Additionally, TSG employs spatial planning tools to optimize the spatial layout of service facilities based on scale and demand differences and deepens the reform of the household registration system to eliminate identity disparities, promoting benefit sharing.

In terms of governance measures, the multifaceted nature of territorial space—encompassing political, economic, and social dimensions—naturally positions spatial governance activities as a complex system. The approach within this system must address the challenges of urban and rural spatial development and use, establishing a comprehensive governance toolkit. This includes territorial spatial planning, institutional design, empowerment reconstruction, and other governance strategies. Territorial spatial planning is fundamental, focusing on guiding spatial layout and making anticipatory arrangements throughout the governance process. Its development and implementation involve multi-agent collaboration, where local governments often integrate land oversight and restrictive strategies to standardize land use patterns and regulate the organizational structure of spatial elements in urban and rural areas [36,37]. For example, national master plans for functional zones apply differentiated spatial zoning controls to clarify development orientations for territorial units, providing strategic guidance for spatial governance [50].

The institutional design ensures governance practices are standardized and effective by formulating and refining relevant policies and regulations, serving as a foundational guarantee for constructing a high-quality spatial governance framework. China's recent land system reforms focus on land property rights, agricultural land empowerment, and land transactions, adjusting the spatial allocation of land, capital, labor, and other elements to achieve coordinated human, land, industry, and production development, thus facilitating rural transformation and enhancing regional functions [51].

The land use control system prioritizes farmland protection to maintain total farmland area stability and food production security, balancing production, living, and ecological spaces by controlling construction land quantities [52]. It sets protection red lines and refines development indicators for various land types, ensuring fair allocation and effective development of land resources, and promoting the stable operation of the territorial space system.

Concurrently, the integration of household registration, education, social security, evaluation, and land systems, along with a mechanism linking land increase and decrease, forms an institutional bundle. This drives TSG toward legalization and refinement, enhancing the authority, humanity, and scientific basis of institutional design. Empowerment reconstruction focuses on the equitable allocation of development rights among multiple entities, establishing responsibilities, and improving the distribution mechanism by adjusting the rules for allocating and redistributing territorial space development rights [53]. Additionally, it explores new methods for utilizing territorial space alongside land system reforms, addressing the disparity in urban and rural market values, exploring new avenues for rural value realization, and achieving equitable urban and rural development.

Regarding governance objectives, the integrated development of urban and rural areas aims to reduce regional development disparities, enhance living standards, promote the free movement of resources, create complementary functions, and achieve equivalent development [54]. This ultimately results in comprehensive, coordinated, and sustainable development for both urban and rural communities. Consequently, TSG pursues the values of rights allocation and spatial equity, focusing on resolving various dilemmas in the distribution pattern, organizational layout, and flow rate of development factors between urban and rural spaces. By reasonably managing the timing and processes of territorial space utilization, it enhances the efficiency and functionality of these spaces, providing fundamental support for integrated urban–rural development. Additionally, this governance approach facilitates ongoing negotiation and reconciliation among various stakeholders, further clarifying the ownership relationships of urban and rural spaces [12], ensuring the efficient utilization of territorial space, and establishing a network of equal and reciprocal social relations.

2.3. Theoretical Rationales on Leveraging Spatial Governance for Advancing Urban–Rural Integration

The analysis above illustrates that TSG transforms the allocation and flow patterns of urban and rural factors, thereby accelerating functional interactions between urban and rural systems and unlocking the potential value in rural areas. This process ultimately reshapes spatial, human–land, and interpersonal relations during URI, enhancing the overall level of integration. The following discussion will further explore the driving mechanisms of TSG on URI, focusing on the empowerment pathway of "factors/functions/values" (Figure 2).

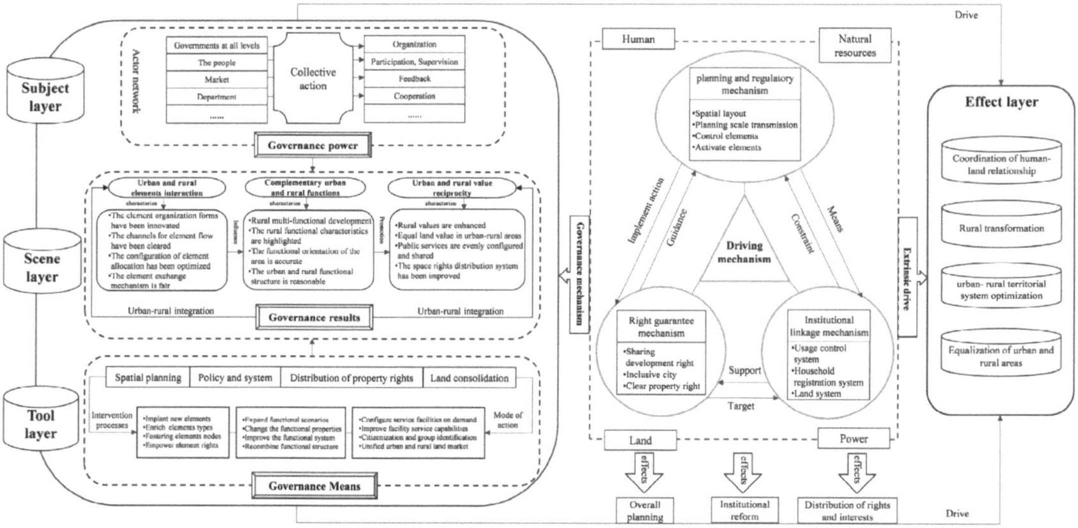

Figure 2. Theoretical rationales on spatial governance to advance URI.

2.3.1. Spatial Governance of Urban–Rural Element Flows

The free and efficient circulation of urban and rural factors is a prerequisite for achieving integrated urban–rural development. However, the unilateral outflow of rural development factors, such as a large number of laborers, has weakened the momentum for rural development. Meanwhile, China's land management system exhibits distinct urban–rural dual characteristics, specifically divided into state ownership of urban land and collective ownership of rural land [55]. This ownership environment creates significant disparities in the integration of urban and rural construction land markets, thereby impeding the efficient flow of key development factors like capital, information, and labor, which obstructs the process of URI [5]. TSG addresses these challenges by utilizing policies to enhance "flow spaces" such as information and economic flows. It establishes a synergistic approach to update resource types, improve flow environments, and upgrade circulation channels. This approach corrects the previous misallocation of urban and rural resources, which favored urban areas, thereby encouraging orderly growth and a virtuous cycle of resource exchange between urban and rural regions.

To address the challenge of unilateral factor circulation in URI, TSG aligns with the developmental needs of multifunctional and global villages, focusing on the modernization of rural economic and social systems. It clarifies the new requirements posed by urban–rural regional linkages and emphasizes the integration of locally rooted factors shaped by rural areas' unique natural environments and cultural heritage. Government departments, in partnership with village collective organizations, employ comprehensive land remediation to transform physical spaces, creating modern consumption scenarios that are functional, distinctive, and culturally grounded [56]. The aim is to enhance rural appeal, thereby attracting industrial and commercial urban capital to support rural development. The infusion of social capital provides essential support for rural areas [57]. The convergence of various capital factors enhances the value of land through market pricing mechanisms, increasing employment opportunities for local farmers and returning laborers. This approach helps construct a new factor pattern of people, land, industry, and capital, thus enriching and improving the types and structures of factors within rural territorial systems through the spillover effects of multiple factors. This strategy addresses the shortcomings of rural development capabilities within the context of URI.

The fundamental driving force behind the dynamic evolution of URI is a fair factor flow mechanism [58]. TSG is dedicated to establishing a multi-scale planning linkage system and empowering factors to enhance the flow environment. This approach facilitates the transformation and upgrading of circulation channels, providing a stable platform for the exchange of material, energy, and information between urban and rural factors. A prominent challenge is the impotent spatial mobility network between urban and rural areas, which hinders the creation of a large integrated market and the coupling of urban and rural factors [13]. The "flow space" concept highlights the advantages of TSG by emphasizing the integration of multiple plans to achieve a balanced allocation of urban and rural elements. It ensures collaborative construction, mutual benefits, interoperability, and complementarity in spatial planning across all levels.

For instance, at national and provincial levels, planning is guided by national territorial spatial frameworks and main functional zones, reinforcing factor planning at the regional and watershed scales based on geographical differentiation features and economic characteristics. This provides strategic guidance for the layout and positioning of urban and rural productivity [13]. At the city and county levels, authorities implement the requirements of higher level plans, focusing on total quantity control, flow scale, and the differentiated index allocation of factors. They are also tasked with relaxing rigid local planning controls and exploring flexible planning schemes tailored to city government needs, optimizing factor input structures at these scales. Achieving a relative balance among governments in factor distribution is essential to address supply shortages and create conditions for a URI model with counties as key carriers [59].

The town–village scale spatial planning approach emphasizes the interaction between elements and optimizes their structure from a micro perspective. It facilitates the rational allocation of resources within rural settlements and villages by implementing practical village planning through classification, zoning, and grading [1,60]. The convergence of advantageous factors drives TSG, which uses detailed rural planning to revitalize idle land and other resources, cultivate factor space nodes, and promote the interaction and coupling of development factors such as land, labor, capital, industry, and technology [61]. This approach improves the efficiency of rural factor allocation and supports the creation of a broader circulation network for factors.

On the other hand, TSG empowers the development of elements within a fair allocation model. With the comprehensive deepening of the land system reform, TSG takes land elements as a starting point. It relies on the reform of rural collective property rights and the homestead system [10]. This strategy addresses issues like limited element circulation freedom and weak bidirectional interaction through cross-regional transaction rules and orderly paid withdrawal policies for land use indicators. Breaking regional spatial restrictions and policy barriers is essential to unify the factor market, establish cross-scale and cross-regional circulation corridors, improve cross-regional transaction rates of land and capital, and eliminate institutional obstacles to integrating urban and rural factors.

2.3.2. Spatial Governance of Urban–Rural Functional Complementarity

Urban–rural functional complementarity (URFC) is a key link for fostering the dynamics of URI and development [62]. It serves as an essential support for establishing economic cycles within the urban–rural territorial system. Villages undergoing rapid urbanization risk functional decline, marginalization, and the homogenization phenomenon known as "a thousand villages in a row". Additionally, the dysfunctional structure and low service levels within the URI system hinder deep integration between urban and rural areas, highlighting the weak interaction effect of the urban–rural system [4]. To address the diverse needs of integrated economic, social, and spatial development in urban and rural areas, TSG aims to optimize the rural functional system and coordinate the organization of urban–rural functional structures. This approach promotes adjustments and mutual symbiotic development within the functional landscape of urban and rural territorial spaces.

The city and the countryside function as a pair of intertwined organisms [63], each with unique and irreplaceable intrinsic values, together forming the space for human socio-economic activities. Historically, the countryside has primarily supported urban development by providing raw materials and labor, but its potential for modernization remains underexplored. There is an urgent need to clarify the countryside's role and status within urban–rural development. Currently, the issue of "rural disease" is a common global challenge, manifested through village depopulation, atomization, cultural erosion, and environmental pollution. These issues have weakened rural functions related to production, life, ecology, culture, and education [64–66].

Rural areas face multiple difficulties such as vague functional positioning, insufficient exploration, and loss of function vitality [67]. TSG aims to enhance the rural functional system by expanding and innovating rural function types and optimizing the urban–rural functional distribution through redefined positioning. This approach seeks to build a sustainable functional system for the countryside. Within a market economy, a sound institutional mechanism for market-based factor allocation is crucial for increasing factor interaction intensity [68]. TSG incorporates new development elements through institutional reforms that facilitate factor flow, leveraging agricultural resources and technological innovation to introduce new capital structures, management modes, and operational methods. It also transforms land use patterns [69], reorganizes idle resources through industrial integration, and promotes structural changes in functional attributes, leading to diversified, digitized, and vernacular rural industries. This aligns with the overall trend of urban–rural functional transformation, breaking the urban–rural functional segmentation and facilitating functional exchange. Under renewed governance focused on updating rural functions, diverse business models such as urban, tourism, and recycling agriculture overlap [70,71]. This creates new consumption spaces that meet the evolving material and spiritual needs of urban and rural residents, establishing favorable conditions for a new order of urban–rural functional development and enhancing inter-functional interaction.

Recognizing the differences between urban and rural functions is essential for the dynamic evolution of URI. The heterogeneous nature of these functions depends on the comprehensive arrangement of TSG across spatial and target dimensions. Currently, the fragmented functional forms, unclear positioning, and disorganized layout of urban and rural areas contribute to unequal value patterns. The countryside has often been viewed as either an appendage of the city or a traditional production unit [72], leading to a low level of functional coupling between urban and rural areas. TSG employs flexible spatial planning through classification, zoning, and use control of urban and rural spaces. This involves delineating three zones and three lines to guide main functions on a macro scale, while village planning refines land use at the micro level.

From the perspective of urban–rural interaction, spatial governance strengthens the unique functional roles of the countryside, such as preserving Chinese farming civilization, ensuring food quality and ecological security, providing spaces for healthy living, and serving as a hinterland for urbanization and industrialization [73]. This establishes an urban–rural functional system characterized by distinct differences, unique features, and synergistic interactions to meet the needs of integrated development. Achieving complementary and symbiotic urban–rural functions requires reconstructing the spatial organization of existing functional structures [74]. TSG coordinates the overarching structural arrangement based on regional perspectives. Studying regional advantages and development plans creates functional nodes of various levels and purposes through township system planning, emphasizing the critical roles of county units in URI and development in terms of their functions of evacuation and urbanization in the vicinity of towns and villages [75]. This facilitates reasonable governance of functional hierarchies. Moreover, TSG uses rural revitalization planning to focus on synergizing residential and industrial systems. It develops revitalization strategies for villages based on location and development level, enhancing the functional structure of rural communities. This approach promotes the articulation of functional systems at county–municipal–territorial scales and supports the complementarity of the urban–rural territorial system [76].

2.3.3. Spatial Governance of Urban–Rural Value Reciprocity

China's long-standing urban–rural dual structure has significantly undermined rural values due to unequal factor mobility and imbalanced functional structures between urban and rural areas. The urban-centric development model has increasingly encroached on rural spaces, gradually leaving rural systems in a precarious position concerning their legitimate rights and interests [77]. This issue is evident in the slow urbanization of the rural population, limited social integration for farmers, and unequal distribution of public service facilities [5,78]. A primary goal of URI is to achieve equitable development between urban and rural areas, essentially fostering positive interactions within the urban–rural territorial system, and emphasizing rural values to ensure equal development rights and opportunities for all entities [79]. TSG aims to rectify the disordered pattern of urban–rural values, supporting spatial justice and embodying the dual value criteria of equitable public services and fair distribution of rights and interests. Accordingly, its core governance objective is rights reconstruction, promoting fair exchange and equivalent value distribution between urban and rural spaces. Urban capital, information, and advanced productivity should be extended to rural areas through coordinated resource planning. Additionally, the quality of rural service facilities should match that of urban areas. Innovative models for allocating development rights should be implemented to ensure fairness and equivalency in urban–rural values [13,80].

Urban–rural value reciprocity (URVR) should prioritize a balanced allocation of service facilities between urban and rural areas, ensuring that the physical space structure aligns with supply and demand to advance the equalization of public services [81]. Utilizing the latest territorial spatial planning across various regions, facilities' spatial layout should be integrated to address the needs of all age groups and the daily activities of urban and rural residents. This involves refining allocation standards through multi-level spatial planning to achieve Pareto improvements in spatial requirements and avoid the uniform distribution of public services. Consequently, the supply of service facilities in urban–rural integrated development will be enhanced, promoting public service integration. Addressing the core demand for urban–rural equal value development [82] and achieving fairness in the distribution of rights and interests is essential for TSG and the deeper integration of urban and rural areas. The challenges facing URI are fundamentally linked to flawed mechanisms for the allocation of development rights and unfair distribution of rights and interests [83,84].

The urban-productivism growth model, predominant during rapid urbanization, has exacerbated the inequality in development rights, diminishing the dynamics of rural territorial systems and hindering the manifestation of rural values. TSG leverages the macro-regulatory role of government, using reforms in household registration and employment systems to enhance citizenship and status for various groups, including the floating and informally employed populations. This approach allows diverse interest groups to express their spatial interests, improving inclusive urban governance and addressing barriers to resources like education, employment, housing, and healthcare [85]. This creates favorable conditions for a people-centered urbanization model. The entrenched urban–rural dual structure results in significant differences in land use, rights allocation, and value assessment, limiting fair and efficient land transfers and reducing land use efficiency [45,68]. This presents a major obstacle to URI in China. To address land value suppression caused by the urban–rural partitioned land system, TSG employs market-based allocation systems for land development rights, ensuring spatial benefit deployment. Enhancing the urban–rural integrated land market mechanism establishes a synergy between rural land value impairment and compensation, addressing land allocation malfunctions and property rights ambiguities. This unleashes the capitalization effect of rural land resources, reshapes land empowerment, and supports URI and development.

3. Materials and Methods

3.1. Study Area

This study focuses on a case located in Jintang County, Chengdu City, Sichuan Province, China. The area is strategically positioned near major transportation routes, including the Chengdu-Bazhong Expressway, Qing Baijiang-Jintang Expressway, and Da Zhou-Chengdu Railway, providing a comprehensive external transportation system and an advantageous geographical location. The whole area is about 29 km^2, including four communities of Shi Ziling, Jiang yuan, Yun Xiu, and Guanyin Mountain, with a registered population of 26,426 and a permanent population of 24,079 in 2021. The cultivated land area in the area is 673.79 ha, mainly irrigated land and paddy fields, with good quality. The garden area is 676.48 ha, mainly distributed in the Shi Ziling community of Guanyin Mountain, with rich forest resources, mainly woodland in the hill area and Linpan of western Sichuan. Agricultural planting in the region is dominated by traditional grain and oil planting, and the per capita cultivated land area is small. There are more than 30 food processing, furniture and building materials, plastic products, brick factories, and other enterprises, and the problem of idle industrial and mining land is prominent. Part of the area has a foundation for the pension industry, and Guanyin Mountain has been rated as the "first batch of forested villages in China", which has a certain potential for the development of agricultural, commercial, cultural, and tourism integration (Figure 3).

During the research process, it was discovered that, in recent years, Jintang County, in response to President Xi Jinping's park city initiative, has explored rural expressions of Chengdu Park City. The county has actively integrated into the synergistic development framework of the European Industrial City and Huizhou New City, forming a dual-city functional linkage. It has established a comprehensive planning strategy focused on "controlling the bottom line, revitalizing the industry, re-regulating, optimizing facilities, and strengthening guarantees". Through collaboration among government bodies, village collectives, villagers, and planners, specific governance actions have been implemented, including village and land renovation, spatial layout optimization, and innovative operational models. This effort has resulted in the construction of 13 new communities, 2 new aquaculture lands, and 16 planting facility construction sites, providing 116.21 ha of land for major infrastructure, transportation, water conservancy, and industrial development projects (Table 1). These projects have enhanced livable and convenient tourism service facilities, contributed to beautiful rural community development, and supported nearby urbanization, making the area a significant growth pole for rural revitalization and URI. Over its historical development, the area has innovated its regional policy through TSG, ensuring abundant land resources via land consolidation. Territorial space planning has facilitated functional optimization and clear area positioning. Institutional optimization has led to the establishment of an asset management team, which has used the land system to foster mutually beneficial value exchanges between urban and rural areas, promoting integrated development. This case exemplifies China's efforts toward URI and represents a common developmental trajectory achieved through TSG measures. Its practical initiatives offer universally applicable insights for exploring URI in similar regions in China and globally (Figure 4).

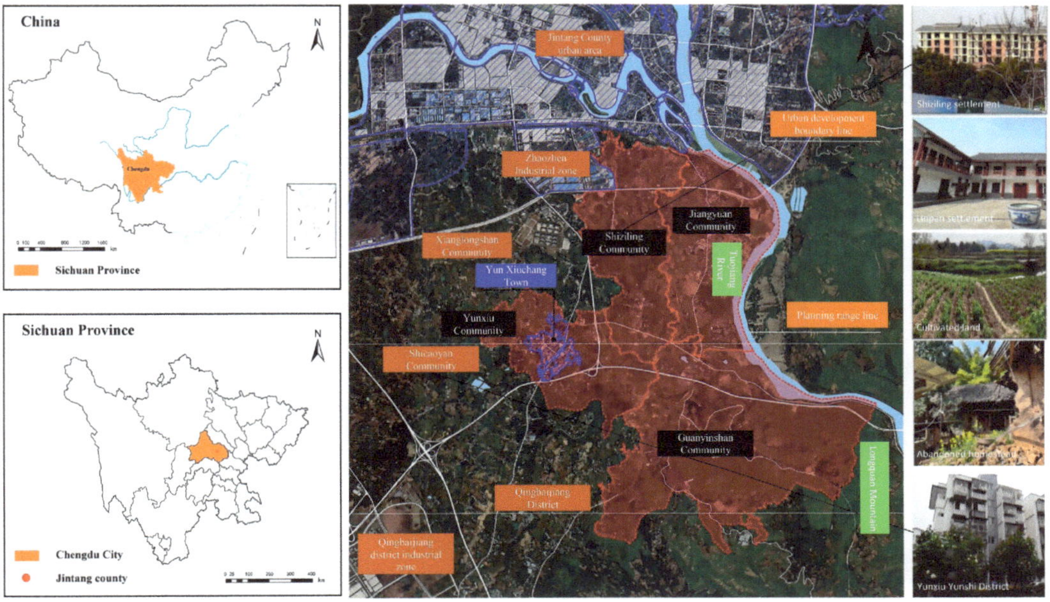

Figure 3. Case study area.

Table 1. Land allocation for various development projects in the study area.

Number	Project Name	Area (ha)
1	Guanyinshan Lake Chengdu Guanyinshan Jinshu Agricultural Science and Technology Development Co., Ltd.	0.28
2	Rural Revitalization Project	0.19
3	Farming Cloud Agriculture Innovation and Incubation Park	0.85
4	Qinghua Reservoir	0.02
5	Shi Ziling Village Collective Peony Industry Park	0.69
6	Sichuan Guofu Investment Co., Ltd.	0.03
7	New Resettlement Site	3.01
8	Yunguan Farm	5.43
9	Yunxiu Flower Field Project	0.02
10	Shangling Aquaculture Cooperative	0.06
11	New Rural Road Construction	0.16
12	S422 Zhaozhen to Huaikou Road Renovation Project	0.12
13	Chengdu Jintang Yangliu 110 kV Substation Expansion Project	0
14	Chengdu to Bazhong Railway, Chengdu to Santai Section	1.2
15	Dacheng Railway Capacity Expansion	6.74
16	Guanyin Mountain to Yunding Mountain Tourist Road Construction Project (Phase 2)	1.55
17	Jintang Gorge Flood Discharge Capacity Enhancement Project	4.16
18	Jintang County Longquanshan Urban Forest Park Water Source Project Phase I (Guanyin Mountain Project)	0.6
19	Jintang County Social Welfare Home Relocation Project	0.25
20	Chengdu Fuxing Elderly Care Service Center	0.52
21	Tuanjie Water Hub–Pihe Diversion Canal	6.77
22	Tuanjie Water Hub–Qimu River Flood Storage Reservoir	15.84
23	Tuanjie Water Hub–Tuojiang River Diversion Canal	47.53

Table 1. *Cont.*

Number	Project Name	Area (ha)
24	Jintang County Reclaimed Water Plant and Supporting Pipeline Project (Phase I)	3.47
25	Jintang County Construction Waste Disposal Project	2.79
26	Jintang County Sewage Treatment Plant Sludge Centralized Deep Processing Center Project (Phase I)	0.04
27	Jintang County Sewage Treatment Plant Sludge Centralized Deep Processing Center Project (Phases II & III)	3.56
28	Longquanshan Urban Forest Park Tourism Loop	3.36
29	Rural New Community Sewage Treatment Facility Construction Project	0.19
30	Pengqing-Huai Expressway	5.22
31	Qingjin Expressway	1.56
	Total	116.21

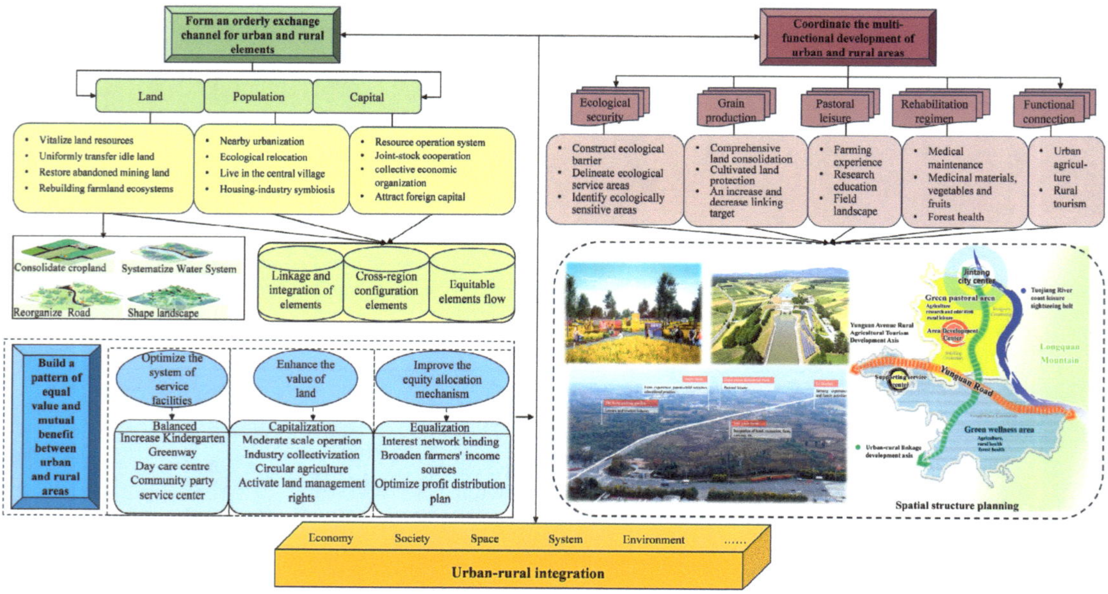

Figure 4. Spatial governance measures are taken in the study area for advancing URI.

3.2. Research Methods

This research adopts field surveys to interview government officials, villagers, community representatives, and other stakeholders. These interviews are supplemented by policy documents, planning texts, and other sources to obtain first-hand information on the current status of URI and the process of TSG in the case area. In March 2022, we conducted in-depth research in four communities, focusing on the spatial layout of key industries, construction, and public service facilities. On 8 April 2022, we engaged in discussions with town leaders and community leaders in Zhaozhen Street to understand some basic regional conditions.

In early 2023, we conducted on-site visits, concentrating on villagers' aspirations, industrial conditions, community public service facilities, municipal infrastructure, and other issues. Villagers' opinions and demands were collected through questionnaires. Additionally, the team also obtained relevant information from the Natural Resources and Planning Bureau of Jintang County and town leaders. We also acquired and thoroughly analyzed key documents, including the "Regulations on the Preparation of URI Development Area Plans in Chengdu" (Implemented), the "Guidance for the Preparation of Village-Level Plans in Sichuan Province", and the "Jintang County Territorial Spatial Planning (2019–2035)". These materials have provided rich, authentic, and accurate support for this study (Table 2).

Table 2. Selected policies and spatial plans in the study area.

Number	Policy/Plan	Relevant Points
1	2021 General Design for Rural Revitalization of Guanyin Mountain Community in Jintang County	Combining the ecological characteristics of "hills, valleys, forests, and fields" on the site, creates a characteristic rural area with "immersive" consumption and in-depth experience, and builds a healthy and recuperative Guanyin Mountain.
2	Conceptual Planning for Rural Revitalization of Shi Ziling Community in Zhaozhen, Jintang County	Promotes the development of characteristic rural tourism, and focuses on creating characteristic rural tourism products with comprehensive functions such as leisure and sightseeing, green ecology, and farming culture experience.
3	Project Plan for Rural Revitalization Demonstration of Yunxiu Huatian and Shi Ziling Community Agricultural Park in Zhaozhen Street	Creates an AAAA rural tourism theme park; Jintang County Rural Revitalization Demonstration Zone is a model area for rural revitalization in the province and city.
4	14th Five-Year Plan for the Development of Modern Urban Agriculture in Jintang County (2021–2025)	Focusing on urban agriculture (efficient ecological agriculture, characteristic high-quality agriculture, leisure, and sightseeing agriculture), creates a national model of urban agriculture with characteristics in hilly areas. The specific content is to build a supply base for characteristic agricultural products, a beautiful health and recuperation resort, and a new model of vacation agriculture.
5	Overall Tourism Development Plan of Jintang County	Focuses on the development of leisure and vacation, health, and recuperation industries.
6	Spatial Planning of Land for Modern Service Industry Area in Tianfu Water City, Jintang County (2021–2035)	Pending further docking, strictly transmits the binding indicators such as the cultivated land reserve and permanent basic farmland in this area. The function is positioned to create a collection of flowers and elderly care.
7	Spatial Planning of Land in Jintang County (2019–2035)	Located in the western optimization area, mainly focuses on stock renovation, and strictly controls the development scale along Longquan Mountain. In terms of industrial function layout, creates a cultural tourism and health care service center and an external exchange center in the coordinated area, and the leading industry is the modern service industry of "culture and tourism +".

4. Case Analysis: Leveraging Spatial Governance for Advancing Urban–Rural Integration

4.1. Forming an Orderly Exchange Channel of Urban–Rural Elements

The case area utilizes innovative policy systems and comprehensive land management to tackle challenges in elements circulation. Through the land transfer system, village collective economic organizations or cooperatives lease, contract, or share idle industrial and residential land, revitalizing land resources, enhancing agricultural advantages, and improving the regional market allocation of land resources. A special resource development and management system attracts investment, facilitates shared cooperation, establishes village-level collective economic entities, and promotes interaction among foreign capital, local resources, and labor. According to field surveys, 26% of villagers express a willingness to move to urban areas, while about 56% prefer to centralize their settlements. This is primarily because the Jiang yuan community, located along the ecologically fragile and flood-prone Tuo Jiang River, necessitates relocating scattered rural residences to higher, safer farm towns and central villages, optimizing the spatial and demographic patterns of villages and towns.

Additionally, the improved facilities in the Yun Xiu community attract former residents to return for work and living, benefiting from urban conveniences. The case area also undertakes comprehensive land improvement and ecological restoration through the following four major strategies: rectifying fields, managing water, improving roads, and shaping landscapes. This includes constructing 153.11 ha of high-standard farmland in the Guanyin Mountain community, reclaiming 94.17 ha of homesteads as arable land, and remediating 12.85 ha of mining wasteland. Furthermore, 11.3 ha of construction land has been revitalized inefficiently, while 29.57 ha has undergone functional replacement, providing ample land elements to advance agricultural modernization, rural development, and industrial space expansion in the region.

4.2. Coordinating Multifunctional Development of Urban and Rural Areas

Leveraging the area's resource endowment and strategic location, the case area integrates its functional development needs through comprehensive management strategies that include master planning and detailed refinement. This approach ensures land availability for rural industrial development and promotes distinctive agriculture, such as recycling farming and the cultivation of green organic fruits and vegetables. These efforts align with regional synergy under the "green industry" theme and the foundation of green agriculture. The industrial chain is extended to create a functional integration pattern that combines idyllic recreation and tourism, agricultural research and education, retirement communities, and health and wellness services.

Firstly, areas such as Tuojiang River, Bihe River, and Alder River are designated as crucial for ecological service functions and identified as ecologically sensitive through territorial spatial planning. This is primarily intended to establish spatial barriers for urban and rural ecological security. Concurrently, comprehensive land remediation projects are utilized to supplement approximately 127 ha of arable land and create an additional 14 ha, fulfilling the goal of arable land protection and ensuring agricultural production and residents' livelihoods.

Secondly, while ensuring food security, the region will establish three supply bases for "grains and oil, flowers and trees, and fruits and vegetables". It is designed to introduce diversified elements such as pastoral sightseeing, farming experience, and cultural innovation, thereby enhancing the agricultural landscape. This will form an urban agricultural demonstration area with local characteristics and a green rural leisure agricultural tourism zone. Specifically, Shi Ziling Area develops modern agriculture relying on the market demand advantage of the proximity of Jintang County to supply fresh ready-to-eat agricultural products to urban areas. It also aims to absorb the leisure demand from nearby cities by establishing a rural tourism park that integrates parent–child education, sports, and summer vacation activities. The area develops various types of rural leisure industries and strengthens functional interactions with Chengdu, Qing Baijiang, and Jintang County. Yunxiu Community opens up the health and pension market and builds a significant pension base in western China by utilizing facilities like the Jiujiu Pension Center, Yunxiu Pension Service Center, and Wanxia Community Pension Service Center. Taking advantage of the superior ecological landscape of the Tuojiang River, Guanyin Mountain develops a mountain-based planting base for medicinal herbs, vegetables, and fruits, enhancing the functions of forest health, forest activities, and medical care.

Finally, the initiative aims to promote the functional transformation of existing industrial land. By leveraging its advantageous location and convenient transportation near Zhaozhen Industrial Park, the Jiangyuan community encourages existing factories to relocate to the park through administrative measures, economic compensation, and tailored "one enterprise, one policy" strategies. This strategy targets gradually achieving land reclamation and greening, fostering a new green industrial base characterized by sustainability, low-carbon emissions, and increased income, hence contributing to the dual-carbon goal.

4.3. Constructing a Pattern of Equal and Reciprocal Value Between Urban and Rural Areas

As one of the pilot areas for the comprehensive reform of URI development in Sichuan Province, the case area advances land market transactions centered on land system reform to ensure that the rights and interests of secondary land development are equitably shared. To balance the interests of the collective economy and social capital, a working group on the reform of village asset demutualization has been established. This group leads a comprehensive verification of operational, non-operational, and resource assets under village collective ownership, along with government inputs. Following the model of "party branch leadership, party member initiative, farmer participation, and cooperative operation", these assets are integrated into the village collective economic organization. Shares of this organization are then quantified and allocated to individuals, enhancing farmers' property rights and increasing collective economic income. The proposal proposes an interest network comprising combinations such as "farmers + companies", "farmers + cooperatives + companies", and "collective economy + companies". It encourages qualified farmers to utilize idle rural settlement units, including homesteads, courtyards, dams, and forest trays. Through leasing, shareholding, and other means, these units are transferred to business experts to develop industries like unique homestays, inns, and bars, along with weaving workshops and product yards. This also attracts folk craftsmen and artisans to start businesses in rural areas, forming industrial clusters and expanding income-generating opportunities for farmers (Table 3).

Table 3. Overview of some collective economic incomes in the case area.

Community Name	Enterprise (Individual Business) Name	Type	Annual Output Value (CNY Ten Thousand)	Number of Employed Individuals
Shi Ziling	Hexiangyuan Restaurant	Catering	4000	35
	Qianting Manor	Catering	500	15
	Strawberry Picking Garden of Sichuan Yihong Agricultural Co., Ltd.	Strawberry Picking	50	3
	Happy Farm	Agricultural Research and Study	5000	15
	Agricultural Cloud—Eastward. Cultivate Learning in the Countryside.	Agricultural Innovation and Entrepreneurship Incubation Park	150	10
	Community Collective Peony Daya Industrial Park	Flower and Fruit Appreciation, Cold Storage, Flower Trading Market, Hot Pot City	100	1
	Subtotal		9800	79
Yunxiu	Xiangweiju	Catering	25	5
	Chuanjia Agriculture	Breeding, Catering	400	10
	Native Chicken Restaurant	Catering	27	4
	Sister Tian's Grocery Store	Supermarket	100	5
	Zhejing Supermarket	Supermarket	200	15
	Baihui Supermarket	Supermarket	150	8
	Qilin Commerce and Trade	Supermarket	180	6
	Subtotal		1082	53
Jiangyuan	Jintang Tuozhiyuan Catering Company	Catering	143	38
Guanyinshan	Tian Tian Le Farmhouse in Zhao Town, Jintang County	Catering	110	20
	Tuba Network Ecological Agricultural Farming Culture Theme Park	Planting and Picking	300	12
	Guanyinshan Fruit King Valley	Planting and Picking		6
	Subtotal		410	38
Total			11,435	208

Simultaneously, the reform of "separation of rights" is actively implemented to revitalize land management rights, attract foreign capital, and promote the ongoing development of advantageous industries and moderate-scale operations in the area. This is intended to achieve rational resource allocation and enhance the benefit distribution mechanism within each community. As a result of the land reform, projects such as Yunguan Farm, Guofu Family, and Zhiheng Picking Garden have been successfully introduced. These projects utilize cultivated land, forest land, and construction land to establish an ecological circular agriculture model, including rice/fish, rice/shrimp, rice/turtle, and rice/vegetable systems, thereby maximizing land value and significantly increasing the value-added income for various stakeholders, such as farmers and enterprises. It strengthens the interest relationships among the government, social capital, and village collectives. In addition, the case area optimizes the residential space structure through unified planning and self-directed construction. It guides villagers from regions prone to natural disasters to relocate to central villages, such as the Guanyin Mountain settlement and the Shi Ziling new community. Coincidentally, new facilities such as kindergartens, urban and rural greenway networks, community party service centers, and daycare centers are being added. These efforts contribute to promoting the equalization of urban and rural infrastructure and public service facilities, ensuring that residents in both areas enjoy fair rights and opportunities in residence, education, retirement, work, and leisure.

5. Discussion

5.1. Policy Implications

To advance URI and build a new urban–rural relationship, it is imperative to identify the structural contradictions that impede URI and elucidate the internal logic of TSG in fostering the multidimensional integration of urban–rural elements, functions, and values. From the theoretical analysis above, the following insights are derived.

TSG is a crucial aspect of the national governance system and serves as an effective institutional means to enhance national governance capabilities. It should capitalize on its strengths in spatial allocation, institutional constraint, and developmental guidance. Firstly, within the logical framework of TSG that facilitates the seamless flow of URI elements, it is important to objectively grasp the specific policy environment and land resource situation, advance land system reforms prudently, manage human–land relationships properly, and create a conducive environment for factor mobility and external growth. Secondly, in the context of functional interaction mechanisms, it is crucial to harness the spatial control efficacy of TSG, formulate spatial access lists tailored to regional specifics, balance the multifunctional development with ecological protection, and utilize unique village resource endowments to develop distinctive business formats. This approach fosters a functional landscape characterized by "one village, one product; one enterprise, one policy".

In the specific implementation process, it is essential to focus on rural areas' resource endowments and development needs, assessing their regional status from a macro perspective to accurately guide rural planning and functional allocation. Additionally, identifying rural characteristics and highlighting rural values based on unique cultural attributes is crucial to prevent homogeneous development. At the rural level, special consideration should be given to their distinct functions and attributes within the context of regional urban–rural integrated development. At the county level, it is necessary to optimize the internal spatial structure and residential–industrial systems of rural areas to achieve functional complementarity and spatial connectivity with neighboring towns and counties. In terms of value reciprocity, TSG should utilize its tools to innovate rural value realization methods and refine value distribution mechanisms. By establishing equitable trading platforms such as "lucid waters and lush mountains are invaluable assets banks" and adopting market-oriented operational modes, we can enhance the service capabilities of ecological products, express diverse values, and ensure a sustainable supply by converting resources into assets and capital values. Simultaneously, reforms should be deepened to enable the market entry of collectively constructed operating land and innovate cross-regional and cross-administrative land circulation schemes, thereby increasing land revenues and expanding farmers' profit opportunities.

5.2. Research Contribution

The existing literature has discussed the concepts and implementation pathways of spatial governance and URI, yet these discussions often present fragmented analytical models that overlook the intrinsic relationship between the two, lacking a comprehensive theoretical framework to examine how TSG empowers URI. To address this gap, this paper integrates the current practical demands of URI, delving into the theoretical connotation and systemic content of TSG. It constructs a framework that bridges the research on URI and TSG. As a related approach to land resource management, TSG serves as an important tool for achieving sustainable protection and utilization of urban and rural territorial spaces. Recent studies have examined the relationship between farmland governance and urbanization, attempting to establish a sustainable farmland utilization evaluation system to clarify future directions for land resource protection, thereby effectively promoting regional sustainable development [22,86]. Meanwhile, the existing literature has analyzed effective strategies for promoting urban–rural integrated development from various dimensions, mainly including comprehensive territorial land remediation projects, the construction of urban–rural data-sharing platforms, and land use transformation [87,88]. Numerous studies have confirmed that multi-scale pathways to achieve urban–rural integrated development involve strengthening URI within urban agglomerations, gradually urbanizing county-level populations, enhancing economic ties between urban and rural areas, and improving agriculture connectivity [89,90]. While these studies provide valuable references for this paper, they often fail to systematically analyze the impact pathways of TSG on URI.

In contrast, our research expands on this by analyzing the relationship between urban–rural integrated development and TSG from a spatial governance perspective, exploring the governance mechanisms of URI. We focus on deconstructing how TSG drives URI, offering a unique theoretical perspective and analytical framework for understanding the enabling process and implementation pathways of TSG oriented towards URI. On the other hand, most existing research on TSG is predominantly fragmented in its theoretical analysis, lacking a cohesive analytical framework based on logical chains, and insufficiently exploring the enabling effects and practical implications. Recent studies have analyzed the connotation and implementation pathways of TSG from a structural functionalist perspective [33], and examined the linkage mechanisms between URI, rural transformation, and rural spatial governance across different historical stages in Yucheng, China [91,92]. While the aforementioned theoretical and case studies contribute to understanding the impact of specific governance approaches, they fail to provide a holistic view of TSG. This study dissects the intrinsic logical relationship between TSG and URI, further elucidating the theoretical mechanisms and procedural dynamics through which TSG facilitates the interaction of urban and rural elements, the complementarity of their functions, and their mutual benefits. Furthermore, grounded on systematically collected data from TSG cases in western China, this study analyzes the practical pathways of TSG in concretely promoting URI, thereby verifying the reliability and validity of the theoretical framework. It also proposes a relatively detailed Chinese-style governance approach for the integrated development of urban and rural areas that could apply to similar regions worldwide.

5.3. Research Prospects

China's URI is at a pivotal historical juncture, prompting the scientific community to address essential questions, such as how to delineate the facilitation mechanisms of TSG for URI, and how to leverage URI's leading role in directing TSG. As China enters the latter phase of its urbanization process, it is crucial to reassess national conditions regarding key elements such as urban–rural population, land, and resources. This reassessment aims to expand the new connotations of TSG in line with China's core values and strategic modernization needs. Clarifying the new requirements for URI development and devising spatial governance strategies adapted to emerging integration trends are also essential. Despite advancements, regional development disparities persist in China, and improving the income and intergenerational equity of urban and rural residents remains urgent. Addressing the call for common prosperity and exploring targeted regulatory pathways for TSG based on current conditions continue to be pressing research topics.

The case study examined in this research represents a microcosm of China's URI. Developing countries and similar regions worldwide face comparable challenges when promoting URI. The most prominent issue is the lagging development in rural areas, where rural values are not effectively transformed into assets. Consequently, farmers struggle to earn substantial local wages, contributing to a global trend of rural decline [93]. In response, China has implemented the Rural Revitalization Strategy, which provides increased policy support, capital investment, and talent inflow to rural areas, aiming to elevate the modernization of rural construction. Simultaneously, land remediation projects and land system reform innovations have been carried out across regions to achieve sustainable development of rural territorial systems by accelerating land use transformation. It is evident that continuously enhancing rural modernization and sustainable development capabilities addresses deficiencies in the rural dimension. Revitalizing rural areas fosters regional economic and social growth, narrows the income gap between urban and rural residents, elevates villagers' status, and thereby supports URI.

While TSG offers unique advantages in driving URI, it also encounters potential risks, such as imbalanced governance structures and lagging governance tools [94]. Future research should not only emphasize the facilitative role of TSG in promoting URI but also be aware of the hindering factors and practical challenges in this process. Moreover, given the pronounced spatial heterogeneity in China's URI development, which exhibits a pattern of higher integration in the east compared with the west and middle [58], it is urgent to identify the main contradictions and obstacles factors in URI development across different geographical regions. Developing a differentiated optimization path system using TSG is essential. Tailored, dynamic, and region-specific governance schemes should be devised based on China's national conditions and integration levels of integration across different historical stages.

Additionally, the advent of globalization, intellectualization, and digitization has profoundly impacted global urban–rural systems and spatial relations, necessitating careful consideration. Digital technology, as a new production factor, is reshaping rural spatial production, lifestyles, and social governance structures, initiating the digital transformation of rural areas [95,96]. This transformation creates favorable conditions for URI from a flow space perspective. Responding to the practical needs of digital urban–rural construction, TSG theories should promptly incorporate digital elements and adjust governance strategies. Future research should clarify the theoretical logic, implementation mechanisms, and pathways by which digital technology can enhance TSG, based on an assessment of its transformation trends. Theoretically, a spatial governance framework under a digital context should be established to deepen the coupling relationship between URI and digital TSG, explore governance technologies in the digital era, and better support URI development.

6. Conclusions

In recent years, China's level of urbanization has significantly increased, with ongoing efforts to promote deep URI to achieve a new model of urbanization that effectively coordinates population and land use. Within the framework of enhancing governance capabilities and systems, the multifunctional and multi-value aspects of TSG play a crucial role. This governance approach is vital in pursuing the strategic goal of URI, serving as an essential systematic tool. Drawing from the literature reviews and policy analyses, this paper constructs a theoretical framework and implementation pathway for using TSG to drive URI. This study finds that TSG is a comprehensive process aimed at optimizing the allocation of elements, structural organization, and overall functions of the urban–rural territorial spatial system. It necessitates collective action from various stakeholders and the coordination of tools such as spatial planning and institutional design. By following the logic of element interaction, functional complementarity, and value reciprocity, it seeks to reconstruct the human–land relationship and spatial form, thereby promoting URI.

A qualitative analysis using Jintang as a case study reveals that the path to achieving URI should begin with TSG and progress through three dimensions: elements, functions, and values. Specifically, the orderly exchange of urban–rural elements can be facilitated through land circulation and management, encouraging villagers to move into towns, and land consolidation. Ensuring food security while promoting urban agriculture, green rural leisure industries, and the transformation of industrial land can support coordinated multifunctional development between rural and urban areas. Establishing village-level asset securitization reform teams can secure land appreciation returns, while optimizing village and town spatial structures and facility layouts can achieve balanced urban–rural development. The research conclusions offer a Chinese approach to URI and spatial governance that can be applied to other similar countries and regions worldwide.

Author Contributions: Data curation, L.W.; Writing—original draft, L.T.; Writing—review & editing, Q.C.; Supervision, L.C. All authors have read and agreed to the published version of the manuscript.

Funding: This research was funded by Observation and Research Station of Land Ecology and Land Use in Chengdu Plain, Ministry of Natural Resources (grant numbers: CDORS-2024-07) and Chengdu philosophy and social science planning project (grant numbers: 2024BS037).

Data Availability Statement: The raw data supporting the conclusions of this article are unavailable due to privacy restrictions, and further inquiries can be directed to the corresponding author.

Acknowledgments: We want to thank all experts and editors who reviewed the paper after its completion. Their opinions are important references for us to polish and improve the paper.

Conflicts of Interest: Author Lan Wang was employed by the company Sichuan University Engineering Design and Research Institute Co. Ltd. The remaining authors declare that the research was conducted in the absence of any commercial or financial relationships that could be construed as a potential conflict of interest.

References

1. Liu, Y.; Li, Y. Revitalize the World's Countryside. *Nature* **2017**, *548*, 275–277. [CrossRef] [PubMed]
2. Cui, Q.; Tan, L.; Ma, H.; Wei, X.; Yi, S.; Zhao, D.; Lu, H.; Lin, P. Effective or Useless? Assessing the Impact of Park Entrance Addition Policy on Green Space Services from the 15-Min City Perspective. *J. Clean. Prod.* **2024**, *467*, 142951. [CrossRef]
3. Liu, Y.; Long, H.; Chen, Y.; Wang, J.; Li, Y.; Li, Y.; Yang, Y.; Zhou, Y. Progress of Research on Urban-Rural Transformation and Rural Development in China in the Past Decade and Future Prospects. *J. Geogr. Sci.* **2016**, *26*, 1117–1132. [CrossRef]
4. Liu, Y.; Dai, L.; Long, H.; Woods, M.; Fois, F. Rural Vitalization Promoted by Industrial Transformation under Globalization: The Case of Tengtou Village in China. *J. Rural Stud.* **2022**, *95*, 241–255. [CrossRef]
5. Ge, D.; Sun, P.; Tang, L. The Logic and Path of Supporting Urban-Rural Integrated Development through Territorial Space Planning. *China Land Sci.* **2023**, *37*, 1–9.
6. Niu, B.; Ge, D.; Sun, J.; Sun, D.; Ma, Y.; Ni, Y.; Lu, Y. Multi-Scales Urban-Rural Integrated Development and Land-Use Transition: The Story of China. *Habitat Int.* **2023**, *132*, 102744. [CrossRef]
7. Long, H.; Zou, J.; Pykett, J.; Li, Y. Analysis of Rural Transformation Development in China since the Turn of the New Millennium. *Appl. Geogr.* **2011**, *31*, 1094–1105. [CrossRef]
8. Laurin, F.; Pronovost, S.; Carrier, M. The End of the Urban-Rural Dichotomy? Towards a New Regional Typology for SME Performance. *J. Rural Stud.* **2020**, *80*, 53–75. [CrossRef]
9. Long, H.; Qu, Y. Land Use Transitions and Land Management: A Mutual Feedback Perspective. *Land Use Policy* **2018**, *74*, 111–120. [CrossRef]
10. Guipeng, Z.; Hualou, L.; Dazhuan, G.E.; Linzeng, X.U.; Sijie, L.I. Study on Three-Dimensional Measurement of Recessive Farmland Use Morphology and Its Regional Types: A Case of the Huang-Huai-Hai Region. *Geogr. Res.* **2024**, *43*, 180–199. [CrossRef]
11. Long, H.; Chen, K. Urban-Rural Integrated Development and Land Use Transitions: A Perspective of Land System Science. *Acta Geogr. Sin.* **2021**, *76*, 295–309.
12. Dazhuan, G.E.; Hualou, L. Rural Spatial Governance and Urban-Rural Integration Development. *Acta Geogr. Sin.* **2020**, *75*, 1272–1286. [CrossRef]

13. Sun, P.; Ge, D.; Yuan, Z.; Lu, Y. Rural Revitalization Mechanism Based on Spatial Governance in China: A Perspective on Development Rights. *Habitat Int.* **2024**, *147*, 103068. [CrossRef]
14. Dazhuan, G.E. The Characteristics and Multi-Scale Governance of Rural Space in the New Era in China. *Acta Geogr. Sin.* **2023**, *78*, 1849–1868. [CrossRef]
15. Zhang, Y.; Long, H. Theory and Practice of County-Level Territorial Space Planning under the Context of Urban-Rural Integrated Development. *China Land Sci.* **2023**, *37*, 1–10. [CrossRef]
16. Chen, K.; Long, H.; Liao, L.; Tu, S.; Li, T. Land Use Transitions and Urban-Rural Integrated Development: Theoretical Framework and China's Evidence. *Land Use Policy* **2020**, *92*, 104465. [CrossRef]
17. Luo, X.; Jin, X.; Liu, X.; Zhang, S.; Ying, S.; Zhou, Y. Mechanism and Model of Comprehensive Land Consolidation Promoting Urban-Rural Integration in Peri-Urban from the Perspective of Symbiosis Theory. *J. Nat. Resour.* **2024**, *39*, 1053. [CrossRef]
18. Song, X.-P.; Hansen, M.C.; Stehman, S.V.; Potapov, P.V.; Tyukavina, A.; Vermote, E.F.; Townshend, J.R. Global Land Change from 1982 to 2016. *Nature* **2018**, *560*, 639–643. [CrossRef]
19. Li, Y.; Westlund, H.; Liu, Y. Why Some Rural Areas Decline While Some Others Not: An Overview of Rural Evolution in the World. *J. Rural Stud.* **2019**, *68*, 135–143. [CrossRef]
20. Yan, J.; Chen, H.; Xia, F. Toward Improved Land Elements for Urban–Rural Integration: A Cell Concept of an Urban–Rural Mixed Community. *Habitat Int.* **2018**, *77*, 110–120. [CrossRef]
21. Han, L.; Qu, Y.; Liang, S.; Shi, L.; Zhang, M.; Jia, H. Spatiotemporal Differentiation of Land Ecological Security and Optimization Based on GeoSOS-FLUS Model: A Case Study of the Yellow River Delta in China Toward Sustainability. *Land* **2024**, *13*, 1870. [CrossRef]
22. Yang, Q.; Pu, L.; Huang, S. Review of Urbanization-Associated Farmland Research in China: A Sustainability Perspective. *Land* **2024**, *13*, 534. [CrossRef]
23. Johansson, J.; Holmquist, M. LEADER and Rural Development Policy—What's the Problem Represented to Be? *J. Rural Stud.* **2024**, *108*, 103287. [CrossRef]
24. Lin, J.; Gao, Y.; Zhao, Y. Exploration of National Territory Spatial Governance from the Perspective of Spatial Development Rights. *J. Nat. Resour.* **2023**, *38*, 1393–1402. [CrossRef]
25. Gui-Peng, Z.; Hua-Lou, L. The Mechanism of Land Use Transitions and Optimization of Territorial Spatial Development Patterns: Analysis Based on the Spatial Functions of Land Use Benefits. *J. Nat. Resour.* **2023**, *38*, 2447–2463. [CrossRef]
26. McCarthy, J. Rural Geography: Multifunctional Rural Geographies-Reactionary or Radical? *Prog. Hum. Geogr.*. **2005**, *29*, 773–782. [CrossRef]
27. Rahman, M.M.; Szabó, G. A Geospatial Approach to Measure Social Benefits in Urban Land Use Optimization Problem. *Land* **2021**, *10*, 1398. [CrossRef]
28. Esparcia, J.; Escribano, J.; Serrano, J.J. From Development to Power Relations and Territorial Governance: Increasing the Leadership Role of LEADER Local Action Groups in Spain. *J. Rural Stud.* **2015**, *42*, 29–42. [CrossRef]
29. Ansell, C.; Gash, A. Collaborative Governance in Theory and Practice. *J. Public Adm. Res. Theory* **2008**, *18*, 543–571. [CrossRef]
30. Provan, K.G.; Kenis, P. Modes of Network Governance: Structure, Management, and Effectiveness. *J. Public Adm. Res. Theory* **2008**, *18*, 229–252. [CrossRef]
31. Dazhuan, G.E.; Yuqi, L.U. Rural Spatial Governance for Territorial Spatial Planning in China: Mechanisms and Path. *Acta Geogr. Sin.* **2021**, *76*, 1422–1437. [CrossRef]
32. Ge, D.; Zhou, G.; Qiao, W.; Yang, M. Land Use Transition and Rural Spatial Governance: Mechanism, Framework and Perspectives. *J. Geogr. Sci.* **2020**, *30*, 1325–1340. [CrossRef]
33. Zhu, C.; Wang, K.; Zhang, J.; Gan, M.; Yuan, S. The Connotation and Realization Path of Territorial Space Governance from the Perspective of "Elements-Structure-Function-Value". *China Land Sci.* **2021**, *36*, 10–18. [CrossRef]
34. Rao, J. Comprehensive Land Consolidation as a Development Policy for Rural Vitalisation: Rural In Situ Urbanisation through Semi Socio-Economic Restructuring in Huai Town. *J. Rural Stud.* **2022**, *93*, 386–397. [CrossRef]
35. Hualou, L. Land Consolidation and Rural Spatial Restructuring. *Acta Geogr. Sin.* **2013**, *68*, 1019–1028.
36. Friedmann, J. Strategic Spatial Planning and the Longer Range. *Plan. Theory Pract.* **2004**, *5*, 49–67. [CrossRef]
37. Liu, Y.; Zhou, Y. Territory Spatial Planning and National Governance System in China. *Land Use Policy* **2021**, *102*, 105288. [CrossRef]
38. Qu, Y.; Jiang, G.; Li, Z.; Tian, Y.; Wei, S. Understanding Rural Land Use Transition and Regional Consolidation Implications in China. *Land Use Policy* **2019**, *82*, 742–753. [CrossRef]
39. Huang, X. The Reform of the Natural Resource Property Rights System and the Innovation in Territorial Spatial Governance. *Urban Plan. Forum* **2021**, *2*, 53–57.
40. Zhao, P. Planning for Social Inclusion: The Impact of Socioeconomic Inequities on the Informal Development of Farmland in Suburban Beijing. *Land Use Policy* **2016**, *57*, 431–443. [CrossRef]
41. Lundsgaard-Hansen, L.M.; Oberlack, C.; Hunt, G.; Schneider, F. The (In)Ability of a Multi-Stakeholder Platform to Address Land Conflicts—Lessons Learnt from an Oil Palm Landscape in Myanmar. *Land* **2022**, *11*, 1348. [CrossRef]
42. Liu, H.; Fan, W.; Zhou, X.; Wang, Y.; Yuan, C.; Liu, L. From Solo to Cluster Governance: An Empirical Study of Transforming Rural Management in Guiyang, China. *Land* **2024**, *13*, 1564. [CrossRef]

43. Wang, Y.; Li, T. Behavioural Selection of Farmer Households for Rural Homestead Use in China: Self-Occupation and Transfer. *Habitat Int.* **2024**, *152*, 103163. [CrossRef]
44. Zhao, Q.; Bao, H.X.H.; Yao, S. Unpacking the Effects of Rural Homestead Development Rights Reform on Rural Revitalization in China. *J. Rural Stud.* **2024**, *108*, 103265. [CrossRef]
45. Kan, K. Creating Land Markets for Rural Revitalization: Land Transfer, Property Rights and Gentrification in China. *J. Rural Stud.* **2021**, *81*, 68–77. [CrossRef]
46. Baffoe, G.; Zhou, X.; Moinuddin, M.; Somanje, A.N.; Kuriyama, A.; Mohan, G.; Saito, O.; Takeuchi, K. Urban–Rural Linkages: Effective Solutions for Achieving Sustainable Development in Ghana from an SDG Interlinkage Perspective. *Sustain. Sci.* **2021**, *16*, 1341–1362. [CrossRef]
47. CPC Central Committee and State Council. Strategic Plan for Rural Revitalization (2018–2022). Available online: https://www.mca.gov.cn/n152/n162/c82617/content.html (accessed on 26 November 2024).
48. Cattaneo, A.; Nelson, A.; McMenomy, T. Global Mapping of Urban–Rural Catchment Areas Reveals Unequal Access to Services. *Proc. Natl. Acad. Sci. USA* **2021**, *118*, e2011990118. [CrossRef]
49. Ji, X.; Ren, J.; Ulgiati, S. Towards Urban-Rural Sustainable Cooperation: Models and Policy Implication. *J. Clean. Prod.* **2019**, *213*, 892–898. [CrossRef]
50. Xia, H.; Zhang, W.; He, L.; Ma, M.; Peng, H.; Li, L.; Ke, Q.; Hang, P.; Wang, X. Assessment on China's Urbanization after the Implementation of Main Functional Areas Planning. *J. Environ. Manag.* **2020**, *264*, 110381. [CrossRef] [PubMed]
51. Kunqiu, C.; Hualou, L.; Li, M.A.; Yingnan, Z. China's Rural Land Reform and Rural Vitalization. *Prog. Geogr.* **2019**, *38*, 1424–1434. [CrossRef]
52. Lin, G.; Jiang, D.; Fu, J.; Zhao, Y. A Review on the Overall Optimization of Production–Living–Ecological Space: Theoretical Basis and Conceptual Framework. *Land* **2022**, *11*, 345. [CrossRef]
53. Wen, L.; Chatalova, L.; Butsic, V.; Hu, F.Z.; Zhang, A. Capitalization of Land Development Rights in Rural China: A Choice Experiment on Individuals' Preferences in Peri-Urban Shanghai. *Land Use Policy* **2020**, *97*, 104803. [CrossRef]
54. Simeng, L.I.; Ren, Y.; Hualou, L. The Theoretical Evolution of Global Rural Development and Its Enlightenment to Rural Modernization in China. *Acta Geogr. Sin.* **2024**, *79*, 854–872. [CrossRef]
55. Xu, J.; Zeng, J.; Hu, Y. Property Rights System and Market Evolution: Plot-Level Evidence from China's Land Titling. *Land Use Policy* **2024**, *145*, 107253. [CrossRef]
56. Yuancheng, L.I.N.; Ren, Y.; Yingxian, D. Urban Consumerization Transformation of Rural Space and Rural Revitalization in China. *Geogr. Res.* **2023**, *42*, 1697–1714. [CrossRef]
57. Li, S.; Yang, R.; Long, H.; Lin, Y.; Ge, Y. Rural Spatial Restructuring in Suburbs under Capital Intervention: Spatial Construction Based on Nature. *Habitat Int.* **2024**, *150*, 103112. [CrossRef]
58. Yang, Y.; Bao, W.; Wang, Y.; Liu, Y. Measurement of Urban-Rural Integration Level and Its Spatial Differentiation in China in the New Century. *Habitat Int.* **2021**, *117*, 102420. [CrossRef]
59. Nilsson, M.; Griggs, D.; Visbeck, M. Policy: Map the Interactions between Sustainable Development Goals. *Nature* **2016**, *534*, 320–322. [CrossRef] [PubMed]
60. Liu, Y.; Zang, Y.; Yang, Y. China's Rural Revitalization and Development: Theory, Technology and Management. *J. Geogr. Sci.* **2020**, *30*, 1923–1942. [CrossRef]
61. Tan, L.; Zhao, S.; Chen, L. The Inner Mechanism and Path Selection of Rural Spatial Governance Enabling Rural Transformation. *Geogr. Res.* **2024**, *43*, 1572–1590.
62. Westlund, H.; Nilsson, P. Agriculture's Transformation and Land-Use Change in a Post-Urban World: A Case Study of the Stockholm Region. *J. Rural Stud.* **2022**, *93*, 345–358. [CrossRef]
63. Qian, H.; Wong, C. Master Planning under Urban–Rural Integration: The Case of Nanjing, China. *Urban Policy Res.* **2012**, *30*, 403–421. [CrossRef]
64. Marsden, T.; Sonnino, R. Rural Development and the Regional State: Denying Multifunctional Agriculture in the UK. *J. Rural Stud.* **2008**, *24*, 422–431. [CrossRef]
65. Wilson, G.A. The Spatiality of Multifunctional Agriculture: A Human Geography Perspective. *Geoforum* **2009**, *40*, 269–280. [CrossRef]
66. Long, H.; Liu, Y.; Li, X.; Chen, Y. Building New Countryside in China: A Geographical Perspective. *Land Use Policy* **2010**, *27*, 457–470. [CrossRef]
67. Nguyen, H.Q.; Warr, P. Land Consolidation as Technical Change: Economic Impacts in Rural Vietnam. *World Dev.* **2020**, *127*, 104750. [CrossRef]
68. Dazhuan, G.E.; Yuqi, L.U.; Pan, S.U.N. The Logic of Rural Spatial Governance and Revitalization. *Acta Geogr. Sin.* **2022**, *77*, 777–794. [CrossRef]
69. Jiang, Y.; Wang, T.; Xu, Y. Deciphering Land Use Transitions in Rural China: A Functional Perspective. *Land* **2024**, *13*, 809. [CrossRef]
70. Kiminami, L.; Furuzawa, S.; Kiminami, A. Social Entrepreneurship and Social Business Associated with Multiple Functions of Urban Agriculture in Japan. *Asia-Pac. J. Reg. Sci.* **2020**, *4*, 521–552. [CrossRef]

71. Rivera, M.; Knickel, K.; de los Rios, I.; Ashkenazy, A.; Pears, D.Q.; Chebach, T.; Šūmane, S. Rethinking the Connections between Agricultural Change and Rural Prosperity: A Discussion of Insights Derived from Case Studies in Seven Countries. *J. Rural Stud.* **2018**, *59*, 242–251. [CrossRef]
72. Mcdonagh, J. Rural Geography I Changing Expectations and Contradictions in the Rural. *Prog. Hum. Geogr.* **2013**, *37*, 712–720. [CrossRef]
73. Yansui, L.I.U. The Basic Theory and Methodology of Rural Revitalization Planning in China. *Acta Geogr. Sin.* **2020**, *75*, 1120–1133. [CrossRef]
74. Jiang, Y.; Long, H.; Ives, C.D.; Deng, W.; Chen, K.; Zhang, Y. Modes and Practices of Rural Vitalisation Promoted by Land Consolidation in a Rapidly Urbanising China: A Perspective of Multifunctionality. *Habitat Int.* **2022**, *121*, 102514. [CrossRef]
75. Cheng, M.; Yin, Z.; Westlund, H. Counterurbanization in China? A Case Study of Counties in Huang-Huai-Hai Area from the Perspective of Urban-Rural Relations. *J. Rural Stud.* **2024**, *110*, 103386. [CrossRef]
76. Yansui, L.I.U. Research on the Urban-Rural Integration and Rural Revitalization in the New Era in China. *Acta Geogr. Sin.* **2018**, *73*, 637–650. [CrossRef]
77. Nordberg, K. Spatial Justice and Local Capability in Rural Areas. *J. Rural Stud.* **2020**, *78*, 47–58. [CrossRef]
78. Liu, Z.; Liu, S.; Jin, H.; Qi, W. Rural Population Change in China: Spatial Differences, Driving Forces and Policy Implications. *J. Rural Stud.* **2017**, *51*, 189–197. [CrossRef]
79. Chen, K.; Long, H. Impacts of Land Market on Urban-Rural Integrated Development in China. *J. Nat. Resour.* **2019**, *34*, 221–235. [CrossRef]
80. Chen, N.; Kong, L. Rural Revitalization in China: Towards Inclusive Geographies of Ruralization. *Dialogues Hum. Geogr.* **2022**, *12*, 213–217. [CrossRef]
81. Cui, X.; Ma, L.; Tao, T.; Zhang, W. Do the Supply of and Demand for Rural Public Service Facilities Match? Assessment Based on the Perspective of Rural Residents. *Sustain. Cities Soc.* **2022**, *82*, 103905. [CrossRef]
82. Liu, Y.; Lu, S.; Chen, Y. Spatio-Temporal Change of Urban–Rural Equalized Development Patterns in China and Its Driving Factors. *J. Rural Stud.* **2013**, *32*, 320–330. [CrossRef]
83. Chari, A.; Liu, E.M.; Wang, S.-Y.; Wang, Y. Property Rights, Land Misallocation, and Agricultural Efficiency in China. *Rev. Econ. Stud.* **2021**, *88*, 1831–1862. [CrossRef]
84. Adam, A.G.; Dadi, T.T. Perspectives for Smooth Bridging of Dichotomized Urban–Rural Land Development in the Peri-urban Areas of Ethiopia: Toward a Continuum Approach. *Reg. Sci. Policy Pract.* **2024**, *16*, 12733. [CrossRef]
85. Wu, Y.; Sun, X.; Sun, L.; Choguill, C.L. Optimizing the Governance Model of Urban Villages Based on Integration of Inclusiveness and Urban Service Boundary (USB): A Chinese Case Study. *Cities* **2020**, *96*, 102427. [CrossRef]
86. Hu, X.; Dong, C.; Zhang, Y. Impacts of Cropland Utilization Patterns on the Sustainable Use Efficiency of Cropland Based on the Human–Land Perspective. *Land* **2024**, *13*, 863. [CrossRef]
87. Chen, K.; Wang, Y.; Li, N.; Xu, Y.; Zheng, Y.; Zhan, X.; Li, Q. The Impact of Farmland Use Transition on Rural Livelihood Transformation in China. *Habitat Int.* **2023**, *135*, 102784. [CrossRef]
88. Rojo-Mendoza, F.; Salinas-Silva, C.; Alvarado-Peterson, V. The End of Indigenous Territory? Projected Counterurbanization in Rural Protected Indigenous Areas in Temuco, Chile. *Geoforum* **2022**, *133*, 66–78. [CrossRef]
89. Boudet, F.; MacDonald, G.K.; Robinson, B.E.; Samberg, L.H. Rural-Urban Connectivity and Agricultural Land Management across the Global South. *Glob. Environ. Change* **2020**, *60*, 101982. [CrossRef]
90. He, Y.; Zhou, G.; Tang, C.; Fan, S.; Guo, X. The Spatial Organization Pattern of Urban-Rural Integration in Urban Agglomerations in China: An Agglomeration-Diffusion Analysis of the Population and Firms. *Habitat Int.* **2019**, *87*, 54–65. [CrossRef]
91. Ge, D.; Lu, Y. A Strategy of the Rural Governance for Territorial Spatial Planning in China. *J. Geogr. Sci.* **2021**, *31*, 1349–1364. [CrossRef]
92. Ge, D.; Long, H.; Qiao, W.; Wang, Z.; Sun, D.; Yang, R. Effects of Rural–Urban Migration on Agricultural Transformation: A Case of Yucheng City, China. *J. Rural Stud.* **2020**, *76*, 85–95. [CrossRef]
93. Long, H.; Liu, Y. Rural Restructuring in China. *J. Rural Stud.* **2016**, *47*, 387–391. [CrossRef]
94. Valderrama, E.-L.; Polanco, J.-A.; Hernandez-Diaz, P.M. Does Collaborative Governance Mediate Rural Tourism to Achieve Sustainable Territory Development? Evidence of Stakeholders' Perceptions in Colombia. *Sustain. Dev.* **2024**, 1–15. [CrossRef]
95. Habibipour, A.; Lindberg, J.; Runardotter, M.; Elmistikawy, Y.; Ståhlbröst, A.; Chronéer, D. Rural Living Labs: Inclusive Digital Transformation in the Countryside. *Technol. Innov. Manag. Rev.* **2022**, *11*, 59–72. [CrossRef]
96. Woods, M.; Fois, F.; Heley, J.; Jones, L.; Onyeahialam, A.; Saville, S.; Welsh, M. Assemblage, Place and Globalisation. *Trans. Inst. Br. Geogr.* **2021**, *46*, 284–298. [CrossRef]

land

MDPI

Article

A Leverage Points Perspective on China's Governance of Marine Protected Areas: Current State and Ways Forward

Jinpeng Wang [1,†], Zhengkai Mao [1,†] and Zhijun Zhang [2,*]

[1] School of Law, Ocean University of China, Qingdao 266100, China; wangjinpeng@ouc.edu.cn (J.W.); maozhengkai@163.com (Z.M.)
[2] Institute of Marine Development, Ocean University of China, Qingdao 266100, China
* Correspondence: lzuzhangzhijun@163.com
† These authors contributed equally to this work and share first authorship.

Abstract: As a key element of spatial governance, marine protected areas (MPAs) have been increasingly established in various countries, with lessons learned from terrestrial environmental protection. Nevertheless, the development of MPAs in China continues to trail behind that of their land-based counterparts. Here, following the leverage points perspective of sustainability interventions, this article presents a systematic analysis of the governance and evolution of China's MPAs, identifying key areas for improvement. The analysis encompasses the number, effectiveness, legal framework, governance structure, value, and paradigm of MPAs, and highlights the associated governance challenges facing China. Drawing on relevant experiences from the United States, Australia, and the European Union, the article offers valuable insights for informing China's future MPA strategies. The study concludes that while China has made significant progress in the development of MPAs, further efforts are needed, including paradigm shifts, refinement of the legal system, optimization of governance structures, and enhancement of MPA effectiveness.

Keywords: marine protected areas; spatial governance; environmental protection; national park; nature reserve; leverage points; China

1. Introduction

The oceans, encompassing nearly three-quarters of the Earth's surface, harbor rich biodiversity and furnish essential ecosystems that yield food and other vital resources for human sustenance. Furthermore, they function as the planet's largest carbon sink, underscoring their critical role in mitigating climate change. In recognition of their importance, the United Nations' Sustainable Development Goal 14 champions the conservation and sustainable utilization of the world's oceans, seas, and marine resources to foster sustainable development. Marine spaces are the maritime regions across the world under national jurisdiction, which are vast in area, abundant in resources, and hold considerable ecological importance. However, past studies on spatial governance have predominantly focused on terrestrial spaces, leaving marine spaces largely underrecognized for the whole spatial sustainability.

In fact, the idea and movement of sustainable development kind of originated from the marine sector, interestingly. Back in the 1950s, with the decline in catch or effort ratios in various fisheries around the world, the need to devise methods to manage international fishery conflicts and protect marine environments at the global level became imperative [1], prompting the development of the concept of protecting specific areas and habitats, and eventually leading to the establishment of marine protected areas (MPAs) as a zoning

 https://doi.org/10.3390/land14020425

management tool to foster marine ecological conservation and sustainable resource utilization. In 2003, the Ad Hoc Technical Expert Group on Marine and Coastal Protected Areas of the Convention on Biological Diversity (CBD) defined a marine and coastal protected area as "any defined area within or adjacent to the marine environment, together with its overlying waters and associated flora, fauna, and historical and cultural features, which has been reserved by legislation or other effective means, including custom, with the effect that its marine and/or coastal biodiversity enjoys a higher level of protection than its surroundings" [2]. In the face of mounting threats from marine pollution and climate change, establishing a robust and well-managed network of MPAs has emerged as a vital strategy for bolstering the conservation of marine ecosystems.

The establishment of MPAs draws on the experience of terrestrial conservation, leveraging management practices and lessons learned from land-based protected areas to inform the governance of MPAs. The concept of protected areas originated in the 19th century in countries such as Australia, Canada, South Africa, and the United States, and by the 20th century, it had gained global traction [3]. For decades, protecting rare and endangered species, as well as preserving their habitats or areas of exceptional natural beauty, has proven to be an effective conservation measure on land. According to Article 2 of the CBD, a protected area is defined as a geographically defined area which is designated or regulated and managed to achieve specific conservation objectives. However, protected areas should not be isolated from their surroundings; instead, they should form part of a network of protected areas that integrate into the broader environment, promoting conservation and sustainable use of land and water resources [4]. The interconnectedness and interdependence of terrestrial and marine ecosystems are underscored by the complex exchanges of matter and energy between them. Notably, a significant portion of marine pollution originates from land-based sources, with an estimated 70% to 80% of plastic pollution in the ocean, by weight, being transported from land to sea through rivers or coastal areas [5]. The effective conservation of coastal protected areas hinges on the integration of MPAs into a broader network of land–sea ecosystems, as the strong land–sea interactions and fluidity of seawater dynamics render isolated conservation efforts inadequate [6].

In particular, the Kunming-Montreal Global Biodiversity Framework (KM-GBF) underscores the importance of establishing MPAs, reflecting the international community's commitment to protecting marine biodiversity. The KM-GBF emphasizes the need to enhance the management effectiveness of MPAs, improve their representativeness and connectivity, and integrate them into broader marine spatial planning. It urges countries to strengthen marine conservation efforts, designating at least 30% of the ocean as protected areas, as well as using Other Effective Area-Based Conservation Measures to address the pressing challenges of marine biodiversity loss. The establishment of MPAs has emerged as a widely accepted consensus among nations in the context of global biodiversity governance and ocean governance. However, despite this broad agreement, the development of MPAs significantly lags behind that of terrestrial protected areas. Statistics reveal that only 8.3% of the world's oceans are designated as MPAs, and most are either protected in name only or so loosely regulated that substantial harmful activities are allowed to continue within them [7], highlighting a pressing need to accelerate progress in marine conservation efforts.

China is home to one of the world's most diverse marine ecosystems, with its waters supporting over 28,000 documented species—representing approximately 11% of global marine biodiversity [8]. However, as marine resources continue to be intensively exploited and utilized, a range of pressing environmental issues has emerged, including accelerating marine environmental degradation, expanding areas of ecosystem decline, and widespread loss of coastal wetlands [9]. Since the 18th National Congress of the Communist Party of China, the country has prioritized the development of a marine ecological civilization and the protection of the marine environment. MPAs have emerged as a crucial tool in managing the marine ecosystem, playing a vital role in safeguarding the health of the marine environment, maintaining ecosystem balance, and promoting the sustainable use of marine resources [10].

Regarding the literature search, by using the terms "Marine Protected Areas" in conjunction with "China" to form the query ["Marine Protected Areas" AND "China"], a total of 416 articles were retrieved from the Web of Science Core Collection. It shows that studies on China's MPAs have gradually developed since the 1990s, with the number of relevant articles increasing annually (Figure 1).

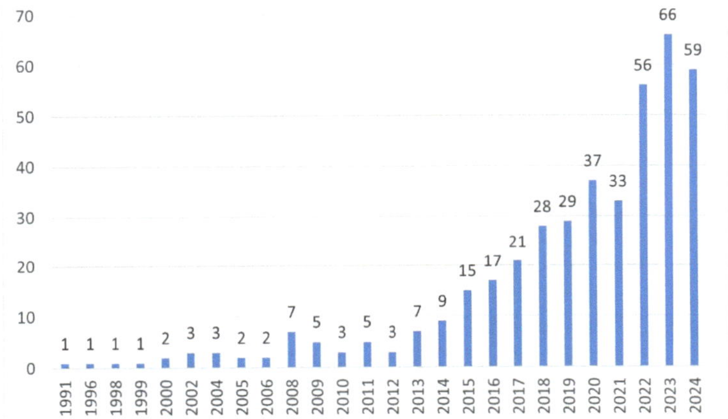

Figure 1. Number of articles on research of China's MPAs.

Additionally, VOSviewer was utilized to construct and visualize keyword co-occurrence networks for both Marine Protected Areas and China, allowing for the deduction of primary research themes from the interlinking patterns of keywords [11]. A comprehensive analysis of 407 articles from 2004 to 2024 reveals that the research focuses on exploring the value and role of MPAs in conserving fish and other biological resources, protecting coral reefs, addressing climate change, and providing ecosystem services. Additionally, the studies also examine the management models and development of MPAs (Figure 2). For example, Qiu et al. analyzed the characteristics of China's system of MPAs, suggesting that it is characterized by decentralized designation and management [12]. Hu et al. analyzed the development process of China's MPAs, arguing that China implemented unified management of MPAs after its institutional reforms in 2018, which made new top-level design possible [13]. Zhao et al. [14] and Zeng et al. [15] indicated that the major issues hampering management effectiveness include multiple management agencies, lack of systematic planning for classification and zoning, and limited survey data. Some studies emphasize the involvement of stakeholders in MPA design, implementation, and management [16–19]. However, in the context of China's efforts to promote integrated land and sea development and build a natural protected area system, research on systematically optimizing the governance of MPAs remains insufficient. This is particularly true regarding the connections and priorities between different elements and leverage points within the governance framework. To address this gap, a deeper understanding of how to intervene effectively in complex systems is essential.

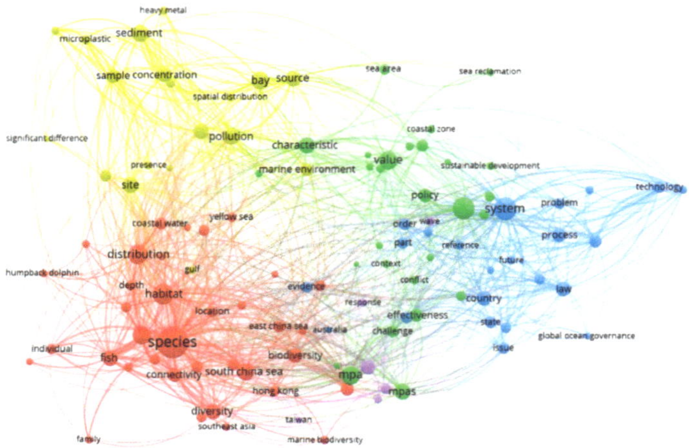

Figure 2. Network analysis of keyword co-occurrence on MPAs and China.

In this regard, Meadows' framework of 12 types of leverage points provides valuable governance insights for understanding where and how to intervene in complex systems to bring about transformative changes [20]. Building on this, Abson et al. propose four realms of leverage—Parameters, Feedback, Design, and Intent—ranging from shallow to deep leverage points. They argue that Design and Intent, as deep leverage points, have great potential but are under-researched [21]. Followingly, Fisher and Riechers expanded this framework into four realms of leverage for sustainability interventions: Material, Processes, Design, and Intent [22]. Tools in the realm of shallow leverage points focus more on past and present system states, while tools in the realm of deep leverage points often use participatory and transdisciplinary approaches across different academic disciplines and sources of knowledge [23]. Guided by this leverage points perspective, this research conducts a comprehensive examination of China's MPA governance, including a case study of the Dalian Spotted Seal National Nature Reserve, an analysis of the Ministry of Ecology and Environment's ecological assessments of 10 national marine nature reserves, and an evaluation of the current status of sea-related national parks, either established or under development. Furthermore, the research undertakes a comparative analysis of China's MPA governance model with those of the United States, Australia, and the European Union, and provides recommendations for the integrated optimization and governance of China's MPAs.

2. Analytical Framework Based on the Leverage Points Perspective

The leverage points perspective offers a valuable framework for understanding and addressing environmental governance challenges by emphasizing systemic and sustainable interventions. By shifting the focus from short-term fixes to transformative changes, this approach provides a powerful tool for tackling complex environmental issues. It enables the examination of interactions between shallow and deep system changes, allowing for a more nuanced understanding of the dynamics at play. Scholars have successfully applied the leverage points perspective to study various issues related to environmental governance [22]. For instance, Bartkowski and Bartke regarded the determinants of farmers' behavior and decision-making regarding soil management as leverage points for soil governance [24]. Building on this foundation, this article explores the development, challenges, and improvement of China's governance of MPAs through the leverage points perspective.

As a crucial tool for managing the marine ecological environment, MPAs play a vital role in conserving marine ecosystems and promoting the sustainable utilization of marine resources. China's marine conservation efforts commenced in 1963 with the establishment of its first MPA on Snake Island in Dalian, followed by the designation of Snake Island and the adjacent Old Iron Mountain as a national-level nature reserve by the State Council in 1980. Over the years, China has made significant strides in developing its MPA network, which now forms a comprehensive system centered on marine ecological environment protection [25]. Drawing on the research experience of existing literature on environmental governance from the leverage points perspective, this article identifies leverage points in MPA governance, ranging from shallow to deep (See Table 1). By examining specific leverage points related to China's MPA governance, the article prioritizes actions that address deeper, systemic issues rather than superficial fixes. The identified leverage points include surface-level factors, such as the number and area of MPAs, as well as their role and effectiveness. Middle-level leverage points comprise the legal system and policies for MPA governance, as well as the protected area governance structure, including government departments and community participation. Deep leverage points primarily involve the value and paradigm of MPA governance, which underpin the overall approach to marine conservation.

Table 1. A leverage points perspective on MPA governance.

Realm of Leverage	Leverage Point	Specific Leverage Points on MPA Governance
Shallow leverage points	Numbers	Number and size of MPAs
	Parameters	Effectiveness of MPAs
Mid-level leverage points	System	Laws and Policies
	Structure	Inter-departmental collaboration
		Community participation
Deep leverage points	Values	Ecological values
	Paradigms	Transformation of governance model

3. The Number and Effectiveness of MPAs in China

3.1. Development Progress and Increased Quantity of MPAs in China

China's Marine Environmental Protection Law has established the legal framework for marine nature reserves and marine special protected areas, defining their respective legal status and establishment conditions. Marine nature reserves prioritize the conservation of rare, natural, and pristine ecosystems, focusing on three main categories: marine biological species, marine and coastal ecosystems, marine natural relics and non-biological resources. In contrast, marine special protected areas strive for a balance between conservation and development, promoting the sustainable use of marine resources while protecting the environment. These areas are categorized into four types: marine parks, marine ecological protection areas, marine resource protection areas, and marine special geographical condition protection areas (See Table 2), with designations at both national and local levels [26,27]. Currently, China has established a comprehensive network of MPAs, effectively covering all coastal provincial administrative units across the mainland [28].

Table 2. Categories of China's MPAs under the Marine Industry Standard of China (HY/T 117-2008) [29].

Category	Protected Targets
Marine special geographical condition protection areas	Sea areas and islands with important marine rights and interests; or with special marine hydrodynamic conditions
Marine ecological protection areas	Areas where rare and endangered species are naturally distributed; areas with concentrated typical ecosystems; and other ecologically sensitive and fragile areas or ecological restoration areas
Marine parks	Areas with special marine ecological landscapes; historical and cultural relics; unique geological landforms and surrounding sea areas; and areas with the function of ecological tourism
Marine resource protection areas	Important marine resource development reserve areas; marine ecological industry zones; and various marine resource development coordination areas

MPAs should be established and managed through a zoning system that reflects the unique characteristics of the marine environment and socio-economic conditions. This will help to create a spatial development framework that achieves a balance between human activity, economic growth, and environmental sustainability [30]. Unfortunately, some MPAs have been established without adequate data, resulting in mismatches between actual sea use and the intended functional zones. This has hindered the enforcement of strict regulations and compromised the effective protection of the primary conservation targets within these areas [31]. Notably, some important marine ecosystems and ecological function areas remain outside of designated protection zones [28].

According to the White Paper on China's Marine Ecological Environment Protection, published in July 2024, China has made significant progress in marine conservation, with the establishment of 352 marine nature reserves covering an impressive 93,300 km^2 of ocean area. Furthermore, five national marine park candidate areas are currently under development, laying the foundation for a robust, well-structured, and functional MPA governance system [32]. However, despite these efforts, China's MPA construction still lags behind its international commitments. The Overall Plan for Ecological Civilization System Reform, issued by the Central Committee of the Communist Party of China and the State Council in 2015, explicitly stated China's intention to "actively participate in global governance", marking a shift towards China taking a proactive role in assuming governance responsibilities and driving global development [33]. As the host of the 15th Conference of the Parties to the CBD [34], China played a key role in promoting the adoption of the KM-GBF. Nevertheless, China's MPAs account for only 4.1% of its total jurisdictional sea area by the end of 2019 [35], falling short of the target set by the KM-GBF, which emphasizes the importance of MPAs and Other Effective Area-Based Conservation Measures in meeting Target 3 criteria [36]. This disparity highlights the need for China to strengthen its planning and implementation of MPAs.

3.2. Case-Based Analysis of the Effectiveness of MPAs in China

MPAs vary in their level of protection, ranging from comprehensive to minimal, with some existing only on paper, and lacking effective implementation [37]. Enhancing the effectiveness of MPAs is a global imperative [38]. Following years of development, China's MPA planning and implementation efforts have yielded significant outcomes, with the construction of marine nature reserves playing a crucial role in promoting the sustainable development of the oceans and conserving biodiversity. This article selects the largest national-level marine nature reserve in China, the Dalian Spotted Seal National Nature Reserve, as a typical case to analyze the effectiveness of the marine reserve and its protection of marine biological species, as well as the challenges it faces. The Dalian Spotted Seal National Nature Reserve is located near Changxing Island in Fuzhou Bay, 20 km northwest of Dalian City in Liaodong Bay, Bohai Sea [39]. It was established in 1992 with the approval of the Dalian Municipal Government and was upgraded to a national-level nature reserve in 1997. The reserve covers an area of 560,000 hectares, with a coastline of approximately 370 km, almost encompassing all the coastal waters of Dalian City within Liaodong Bay. Its primary conservation target is the spotted seal and its associated ecological habitat. Notably, the reserve was designated as a Wetland of International Importance under the Ramsar Convention in January 2002 [40].

Historically, the population of spotted seals in Liaodong Bay was estimated to be approximately 7000–8000 individuals in the 1930s. However, excessive hunting between 1940 and 1970 led to a precipitous decline, with only 2269 seals remaining by 1979 [41]. In response to this alarming trend, Liaoning Province enacted legislation in 1983 that strictly prohibited the hunting of spotted seals. Despite this conservation effort, the population continued to decline, and by 1992, there were fewer than 1000 spotted seals remaining [40]. The establishment of the Dalian Spotted Seal Nature Reserve in 1992 marked a pivotal moment in conservation efforts, and when combined with successful artificial breeding programs, the reserve contributed to the revival of the population. By 2006, the population of spotted seals had increased to an estimated 2000 individuals in the waters under China's jurisdiction [42]. Subsequently, the population stabilized and exhibited signs of growth. A 2021 survey conducted by the National Forestry and Grassland Administration revealed a notable sighting of over 200 spotted seals at Ant Island's reef and adjacent waters, representing the largest single observation of its kind in recent years for a concentrated haul-out site in Dalian waters [40].

Institutional design-level measures played a crucial role in the recovery of the spotted seal population, driven by two key factors. Firstly, the development of a comprehensive legal and regulatory framework has provided a solid foundation for conservation efforts. The State Council's issuance of the "Outline of China Aquatic Biological Resources Conservation Action" (2006–2020) and the "China Biodiversity Conservation Strategy and Action Plan" (2011–2030) has been instrumental in this regard. Furthermore, the Liaoning Provincial People's Government has introduced a series of local regulations, including the "Regulation of Liaoning Province on the Protection of Spotted Seals" (2007), the "Implementation Measures of Liaoning Province on the Protection of Key Protected Terrestrial Wild Animal Spotted Seals" (2018), and the "Plan of Liaoning Province on the Protection of Spotted Seals' Population and Habitat" (2019). These regulations have effectively promoted the conservation work of the Dalian Spotted Seal National Nature Reserve, providing a robust legal framework for protection. Secondly, the Dalian Spotted Seal National Nature Reserve's effective conservation measures have played a crucial role in the recovery of the spotted seal population. The reserve's management agency has enforced stringent protection of the spotted seals' habitats, including beaches and reefs, and implemented coastal zone and island vegetation restoration projects. Additionally, a comprehensive monitoring system has been established to track the spotted seal population's size, distribution, and breeding patterns on a regular basis, providing valuable insights for conservation efforts [43].

The overarching objective of China's MPA governance is to integrate land and sea management mechanisms through the enhancement and implementation of protected area systems, thereby continually improving the efficacy of marine ecological environment governance [32]. Nevertheless, the current governance effectiveness of China's MPAs remains in need of improvement. A comprehensive assessment of the ecological environment of 10 national marine nature reserves, conducted by China's Ministry of Ecology and Environment in 2023, underscores this imperative. The results revealed that five reserves—Dalian Spotted Seal National Nature Reserve in Liaoning, Yellow River Delta National Nature Reserve in Shandong, Huidong Sea Turtle National Nature Reserve in Guangdong, Zhanjiang Mangrove National Nature Reserve in Guangdong, and Hepu National Nature Reserve in Guangxi—achieved a Grade I rating, signifying excellent ecological conditions. The remaining five reserves, including Yancheng Wetland Rare Bird National Nature Reserve in Jiangsu, Jiuduansha Wetland National Nature Reserve in Shanghai, Xuwen Coral Reef National Nature Reserve in Guangdong, Shankou Mangrove Ecological National Nature Reserve in Guangxi, and Beilunhe Estuary National Nature Reserve in Guangxi, received a Grade II rating, indicating satisfactory but not outstanding conditions (See Table 3) [44].

Table 3. National marine nature reserves with Grade II rating in China's ecological environment quality assessment in 2023 [44].

National Marine Nature Reserve	Ecological Environment Quality	Monitoring Results
Yancheng Wetland Rare Bird National Nature Reserve in Jiangsu	Grade II	There has been a reduction in the area occupied by Spartina alterniflora compared to the previous year
Jiuduansha Wetland National Nature Reserve in Shanghai	Grade II	Spartina alterniflora occupies around 16.4% of the total area
Xuwen Coral Reef National Nature Reserve in Guangdong	Grade II	The average coverage of living coral has decreased year-on-year
Shankou Mangrove Ecological National Nature Reserve in Guangxi	Grade II	The area of mangrove forest and the average density of mangroves have both increased year-on-year
Beilunhe Estuary National Nature Reserve in Guangxi	Grade II	The average mangrove density has decreased year-on-year

As revealed in the "Report on the Ecological Environment of China's Oceans in 2023", released by the Ministry of Ecology and Environment, the primary challenges confronting the ecological environment of marine nature reserves include the incursion of the non-native species Spartina alterniflora, in addition to the diminution of live coral and mangrove populations. Originally native to the eastern seaboard of North America, Spartina alterniflora, a perennial grass species, was intentionally introduced to China in 1979 with the goals of enhancing sedimentation, conserving coastlines, and reducing wave-induced erosion [45]. However, Spartina alterniflora possesses high adaptability, strong interspecific competitiveness, and rapid reproduction characteristics [46], which have led to a high overlap with mangrove habitats in muddy tidal flats and wetlands. Not only does Spartina alterniflora compete with mangroves for survival space, but it also reduces the quality of microhabitats and the biodiversity of large benthic animals in mangrove forests [47].

The unique characteristics of MPAs, particularly their remoteness from land, exacerbate the challenges associated with monitoring and enforcement. Consequently, the level of marine environmental and resource surveying, monitoring, and research in these areas is inadequate, hindering the provision of stable, long-term technical support for marine ecological restoration and conservation management [48]. Furthermore, the coverage of protected areas in fragile ecosystems, such as coasts, coastal wetlands, and river estuaries, is insufficient due to competing economic development interests, resulting in inadequate protection efforts in these regions. Moreover, some national-level MPAs lack comprehensive management plans, and unresolved conflicts over sea use and resources among stakeholders persist. The effectiveness of law enforcement agencies is also compromised by the absence of specific legal authority to enforce laws and regulations [25].

4. Systems and Structure of China's MPA Governance

4.1. Regulatory Framework for China's Governance of MPAs

The legal framework and administrative measures governing national parks, nature reserves, marine special protected areas, geoparks, marine parks, and forest parks in China are multifaceted (See Table 4). To promote the development and governance of MPAs, China has established a comprehensive suite of laws and regulations, building on the foundation of the "Regulations of the People's Republic of China on Nature Reserves". Key legislation and administrative measures includes the "Marine Environment Protection Law of the People's Republic of China", "Measures for the Management of Marine Nature Reserves", "Measures for the Management of Nature Reserves for Aquatic Animals and Plants", "Law of the People's Republic of China on the Management of Sea Area Use", "Law of the People's Republic of China on the Protection of Islands", and "Measures for the Management of Marine Special Protected Areas", which collectively provide a regulatory framework for the establishment and management of marine nature reserves, nature reserves for aquatic plants and animals, and marine special protected areas, as well as the use and management of sea areas and islands [12]. Furthermore, laws and regulations such as the "Fisheries Law of the People's Republic of China", "Regulations on the Protection of Aquatic Wild Animals", "Regulations on the Propagation and Protection of Aquatic Resources", "Cultural Relics Protection Law of the People's Republic of China", and "Regulations on the Prevention of Pollution Damage to the Marine Environment by Land-based Pollutants" have contributed to the development of MPAs from diverse perspectives.

Table 4. Basic management system and institutions for various types of natural protected areas in China.

Type	Main Laws and Regulations	Management System	Zoning Measures	Measures to Prohibit or Restrict Human Activities
National Park	Interim Measures for the Management of National Parks	The State Forestry and Grassland Administration (National Park Management Bureau) is responsible for supervision and management, while each national park management institution is responsible for specific management.	Core protection areas and general control areas	The core protection area prohibits human activities in principle. The general control zone prohibits development and production activities, but allows limited human activities.
Nature Reserve	Regulations on Nature Reserves	The State Council's ecological environment administrative department is responsible for the comprehensive management of national nature reserves. Relevant administrative departments are responsible for managing nature reserves within their respective scopes of responsibility.	Core areas and buffer zones	No one is allowed to enter the core area of the nature reserve. Tourism and production activities are prohibited in the buffer zone of the nature reserve.

Table 4. *Cont.*

Type	Main Laws and Regulations	Management System	Zoning Measures	Measures to Prohibit or Restrict Human Activities
Marine Special Protected Area	Measures for the Management of Marine Special Protected Areas	The State Oceanic Administration is responsible for the supervision and management of national marine special protected areas, and collaborates with coastal provinces, autonomous regions, and municipalities to develop and implement construction and development plans for national marine special protected areas, while guiding the development of local marine special protected areas.	Key protection areas, moderate utilization areas, ecological and resource restoration areas, and reserve areas	Within the key protection area, all construction activities unrelated to protection are prohibited. Within the moderate utilization area, moderate use of marine resources is permitted. Within the ecological and resource restoration area, appropriate artificial ecological restoration and repair measures can be taken. Within the reserve area, human disturbance is strictly controlled.
Geopark	Measures for the Management of World Geoparks	The State Forestry and Grassland Administration is responsible for the organization, coordination, supervision, and management of national geoparks, as well as for providing technical guidance.	Areas overlapping with nature reserves must comply with the relevant management and control requirements of the nature reserves.	Management of global geoparks shall be carried out in accordance with the relevant provisions of UNESCO and national regulations, ensuring effective resource protection and regional sustainable development.
Forest Park	Notice of the State Forestry and Grassland Administration on Further Strengthening the Management of National Forest Parks	The forestry administrative department is responsible for supervision and management.	No specific zoning measures mentioned	Any development activities and behaviors that do not conform to the main function of national forest parks are strictly prohibited.

4.2. The Structure of China's Governance of MPAs

In 2013, the Third Plenary Session of the 18th Central Committee of the Communist Party of China marked a milestone in ecological conservation by introducing the concept of deepening ecological civilization reform and accelerating the establishment of an ecological civilization system [49]. Prior to these reforms, China's MPAs were governed under a complex management framework that encompassed both departmental and hierarchical structures [50]. Specifically, marine special protected areas fell under the unified management of the former State Oceanic Administration, whereas marine nature reserves were overseen by the former Ministry of Environmental Protection. Meanwhile, the forestry, ocean, environment, agriculture, and land management departments each assumed sector-specific supervisory responsibilities. Following the release of the "Integrated Reform Plan for Promoting Ecological Progress" in 2015, the Communist Party of China Central Committee and the State Council jointly issued the "Plan for Deepening the Reform of Party and State Institutions" in 2018, thus initiating a comprehensive overhaul of China's protected area management system and effectively promoting the development of MPAs. The unification of MPA management under the National Forestry and Grassland Administration represents a significant step toward overcoming the challenges posed by fragmented management structures [13]. Nonetheless, China continues to draw on a hierarchical management approach for MPAs, categorizing protected areas into national and local designations based on their significance, representativeness, and distinctiveness. The local designation is further stratified into three levels: provincial, municipal, and county, to foster a nuanced approach to marine conservation [28].

Prior to the 2018 State Council institutional reform, the government departments responsible for managing MPAs in China included the following: the Ministry of Environmental Protection, the Ministry of Land and Resources, the State Oceanic Administration, the Ministry of Agriculture, the State Forestry Administration, and various levels of local governments. Marine nature reserves were established with the approval of the State Council and different levels of local governments, while marine special protected areas were mainly established with the approval of the State Oceanic Administration. The establishment of nature reserves in China has proceeded without unified, coordinated top-level planning. Instead, various government departments have created their own reserves under their respective mandates, resulting in a national nature reserve classification system lacking both scientific rigor and systematic coherence. This fragmented approach has produced ambiguous and uncoordinated functional designations. Moreover, the piecemeal establishment of reserves by different departments has led to significant spatial overlap and redundancy, with approximately 49.8% of the country's nature reserves exhibiting overlapping boundaries [51].

The 19th National Congress of the Communist Party of China highlighted the need of reforming the ecological and environmental regulatory framework, developing a spatial planning system for land use and conservation, strengthening policies aligned with functional zoning, establishing a natural protection area system centered on national parks, and strictly preventing and penalizing activities that harm the ecological environment [52]. In March 2018, the State Council implemented institutional reforms by creating the Ministry of Natural Resources and assigning oversight of all marine and terrestrial protected areas to the new ministry and its subordinate agency, the National Forestry and Grassland Administration, also known as the National Park Administration. Concurrently, the Ministry of Ecology and Environment was formed, assuming the marine environmental protection responsibilities previously held by the State Oceanic Administration [53]. In June 2019, the General Office of the Communist Party of China Central Committee and the General Office of the State Council issued the "Guiding Opinions on Establishing a System of Natural Protected Areas with National Parks as the Mainstay" [49]. This policy formally integrated MPAs into the overarching national system of natural protected areas, designating the National Forestry and Grassland Administration as the supervisory and management authority for all types of natural protected areas, including MPAs. The document underscores the importance of "establishing a system of natural protected areas with national parks as the main body, categorized scientifically, laid out reasonably, protected effectively, and managed efficiently". It further requires "conducting a comprehensive evaluation of existing natural protected areas, such as nature reserves, scenic spots, geological parks, forest parks, marine parks, wetland parks, glacier parks, grassland parks, desert parks, grassland scenic areas, and adjusting and classifying them according to their natural attributes, ecological values, and management goals, to gradually form a classification system of natural protected areas with national parks as the main body, nature reserves as the foundation, and various natural parks as supplements". In 2022, China released its first national five-year plan for marine ecological environment protection, proposing comparable objectives and reflecting how the principles of ecological priority and green development now permeate the entire marine sector [35].

In addition to the government-led "top-down" governance approach, public participation in protected areas is also crucial. As a part of integrated coastal management, the promotion of MPAs should consider not only the protection of marine biodiversity and ecosystem functions, but also the sustainable use of marine resources and public interests [54]. Many of China's MPAs have resident populations or are centers of intensive human activity. For instance, the Changdao National Marine Park includes fishing villages and communities within its boundaries; in 2024 alone, 5520 vessels were involved in fishing, aquaculture, tourism, and shipping within the park's waters [55]. At present, community participation in MPA governance largely takes the form of public hearings and official announcements during the establishment of MPAs. However, mechanisms to ensure public involvement in the early stages, such as site application and selection, remain inadequate. The 2024 draft edition of China's National Park Law similarly highlights the need for a community governance system in national parks, encouraging the inclusion of local residents, experts, scholars, and social organizations in the establishment, planning, management, and operation of these areas.

5. The Value and Paradigms of China's MPA Governance

5.1. The Value of China's MPA Governance

By prioritizing eco-environmental conservation, China has integrated ecological progress into its overall plan for marine development, adopting a well-conceived and sustainable approach to utilizing marine resources. Marine ecological civilization plays a crucial role in achieving China's broader ecological civilization objectives, and the values underlying China's MPA governance reflect the nation's overarching environmental and developmental goals. China's MPAs aim to protect critical marine habitats, endangered species, and unique ecosystems, which serve as sanctuaries that promote biodiversity and ecological balance. By safeguarding habitats such as coral reefs, mangroves, and seagrass beds, these MPAs enhance marine ecosystem resilience against both climate change and human activities. Despite China's rapid economic growth and substantial government investments in biodiversity conservation, the country's extensive territory, rich biodiversity, and sizeable conservation workload have led to insufficient protected area development and management, as well as funding shortages. Critical areas such as biodiversity surveys and monitoring, in-situ conservation, and ecosystem restoration urgently require more attention and support [56]. Furthermore, although the importance of biodiversity is increasingly acknowledged in China, evaluating the effectiveness of conservation efforts remains challenging due to the extended timeframe needed to observe tangible outcomes and the absence of standardized metrics. Beyond monitoring species populations statistically, there is a notable lack of consistent methods and criteria to assess other pivotal aspects of biodiversity conservation [57].

China prioritizes the integrated governance of the mountain–water–forest–farmland–lake–grassland–sand ecosystem, recognizing the interconnected and systematic nature of ecological systems. This holistic ecological perspective underscores the universal connections within ecosystems, highlighting the intrinsic relationships among their various components. At its core, this approach emphasizes the principle of universal connection rather than a whole-centered perspective [58]. Similarly, the ocean ecosystem constitutes a complex web of interactions between marine life and the marine environment, which mutually influence one another to maintain a dynamic equilibrium over time [59]. Consequently, the development of MPAs should be grounded in ecosystems or biogeographic regions [60], adopting an ecosystem-based approach that fosters cross-sectoral and interdisciplinary integrated marine management [61]. In 2000, the 5th Conference of the Parties to the CBD adopted Decision V/6, outlining 12 key principles for applying an ecosystem-based approach to conservation and management [62]. Subsequently, the 2006 annual report on Oceans and the Law of the Sea, submitted by the UN Secretary-General, provided a comprehensive overview of the ecosystem approach [63]. While noting the lack of an internationally agreed-upon definition, it describes the ecosystem approach as managing ecosystems based on the best available understanding of ecological interactions and processes, aiming to preserve their structure and function. Adopting this approach is essential for the effective management of MPAs and the development of MPA networks. Under its guidance, China's MPAs must be further integrated and optimized to ensure sustainable and holistic conservation outcomes.

5.2. The Paradigms of China's MPA Governance

A pressing challenge facing China's governance paradigms for natural protected areas is the need for systematic integration of multiple, often fragmented areas. For example, Hainan Tropical Rainforest National Park integrates Diaoluoshan National Forest Park, Jianfengling National Nature Reserve, and Limushan Provincial Nature Reserve, among others [64], covering a total area of 4269 km^2. Of this total, 2331 km^2 comprise the core protected area, while 1938 km^2 fall under the general control area [65]. A similar need for integration exists in MPAs. China is currently planning five national park candidates related to coastal and marine environments, including Tropical Ocean, Yellow River Estuary, Liao River Estuary, Chang Island, and Nanji Islands [66]. For instance, the Yellow River Estuary National Park spans 352,291 hectares. Its main body incorporates eight natural protected areas, including the Shandong Yellow River Delta National Nature Reserve, Shandong Yellow River Delta National Geopark, and Shandong Yellow River Estuary National Forest Park; it also encompasses areas of ecological significance, such as spawning and nursery grounds for marine species. This park is divided into a core protected area of 184,103 hectares—all state-owned—and a general control area of 168,188 hectares containing agricultural land, scattered residential locations, educational and research facilities, and recreation sites [67]. Similarly, the Liao River Estuary National Park was formed by merging and optimizing the original Liao River Estuary National Nature Reserve with seven other protected areas, creating a more manageable and effective conservation region [68]. In another example, the Changdao National Park candidate area centers on the Changdao Marine Ecological Civilization Comprehensive Experimental Zone, encompassing 151 islands and their surrounding waters. This area includes nine existing protected sites—such as the Shandong Changdao National Nature Reserve, Changdao National Marine Park, Changdao National Geopark, and Changdao National Forest Park—along with the Changshanwei Marine Geological Relics Provincial Marine Special Protected Area. Moreover, it covers critical migration routes and habitats of species such as the spotted seal and the East Asian finless porpoise [69].

The development of national marine parks faces the dual challenge of coordinating both spatial planning and administrative authority. This process requires integrating management frameworks and institutional arrangements for various types of natural protected areas within park boundaries, such as marine parks, forest parks, and geological parks. Previously, these areas were managed by different departments according to specific functional requirements. Once incorporated into the natural protected area system, they must continue fulfilling their original functions [70]. In 2022, the Chinese National Forestry and Grassland Administration, National Development and Reform Commission, Ministry of Finance, Ministry of Natural Resources, and Ministry of Agriculture and Rural Affairs jointly issued the "Plan for Major Projects on the Construction of National Parks and Other Natural Protected Areas and the Protection of Wild Animals and Plants (2021–2035)". The plan proposes completing the integration, consolidation, and optimization of natural protected areas and implementing a unified registration system for natural resources by 2025. By 2035, the spatial layout of natural protected areas is expected to be further optimized.

China has also introduced an ecological conservation red line system, which identifies areas with critically important ecological functions requiring strict, mandatory protection [71]. All marine natural protected areas and marine natural parks fall under this framework [72]. According to the management guidelines of the marine ecological conservation red line, core areas of marine nature reserves are strictly off-limits to human activities, mirroring the requirements for core areas under the marine nature reserve system. The general control areas of marine nature reserves and the entirety of marine nature parks permit only limited human activities that do not harm ecological functions, while prohibiting developmental and constructive uses. This further refines the management requirements for general control areas under the marine nature reserve system and the management requirements for marine nature parks [73]. Consequently, in terms of institutional alignment, the core areas of marine nature reserves are subject to both the marine nature reserve system and the marine ecological conservation red line system, ensuring stringent and mandatory protection. The general control areas of marine nature reserves and the entirety of marine nature parks, while adhering to their respective management requirements, must also ensure that human activities are kept within the limits that the ecosystem can sustainably withstand. Specifically, limited human activities are only permitted if they do not encroach upon the marine ecological conservation red line or cause harm to the marine ecosystem [74].

6. Ways Forward for Advancing China's Governance of MPAs

6.1. Improving the Paradigms of China's MPA Governance

Deep leverage points are the key to solving complex, entrenched problems and transforming systems over time, as they address the root causes of issues and can shift the dynamics of the entire system. Improving the paradigms of China's MPA governance could focus on adopting a more comprehensive and holistic approach. Marine management in China is characterized by fragmentation, with multiple administrative bodies developing their own plans and standards for marine industry development and environmental protection. This lack of effective coordination hampers efforts to address the complex issues facing the marine environment. A case in point is the management of land-based pollutants, which are a major source of marine pollution. Despite the fact that these pollutants often originate from river basins that span multiple provinces, governance efforts tend to focus on enforcement by marine authorities at the estuaries, rather than adopting a more holistic approach. Furthermore, in coastal areas with high population and industry density, the relentless pursuit of large-scale industrial expansion has led to the proliferation of heavy and chemical industries near ports and coastlines, posing a significant threat to the ecological integrity of MPAs [75].

To address these challenges, a shift toward land–sea integration seems necessary and promising. Land–sea integration is a holistic approach that recognizes the land and sea as two interconnected systems. Effective planning under this framework should transcend the narrow confines of the coastal zone and encompass a wider range of resources [76]. By adopting this more comprehensive approach, land–sea integration can unlock new opportunities for sustainable economic development and environmental stewardship. The Ministry of Natural Resources is promoting a unified planning system for national territorial space, including marine functional zoning. In addition, to strike a balance between conservation and development, it is essential to adopt diverse protection strategies tailored to different types of MPAs. Areas of high ecological significance should be strictly protected, while others may combine conservation with sustainable resource use or focus on tourism development. By assessing ecosystem resources and environmental capacity based on natural growth, biodiversity, and conservation targets, a more flexible MPA system can be established.

6.2. Developing and Refining the Legal System for MPAs

A sound legal system is of utmost significance to the effective governance of MPAs. For example, the National Marine Sanctuaries Act of the United States provides a solid legal foundation for MPA-related work in the United States [77]. In 2000, Executive Order 13158 was issued [78], which further details the organizational structure of the management system of MPAs, the responsibilities of each participating party, the operational mechanism, and the evaluation criteria, laying the institutional foundation for establishing a comprehensive and coordinated national network of MPAs. The legal basis for the establishment of MPAs in the European Union includes the Natura 2000 network of protected areas established under the Birds Directive and the Habitats Directive, MPAs designated under regional marine conventions, and MPAs designated under national legislation [79].

By developing and refining legislation, China's comprehensive natural protected area system in the marine sector can be developed, characterized by scientific classification, rational layout, effective protection, and efficient management, with national parks as the core. Targeted supporting and implementing regulations tailored to the distinct types, features, and functions of natural protected areas should be developed while prioritizing the development of technical standards. This will create a cohesive natural protected area legal framework consisting of basic law, specialized protection laws, and technical standards [80]. China should expedite the enactment of its National Park Law and revise relevant regulations governing nature reserves, marine special protected areas, and world geoparks. Furthermore, it is essential to closely integrate the development of MPA legislation with the construction of the national ecological compensation system. When formulating laws and regulations on ecological compensation, specialized research and surveys of MPAs should be carried out, and specific marine ecological compensation standards and guidelines should be established. Fines should be imposed on behaviors that damage the marine environment and resources, raising the cost of unlawful actions, reducing negative externalities, and ultimately safeguarding marine ecological resources [81].

6.3. Optimizing Structure of China's Governance of MPAs

Optimizing MPA governance structures is vital for effective management and conservation, and the experiences of the United States and Australia can offer valuable insights for China. The United States has a total of 1031 MPAs, covering 12,205,915 km^2 of ocean area [82]. The structure of MPA governance in the United States is characterized by diversity and collaboration, forming a national management system that involves the federal government, state governments, and the public. Within this framework, the National MPA Center, which is a partnership between the National Oceanic and Atmospheric Administration and the Department of the Interior that supports federal, state, territorial, and tribal programs responsible for the nation's ocean health, plays a crucial role in coordinating various parties and strengthening the governance system. The National MPA Center gathers broad input from federal departments, state-level management agencies, regional fisheries management councils, the public, and the MPA Federal Advisory Committee, while maintaining two-way communication and cooperation with management entities at all levels. Federal departments provide assistance, management, and support for MPA governance; management agencies at state levels are directly responsible for the establishment and daily management of MPAs; the public and stakeholders promote the transparency and improvement of the management of MPAs by providing opinions, suggestions, and supervision;

and the MPA Federal Advisory Committee provides independent and professional advice for management decisions while being subject to public supervision.

Australia's MPA governance structure also serves as a valuable reference, especially through its emphasis on safeguarding marine habitats and species. Australia has 2597 MPAs, covering 8,994,340 km^2 of ocean area [83]. Among them, the Royal National Park near Sydney was formally proclaimed in 1879, and in 1975, Australia enacted the Great Barrier Reef Marine Park Act, establishing the Great Barrier Reef Marine Park (GBRMP) to protect a large part of Australia's Great Barrier Reef from damaging activities [84]. The GBRMP in Australia employs a management system in which government control is the primary approach. The Great Barrier Reef Marine Park Authority (GBRMPA) is largely responsible for managing the GBRMP and receives funding from both the federal government and Queensland's state government. Since the park's establishment, numerous plans have been enacted, most notably a 25-year conservation plan developed by the GBRMPA in 1994. Together, the Great Barrier Reef Marine Park Act, along with relevant policies and plans, form the legal basis and guidelines for managing the GBRMP. The GBRMPA has established various committees, and its Board must include one member nominated by the Queensland Government, one member be Indigenous with knowledge of Indigenous issues in the Marine Park, and a member with tourism industry expertise. The GBRMP has been divided into several zones, and corresponding policies and legal guarantees have been put in place for each zone [85]. Public participation and economic impact assessments are also well-integrated into MPA planning in Australia, further enhancing the effectiveness of its governance framework [86].

China can glean valuable insights from countries such as Australia and the United States to enhance its MPA governance structure. Both demonstrate how unified management bodies, clear regulatory mechanisms, and active stakeholder engagement can help streamline MPA supervision, enforcement, and conservation efforts. Since the release of the "Guiding Opinions on Establishing a System of Natural Protected Areas with National Parks as the Mainstay", the Ministry of Natural Resources has assumed responsibility for the unified management of MPAs nationwide, while local governments oversee MPAs within their respective administrative regions. In particular, the National Forestry and Grassland Administration is now in charge of unified supervision and management of national parks, nature reserves, scenic areas, marine special protected areas, natural heritage sites, and geological parks. This change represents a marked shift from the previous system, where various departments managed MPAs under their own distinct mandates. By centralizing oversight within the National Forestry and Grassland Administration of the Ministry of Natural Resources, MPA management can operate in a more coordinated and streamlined manner.

To support this new governance framework, China has introduced several regulatory and legislative measures. The 2022 "Regulations on Nature Reserves (Draft Revision)" stipulate that the State Council's forestry and grassland administrative department shall oversee national nature reserves, including supervision and management responsibilities, while other relevant State Council departments assume related duties in accordance with their areas of responsibility [87]. Likewise, the draft National Park Law mandates the establishment of an interdepartmental coordination mechanism for national parks, led by the State Council's forestry and grassland department and involving other departments such as development and reform, finance, natural resources, and ecological environment, to facilitate and advance national park development [88]. This exemplifies the new management framework for natural protected areas, wherein the forestry and grassland administrative department serves as the central authority, with other competent departments contributing in line with their respective mandates.

6.4. Enhancing the Effectiveness of MPAs in China

Certain gaps remain in need of further measures to ensure the effectiveness of China's MPAs, and the EU's experience can serve as a valuable reference in this endeavor. Within the EU's Natura 2000 network, 3075 MPAs span approximately 451,538 km^2 by 2022 [89]. Although Natura 2000 functions as a unified biodiversity conservation network, its large number, broad area, and extensive distribution make centralized management by the European Commission challenging. Consequently, the EU delegates the selection and management of areas to its member states, each of which proposes potential areas and submits them to the European Commission for review and approval [90]. In 2020, the European Commission adopted the 2030 Biodiversity Strategy as part of the European Green Deal, setting legally binding targets for the EU and its member states. These include establishing at least 30% of land and sea as protected areas, and restoring at least 30% of degraded ecosystems by 2030 [91].

Nevertheless, despite these ambitious goals, MPA management within the EU faces ongoing challenges. According to a 2019 assessment report by the World Wildlife Fund, while 12.4% of the EU's marine waters are designated as MPAs, only 1.8% of these areas have a management plan in place. In practice, even less than 1.8% of the marine waters receive effective management and monitoring. As a result, it is currently impossible to determine which areas genuinely benefit from effective MPA management and successfully conserve biodiversity [92]. The European Parliament resolution of 16 January 2020 on the 15th meeting of the Conference of Parties to the CBD stresses that in the light of the recent IPCC report on the ocean and cryosphere in a changing climate, a comprehensive assessment and significant increase in EU coastal and marine protected areas and their governance is needed. The resolution recommends extending EU MPAs to include more offshore waters, while underscoring that besides the quantity, the quality of protected areas is essential to preventing biodiversity loss. Consequently, it calls for more emphasis on good and sustainable management [93]. Juliette Aminian-Biquet et al.'s research on EU MPAs in 2022 found that 86% of the 11.4% of EU waters covered by MPAs showed light, minimal, or no protection from the most harmful human activities, such as dredging, mining, or the most damaging fishing gears [94]. To address this challenge, the 2030 EU Biodiversity Strategy states that the EU will ensure better implementation of relevant environmental legislation affecting biodiversity, and review and revise it where necessary [95]. The European Commission will pay particular attention to areas with high biodiversity value or potential, and at least one-third of protected areas will be strictly protected [96].

In China, while MPAs are now under the unified management of the Forestry and Grassland Administration, the original administrative departments that previously managed these areas have established preliminary mechanisms that have shown some success. These departments, despite varying management measures, offer valuable lessons that can be built upon. As an example, in the area of ecological environment supervision and law enforcement in natural protected areas, the Ministry of Ecology and Environment promulgated the "Interim Measures for the Supervision and Management of Ecological and Environmental Protection in Natural Protected Areas" in 2020 [97]. These measures explicitly state that the ecological environment department retains responsibility for overseeing and managing the ecological environment of natural protected areas at all levels and categories. They also mandate that the Ministry of Ecology and Environment shall establish an ecological environment monitoring system for natural protected areas, formulate relevant standards and technical guidelines, construct a national ecological environment monitoring network system that integrates air, land, and space-based monitoring for natural protected areas, and conduct evaluations of the effectiveness of ecological environment protection in national-level natural protected areas. This suggests that the forestry and grassland administrative department, natural protected area management agencies, and ecological environment departments will need to maintain a collaborative approach to supervision and law enforcement in MPAs going forward.

To further enhance MPA governance, China should explore the establishment of a comprehensive maritime law enforcement mechanism, including setting up joint maritime law enforcement cooperation platforms and mechanisms. Coastal provinces and regions should collaborate to strengthen law enforcement and supervision in adjacent waters, establish a unified information-sharing mechanism, and streamline the case-transfer process. Such measures would effectively mitigate maritime environmental risks, while also constantly improving the emergency management framework for marine environmental incidents. Finally, China should implement scientifically grounded zoning and develop specific management regulations to guide community residents and businesses in lawful, orderly fishing, aquaculture, and eco-tourism practices, thus ensuring the community's basic livelihoods, protecting enterprises' legitimate interests, and promoting effective MPA governance.

7. Concluding Notes

Our research adopts the leverage points perspective of sustainability interventions to investigate the evolution of MPA governance in China, examining the number, effectiveness, legal framework, governance structure, value, and paradigm of MPAs, as well as highlighting the challenges facing their governance. Over time, China has established a network of MPAs focused on marine ecological protection, combining marine nature reserves and special protection areas. Case studies, including conservation efforts for marine species at the Dalian Spotted Seal National Nature Reserve, ecological status evaluations of ten national marine nature reserves by China's Ministry of Ecology and Environment, and the development of sea-related national parks, demonstrate the significant contributions of China's MPA system to biodiversity conservation. Nevertheless, several challenges persist, and to establish a scientifically classified, rationally distributed, and effectively managed natural protected areas system with national parks at its core, China's MPA governance requires further strengthening, including paradigm improvements, legal system refinement, governance structure optimization, and effectiveness enhancement.

Beyond the empirical scope of this study, we argue that achieving the Sustainable Development Goals and enabling sustainability transitions are fundamentally governance challenges. These challenges extend beyond sustainability envisioning—such as land use and landscape planning—to producing actionable knowledge that transforms envisioned sustainable futures into reality, such as land management. In this vein, a promising research avenue is to empirically investigate how real-world governance regimes manifest across the realms of leverage in domains like MPA governance and spatial governance at large. While theories provide elegant frameworks, practical success requires nuanced considerations tailored to specific contexts. Equally important, we contend that the conventional understanding of "land" as strictly terrestrial must be expanded to a more holistic and inclusive conceptualization that encompasses both terrestrial and marine environments. Recognizing this interconnectedness is crucial for advancing integrated spatial governance and sustainability efforts. In this regard, systematic and comparative studies across diverse national and regional contexts are particularly needed to inform integrated spatial governance of land and sea. By embracing this broader framing, land-related users, professionals, scholars, policymakers, and other stakeholders can better recognize that there is so much we can do to advance sustainability through land-related approaches. Let us, therefore, redefine "land" to include the vast, interconnected realms of terrestrial and marine environments and foster interdisciplinary, cross-sectoral, and transboundary collaborations to build a more sustainable future.

Author Contributions: Conceptualization, J.W., Z.M. and Z.Z.; methodology, J.W. and Z.M.; formal analysis, J.W. and Z.Z.; investigation, J.W. and Z.M.; data curation, J.W., Z.M. and Z.Z.; writing—original draft preparation, J.W., Z.M. and Z.Z.; writing—review and editing, J.W. and Z.Z.; visualization, Z.M.; supervision, J.W., Z.M. and Z.Z.; project administration, Z.M. and Z.Z.; funding acquisition, J.W. All authors have read and agreed to the published version of the manuscript.

Funding: This research was funded by a major project of the Key Research Base of Humanities and Social Sciences of the Ministry of Education, grant number 23JJD820014.

Data Availability Statement: Data will be available upon reasonable request.

Conflicts of Interest: The authors declare no conflicts of interest.

References

1. Ocean Studies Board, National Research Council; Committee on the Evaluation, Design, and Monitoring of Marine Reserves and Protected Areas in the United States. *Marine Protected Areas: Tools for Sustaining Ocean Ecosystems*; National Academy Press: Washington, DC, USA, 2001; pp. 146–147.
2. Ad Hoc Technical Expert Group on Marine and Coastal Protected Areas. Marine and Coastal Biodiversity: Review, Further Elaboration and Refinement of the Programme of Work. In Proceedings of the Report of the 8th Meeting of the Subsidiary Body on Scientific, Technical and Technological Advice, UNEP/CBD/SBSTTA/8/9/Add.1, Montreal, QC, Canada, 10–14 March 2003.
3. Eagles, P.F.J.; McCool, S.F.; Haynes, C.D.A. *Sustainable Tourism in Protected Areas: Guidelines for Planning and Management*; IUCN: Gland, Switzerland; Cambridge, UK, 2002; p. 8.
4. Dudley, N. *Guidelines for Applying Protected Area Management Categories*; IUCN: Gland, Switzerland, 2008; p. 10.
5. Ritchie, H. Where Does the Plastic in Our Oceans Come From? OurWorldInData.org. 2021. Available online: https://ourworldindata.org/ocean-plastics (accessed on 19 August 2024).
6. Rees, S.E.; Foster, N.L.; Langmead, O.; Pittman, S.; Johnson, D.E. Defining the qualitative elements of Aichi Biodiversity Target 11 with regard to the marine and coastal environment in order to strengthen global efforts for marine biodiversity conservation outlined in the United Nations Sustainable Development Goal 14. *Mar. Policy* **2018**, *93*, 241–250. [CrossRef]
7. Bloomberg Philanthropies. *Just 2.8% of the World's Ocean is Protected "Effectively"*; Bloomberg Philanthropies Press: New York, NY, USA, 2024. Available online: https://www.bloomberg.org/press/just-2-8-of-the-worlds-ocean-is-protected-effectively/ (accessed on 19 August 2024).
8. Zheng, M. How to protect Marine biodiversity after. *China Natural Resources News*, 23 April 2020; p. A5.
9. Lin, X.; Li, Y.; Wang, L.; Wu, S.; Wang, H.; Shi, Z.; Chen, Z.; Zhang, X.; Zhang, Z.; Liu, X. Research and prospects for China's deep participation in global marine biodiversity conservation. *Chin. Sci. Bull.* **2024**, *12*, 1598–1612.
10. Wang, B. The Protection and Management Strategies for Marine Biodiversity in China. *Biodiv. Sci.* **1999**, *4*, 347–350.
11. Yang, Y.; Dong, Z.; Zhou, B.-B.; Liu, Y. Smart Growth and Smart Shrinkage: A Comparative Review for Advancing Urban Sustainability. *Land* **2024**, *13*, 660. [CrossRef]
12. Qiu, W.; Wang, B.; Jones, P.J.; Axmacher, J.C. Challenges in developing China's marine protected area system. *Mar. Policy* **2009**, *33*, 599–605. [CrossRef]
13. Hu, W.; Liu, J.; Ma, Z.; Wang, Y.; Zhang, D.; Yu, W.; Chen, B. China's marine protected area system: Evolution, challenges, and new prospects. *Mar. Policy* **2020**, *115*, 103780. [CrossRef]
14. Zhao, Y.; Pikitch, E.K.; Xu, X.; Frankstone, T.; Bohorquez, J.; Fang, X.; Zheng, J.; Li, Y.; Chen, Z.; Lin, W.; et al. An evaluation of management effectiveness of China's marine protected areas and implications of the 2018 Reform. *Mar. Policy* **2022**, *139*, 105040. [CrossRef]
15. Zeng, X.; Chen, M.; Zeng, C.; Cheng, S.; Wang, Z.; Liu, S.; Zou, C.; Ye, S.; Zhu, Z.; Cao, L. Assessing the management effectiveness of China's marine protected areas: Challenges and recommendations. *Ocean Coast. Manag.* **2022**, *224*, 106172. [CrossRef]
16. Di Cintio, A.; Niccolini, F.; Scipioni, S.; Bulleri, F. Avoiding "Paper Parks": A Global Literature Review on Socioeconomic Factors Underpinning the Effectiveness of Marine Protected Areas. *Sustainability* **2023**, *15*, 4464. [CrossRef]
17. Jones, P.J.S.; Long, S.D. Analysis and discussion of 28 recent marine protected area governance (MPAG) case studies: Challenges of decentralisation in the shadow of hierarchy. *Mar. Policy* **2021**, *127*, 104362. [CrossRef]
18. Fidler, R.Y.; Ahmadia, G.N.; Amkieltiela; Awaludinnoer; Cox, C.; Estradivari; Glew, L.; Handayani, C.; Mahajan, S.L.; Mascia, M.B.; et al. Participation, not penalties: Community involvement and equitable governance contribute to more effective multiuse protected areas. *Sci. Adv.* **2022**, *8*, eabl8929. [CrossRef] [PubMed]

19. Giglio, V.J.; Moura, R.L.; Gibran, F.Z.; Rossi, L.C.; Banzato, B.M.; Corsso, J.T.; Pereira-Filho, G.H.; Motta, F.S. Do managers and stakeholders have congruent perceptions on marine protected area management effectiveness? *Ocean Coast. Manag.* **2019**, *179*, 104865. [CrossRef]
20. Meadows, D. *Leverage Points: Places to Intervene in a System*; The Sustainability Institute: Hartland, WI, USA, 1999.
21. Abson, D.J.; Fischer, J.; Leventon, J.; Newig, J.; Schomerus, T.; Vilsmaier, U.; Von Wehrden, H.; Abernethy, P.; Ives, C.D.; Jager, N.W.; et al. Leverage points for sustainability transformation. *Ambio* **2017**, *46*, 30–39. [CrossRef]
22. Fischer, J.; Riechers, M. A leverage points perspective on sustainability. *People Nat.* **2019**, *1*, 115–120. [CrossRef]
23. Riechers, M.; Fischer, J.; Manlosa, A.O.; Ortiz-Przychodzka, S.; Sala, J.E. Operationalising the leverage points perspective for empirical research. *Curr. Opin. Environ. Sustain.* **2022**, *57*, 101206. [CrossRef]
24. Bartkowski, B.; Bartke, S. Leverage Points for Governing Agricultural Soils: A Review of Empirical Studies of European Farmers' Decision-Making. *Sustainability* **2018**, *10*, 3179. [CrossRef]
25. Tan, X.; Su, Y. A Research on the Legal System of Marine Protected Areas under Land-Sea Coordination. *J. Ocean Univ. China (Soc. Sci.)* **2021**, *1*, 79–89.
26. Cai, H.; Zhang, H.; Xiang, Y. Discussion on the Issues of Construction and Management of Marine Special Protection Areas. *Ocean Dev. Manag.* **2005**, *3*, 55–57.
27. Yuan, K.; Teng, F.; Zhong, S.; Ding, Y.; Tan, J. Research on the Problems and Countermeasures in the Development of National Marine Protected Areas in the Beihai Region. *Ocean Dev. Manag.* **2015**, *11*, 37–41.
28. Zhao, L.; Cheng, M.; Ying, P.; Qu, F.; Zhang, Z. Current Status, Issues, and Development Countermeasures of Marine Protected Areas in China. *Ocean Dev. Manag.* **2019**, *5*, 3–7.
29. *HY/T 117-2010*; Classification and Grading Standards for Marine Special Protected Areas. People's Republic of China Marine Industry Standard: Beijing, China, 2010.
30. *HY/T118-2010*; State Oceanic Administration of China. Technical Guidelines for Functional Zoning and the Overall Plan Compiling of Special Marine Protection Area. Standards Press of China: Beijing, China, 2010.
31. Li, Y.; Sun, M.; Ren, Y.; Chen, Y. The Implication of Systematic Conservation Planning on China's Marine Protected Area Planning System. *Ocean Dev. Manag.* **2020**, *2*, 41–47.
32. The State Council Information Office of the People's Republic of China. *China's Marine Ecological Environment Protection*; The State Council Information Office: Beijing, China, 2024. Available online: https://www.gov.cn/zhengce/202407/content_6962503.htm (accessed on 19 August 2024).
33. The State Council of the People's Republic of China. The Overall Plan for Ecological Civilization System Reform. China Government Network. 21 September 2015. Available online: https://www.gov.cn/guowuyuan/2015-09/21/content_2936327.htm (accessed on 20 August 2024).
34. China Awarded the Hosting Rights for the 15th Meeting of the Conference of the Parties to the Convention on Biological Diversity. Available online: http://www.gov.cn/xinwen/2016-12/10/content_5146242.htm (accessed on 19 August 2024).
35. China Marine Conservation Industry Report: China Has Established 271 Marine Protected Areas. Available online: http://env.people.com.cn/n1/2020/1014/c1010-31892097.html (accessed on 15 August 2024).
36. Li, Y.; Ma, J.; Costigan, A.; Yang, X.; Pikitch, E.; Chen, Y. Reconciling China's domestic marine conservation agenda with the global 30×30 initiative. *Mar. Policy* **2023**, *156*, 105790. [CrossRef]
37. Grorud-Colvert, K.; Sullivan-Stack, J.; Roberts, C.; Constant, V.; Horta e Costa, B.; Pike, E.P.; Kingston, N.; Laffoley, D.; Sala, E.; Claudet, J.; et al. The MPA Guide: A framework to achieve global goals for the ocean. *Science* **2021**, *373*, eabf0861. [CrossRef] [PubMed]
38. Zeng, X.; Liang, J.; Zeng, J.; Chen, M.; Zeng, C.; Mazur, M.; Li, S.; Zhou, Z.; Ding, W.; Ding, P.; et al. Evaluating the effectiveness of three national marine protected areas in the Yangtze River Delta, China. *Front. Mar. Sci.* **2022**, *9*, 911880. [CrossRef]
39. Li, Y.; Wang, Z.; Lv, C.; Hu, D.; Li, Q.; Wang, X. Effectiveness Analysis and Countermeasure Suggestions of Planning Environmental Assessment in Marine Nature Reserves: A Case Study of the Dalian Spot-Seal National Nature Reserve. In Proceedings of the Chinese Society for Environmental Sciences, 2019 Annual Conference of the Chinese Society for Environmental Sciences, Xi'an, China, 23–25 August 2019; Volume 5.
40. National Forestry and Grassland Administration. Internationally Important Wetlands | Liaoning Dalian Spotted Seal Internationally Important Wetland: Guardian of the "Spirits" of the Liaodong Bay. Available online: https://www.forestry.gov.cn/main/6228/20220926/164135923958390.html (accessed on 19 December 2024).
41. Dong, J.; Shen, F. Estimation of Historical Population Size of Harbour Seal (Phoca Largha) in the Liaodong Bay. *Mar. Sci.* **1991**, *3*, 26–31.
42. Xinhuanet. *This Adorable 'Chubby Baby' Must Be Protected!* Available online: https://app.xinhuanet.com/news/article.html?articleId=70736b3fce6ee2107cd1c9b7ec898048 (accessed on 19 August 2024).

43. Ministry of Agriculture and Rural Affairs of the People's Republic of China. Action Plan for the Protection of the Spotted Seal (2017–2026). Available online: http://www.moa.gov.cn/nybgb/2017/djq/201802/P020180202505868637696.pdf (accessed on 17 December 2024).
44. 2023 Bulletin of Marine Ecology and Environment Status of China. Available online: https://www.mee.gov.cn/hjzl/sthjzk/jagb/202405/P020240522601361012621.pdf (accessed on 19 August 2024).
45. Bernik, B.M.; Li, H.; Blum, M.J. Genetic variation of Spartina alterniflora intentionally introduced to China. *Biol. Invasions* **2016**, *18*, 1485–1498. [CrossRef]
46. Gong, L.; Hu, Y.; Li, B.; Song, Y.; Li, Z.; Li, T.; Li, X. Effects of Spartina alterniflora eradication project on macrobenthos community structure: A case study of Nanhui tidal flat in Shanghai. *Chin. J. Ecol.* **2024**, *43*, 2892–2900.
47. Hu, A.; Yang, J.; Liu, B.; Zou, Y. Prediction on the Changes in Potential Suitable Areas for Mangroves Along the Coast of Guangxi and the Threat from Spartina Alterniflora Invasion. *Chin. J. Appl. Ecol.* **2024**, *3*, 669–677.
48. Mao, M.; Cai, H.; Qian, W. Challenges and Countermeasures of Marine Protected Areas in China under the New Nature Reserve System. *J. Anhui Agric. Sci.* **2023**, *20*, 78–81.
49. CPC Central Committee. The Communique of the Third Plenary Session of the 18th Central Committee of the Communist Party of China. Available online: http://www.ce.cn/xwzx/gnsz/szyw/201311/18/t20131118_1767104.shtml (accessed on 31 October 2024).
50. Liu, H.; Liu, Z. Current Status, Issues, and Countermeasures of Marine Protected Areas in China. *J. Mar. Inform. Technol. Appl.* **2015**, *1*, 36–41.
51. National Park and Other Nature Reserves Construction and Wildlife Protection Major Engineering Planning (2021–2035). Available online: https://www.bypc.gov.cn/upfiles/file/20220317/1647507018897358.pdf (accessed on 15 August 2024).
52. Achieving a Comprehensive Well-off Society and Securing Great Victories in the New Era of Socialism with Chinese Characteristics. Available online: http://www.xinhuanet.com/politics/19cpcnc/2017-10/27/c_1121867529.htm (accessed on 15 August 2024).
53. The Central People's Government of the People's Republic of China. The State Council Institutional Reform Plan. Available online: https://www.gov.cn/guowuyuan/2018-03/17/content_5275116.htm (accessed on 1 October 2024).
54. Nathan, J.B.; Philip, D. Why local people do not support conservation: Community perceptions of marine protected area livelihood impacts, governance and management in Thailand. *Mar. Policy* **2014**, *44*, 107–116.
55. Changdao National Marine Park Management Center. 2024 Annual Work Summary. Available online: https://www.changdao.gov.cn/art/2024/12/10/art_30545_2950311.html?xxgkhide=1 (accessed on 1 October 2024).
56. Qin, T. The Process and Challenges of China's Implementation of the Convention on Biological Diversity. *J. Wuhan Univ. (Philosoph. Sci. Ed.)* **2021**, *74*, 95–107.
57. Xue, D.; Zhang, Y. Biodiversity Conservation Achievements and Prospects in China. *Environ. Prot.* **2019**, *47*, 38–42.
58. Bai, Y.; Yang, X. On the Ecological Holism Implications of Environmental Law and Its Implementation Pathways. *J. Shandong Univ. Technol. (Soc. Sci. Ed.)* **2019**, *1*, 41–47.
59. Encyclopedia of China Editorial Board, Environmental Science Committee. *Encyclopedia of China, Environmental Science*; Encyclopedia of China Publishing House: Beijing, China, 2002.
60. UNEP/CBD/COP/DEC/VII/28, para.8. Available online: https://undocs.org/Home/Mobile?FinalSymbol=UNEP/CBD/COP/DEC/VII/28&Language=E&DeviceType=Desktop&LangRequested=False (accessed on 19 August 2024).
61. UNEP/CBD/COP/DEC/VII/5, Operational Objective 1.1. Available online: https://www.cbd.int/doc/decisions/cop-07/cop-07-dec-05-en.pdf (accessed on 19 August 2024).
62. COP 5 Decision V/6. Available online: https://www.cbd.int/decisions/cop/5/6 (accessed on 19 August 2024).
63. Report of the Secretary General. UN General Assembly Doc. A/61/63. Available online: https://documents.un.org/doc/undoc/gen/n06/473/29/pdf/n0647329.pdf (accessed on 19 August 2024).
64. Pilot Scheme for the System of Hainan Tropical Rainforest National Park. Available online: https://www.hainan.gov.cn/hainan/zchbbwwj/202008/f0a42020ac1547098d502acd161119cf.shtml (accessed on 19 August 2024).
65. Hainan Tropical Rainforest National Park Overview. Available online: https://www.hntrnp.com/news/list-276.html (accessed on 19 August 2024).
66. The Second National Parks Forum Focuses on Biodiversity Conservation. Available online: https://www.forestry.gov.cn/c/www/lcdt/518703.jhtml (accessed on 19 August 2024).
67. The Yellow River Estuary Selected as a Key Area for 'National Park' Creation: Overall Planning Map Released. Available online: https://new.qq.com/rain/a/20211021A0ARPW00 (accessed on 19 August 2024).
68. Liaoning Province Actively Establishes Liaohe Estuary National Park. Available online: https://www.ln.gov.cn/web/ywdt/ggaq/hjbh/406F98361C56454B891548E918E8BF66/index.shtml (accessed on 19 August 2024).
69. Changdao Included in National Park Spatial Planning Scheme. Available online: https://www.changdao.gov.cn/art/2022/11/11/art_64836_2935441.html (accessed on 19 August 2024).

70. Peng, X.; Jin, Y. Planning Compilation Techniques and Integrated Optimization Methods of Natural Protected Area System, National Park and Scenic Spot: Based on Landscaping Governance under Policy Guidelines. *Urban Rural Plan.* **2023**, *6*, 81–90.
71. Several Opinions on Delimiting and Strictly Observing the Ecological Protection Red Line. Available online: https://www.gov.cn/zhengce/2017-02/07/content_5166291.htm?trs=1 (accessed on 19 August 2024).
72. Shi, X.; Shen, W.; Ma, W.; Jiang, J. Suggestions for Biodiversity Conservation Based on Ecological Environment Zoning Management. *Environ. Impact Assess. Rev.* **2023**, *6*, 7–12.
73. Mei, H.; Xu, X. Research on the Institutional Evolution of China's Marine Special Protection Zones. *Ocean Dev. Manag.* **2023**, *12*, 16–27.
74. Wu, B.; Xie, Y.; Li, Y.; Cong, L. How to delineate and zone protected areas under the scope of ecological conservation redline strategy. *Biodiv. Sci.* **2022**, *4*, 21–26. [CrossRef]
75. Cao, Z.; Gao, G. The Connotation of and Policy Recommendation for Overall Planning Development of Land and Sea in China. *China Soft Sci.* **2015**, *2*, 1–12.
76. Han, Z.; Di, Q.; Zhou, L. Discussion on the Connotation and Target of Sea-Land Coordination. *Mar. Econ.* **2012**, *1*, 10–15.
77. National Marine Protected Areas Center. National system of Marine Protected Areas. Available online: https://marineprotectedareas.noaa.gov/ (accessed on 1 October 2024).
78. Executive Order 13158. 26 May 2000. p. 34909. Available online: https://www.federalregister.gov/d/00-13830/page-34909 (accessed on 1 October 2024).
79. Greenpeace. The Theory and Practice of Marine Protected Areas in the European Union. Available online: https://www.greenpeace.org.cn/report-list/eu-mpas-report/ (accessed on 10 October 2024).
80. Lv, Z. Systematic legislation for nature conservation with national parks as the main body. *Biodiv. Sci.* **2019**, *2*, 128–136.
81. Mao, X.; Zhong, Y. Conception, Theory and Mechanism of Eco-Compensation. China Popul. *Resour. Environ.* **2004**, *4*, 38–41.
82. Protected Marine Areas. Detailed Report on Marine Protections in the United States. Available online: https://map.navigatormap.org/countries/United%20States (accessed on 19 August 2024).
83. Protected Marine Areas. Detailed Report on Marine Protections in Australia. Available online: https://map.navigatormap.org/countries/Australia (accessed on 10 October 2024).
84. Fitzsimons, J.A.; Wescott, G. Large-scale expansion of marine protected area networks: Lessons from Australia. *Parks* **2020**, *24*, 19–34. [CrossRef]
85. Maestro, M.; Pérez-Cayeiro, M.L.; Chica-Ruiz, J.A.; Reyes, H. Marine protected areas in the 21st century: Current situation and trends. *Ocean Coast. Manag.* **2019**, *171*, 28–36. [CrossRef]
86. Voyer, M.; Gladstone, W.; Goodall, H. Methods of social assessment in Marine Protected Area planning: Is public participation enough? *Mar. Policy* **2012**, *36*, 432–439. [CrossRef]
87. Regulations on Nature Reserves (Revised Draft) (Solicitation of Opinions). Available online: http://www.forestry.gov.cn/html/main/main_4461/20220826163725291622669/file/20220826163808279973492.pdf (accessed on 19 August 2024).
88. National Park Law (Draft) (Solicitation of Opinions). Available online: https://www.forestry.gov.cn/html/main/main_153/20220819150732442216001/file/20220819170240697255827.pdf (accessed on 19 August 2024).
89. European Environment Agency. Natura 2000 Coverage in Europe's Seas. Available online: https://www.eea.europa.eu/themes/biodiversity/natura-2000/natura-2000-coverage-in-european-seas-4 (accessed on 10 December 2024).
90. European Environment Agency. Spatial Analysis of Marine Protected Area Networks in Europe's Seas. Available online: https://www.eea.europa.eu/publications/spatial-analysis-of-marine-protected (accessed on 10 October 2024).
91. European Commission. EU Biodiversity Strategy for 2030: Bringing Nature Back into Our Lives. 2020. Available online: https://eur-lex.europa.eu/legal-content/EN/TXT/?qid=1590574123338&uri=CELEX:52020DC0380 (accessed on 10 October 2024).
92. WWF. Protecting Our Ocean: Europe's Challenges to Meet the 2020 Deadline. Available online: https://d2ouvy59p0dg6k.cloudfront.net/downloads/protecting_our_ocean.pdf (accessed on 10 October 2024).
93. European Parliament Resolution of 16 January 2020 on the 15th Meeting of the Conference of Parties (COP15) to the Convention on Biological Diversity. Available online: https://www.europarl.europa.eu/doceo/document/TA-9-2020-0015_EN.pdf (accessed on 10 October 2024).
94. Aminian-Biquet, J.; Gorjanc, S.; Sletten, J.; Vincent, T.; Laznya, A.; Vaidianu, N.; Claudet, J.; Young, J.; e Costa, B.H. Over 80% of the European Union's Marine Protected Area Only Marginally Regulates Human Activities. *One Earth* **2024**, *7*, 1614–1629. [CrossRef]
95. European Commission. *Communication from the Commission to the European Parliament, the Council, The European Economic and Social Committee of the Regions: EU Biodiversity Strategy—Bringing Nature Back into Our Lives*; European Commission: Brussels, Belgium, 2020.

96. The Blue Leaders—30x30—Highly and Fully Protected, Health Food Chain Safety Environment. Available online: https://www.health.belgium.be/en/blue-leaders-30x30-highly-and-fully-protected (accessed on 10 October 2024).
97. Interim Measures for the Supervision of Ecological and Environmental Protection in Natural Protected Areas. Available online: https://www.gov.cn/zhengce/zhengceku/2020-12/24/content_5572948.htm (accessed on 19 August 2024).

MDPI AG
Grosspeteranlage 5
4052 Basel
Switzerland
Tel.: +41 61 683 77 34

Land Editorial Office
E-mail: land@mdpi.com
www.mdpi.com/journal/land